Lecture Notes in Physics

Volume 956

The series Lecture Notes in Physics (LNP), founded in 1969, reports new developments in physics research and teaching - quickly and informally, but with a high quality and the explicit aim to summarize and communicate current knowledge in an accessible way. Books published in this series are conceived as bridging material between advanced graduate textbooks and the forefront of research and to serve three purposes:

- to be a compact and modern up-to-date source of reference on a well-defined topic;
- to serve as an accessible introduction to the field to postgraduate students and non-specialist researchers from related areas;
- to be a source of advanced teaching material for specialized seminars, courses and schools.

Both monographs and multi-author volumes will be considered for publication. Edited volumes should however consist of a very limited number of contributions only. Proceedings will not be considered for LNP.

Volumes published in LNP are disseminated both in print and in electronic formats, the electronic archive being available at springerlink.com. The series content is indexed, abstracted and referenced by many abstracting and information services, bibliographic networks, subscription agencies, library networks, and consortia.

Proposals should be sent to a member of the Editorial Board, or directly to the responsible editor at Springer:

Dr Lisa Scalone
Springer Nature
Physics
Tiergartenstrasse 17
69121 Heidelberg, Germany
lisa.scalone@springernature.com

More information about this series at http://www.springer.com/series/5304

Massimo Meneghetti

Introduction to Gravitational Lensing

With Python Examples

 Springer

Massimo Meneghetti
Osservatorio di Astrofisica e Scienza dello Spazio
Istituto Nazionale di Astrofisica
Bologna, Italy

ISSN 0075-8450   ISSN 1616-6361   (electronic)
Lecture Notes in Physics
ISBN 978-3-030-73581-4   ISBN 978-3-030-73582-1   (eBook)
https://doi.org/10.1007/978-3-030-73582-1

This Springer imprint is published by the registered company Springer Nature Switzerland AG.
The registered company address is: Gewerbestrasse 11, 6330 Cham, Switzerland

Acknowledgments

This book emerged from a course in gravitational lensing that I had the privilege to teach for about ten years. I am grateful to the University of Bologna for this opportunity that gave me the possibility to meet so many talented students. In particular, I thank my friend Lauro Moscardini, who pushed me into this venture.

I first developed my lecture notes starting from some scripts that Matthias Bartelmann kindly shared with me. I wish to thank him for his support and the many encouragements he gave me to write this book.

I am grateful to my many colleagues and collaborators for the insightful discussions that helped shape this manuscript and Lisa Scalone, an amazingly efficient publishing editor.

Finally, my most enormous thanks go to my family, particularly my wife Enrica and my kids Eleonora and Paolo. Thank you for your support, patience, and love!

Contents

Part III   Appendixes

About the Author

Massimo Meneghetti is a researcher at INAF, Observatory of Astrophysics and Space Science of Bologna, Italy. He obtained his Ph.D. in astronomy from the University of Padova and worked at the Institute for Theoretical Astrophysics, University of Heidelberg (Germany), and at the Jet Propulsion Laboratory in Pasadena (USA). His research focuses on cosmology and structure formation to understand the nature of dark matter and dark energy, the major components of the universe, topics on which he has authored over 160 papers. In particular, he has taken major roles in several international projects that use gravitational lensing as a tool to investigate the matter distribution in cosmic structures like galaxies and galaxy clusters and to explore the distant universe. He taught gravitational lensing to master's students in astrophysics and cosmology at the University of Bologna for more than ten years. During this period, he developed the lecture notes contained in this book.

Part I
Generalities

A Brief History of Gravitational Lensing

1

In this chapter, we briefly summarize the discoveries that contributed to making gravitational lensing a popular tool for studying the universe's structure and evolution. The next chapters will deal with the subject much more rigorously.

1.1 Corpuscular Theory of Light

The idea that light can be deflected due to gravity has ancient origins. In the eighteenth century, the debate over the true nature of light was intense. According to the corpuscular theory, formulated by Sir Isaac Newton (1642–1726), light is composed of particles which, emitted by a luminous body, propagate in space along rectilinear paths, according to the laws of classical mechanics (Newton 1704). If massive, these particles can undergo the gravitational attraction of other bodies. The intensity of this attraction decreases with increasing distance, as required by the Law of Gravity. This theory contrasted with the wave theory of Christian Huygens (1629–1695), according to which light propagates like a wave in a medium called ether (Huygens 1690). The debate continued until the early 1800s when Thomas Young 1802, and later Augustin-Jean Fresnel 1819 resolved the question in favor of the wave theory through their experiments of interference and diffraction.

In the meantime, however, Newton's idea had sparked the interest of several scholars of the time. John Michell 1784, in a letter sent to the Scottish chemist and physicist Henry Cavendish, hypothesized that he could estimate the mass of a star by measuring how the speed of light would slow down due to the gravity of the star itself. The same idea was later re-proposed independently by Pierre Simon de Laplace 1796. It was probably this exchange of letters that prompted Cavendish to calculate how the trajectory of a particle of light would have changed following its close encounter with a star. This calculation, performed again using the principles

© Springer Nature Switzerland AG 2021
M. Meneghetti, *Introduction to Gravitational Lensing*, Lecture Notes
in Physics 956, https://doi.org/10.1007/978-3-030-73582-1_1

of classical mechanics, led to the following result: a ray of light, which passes at a distance R from a mass M, undergoes a deflection equal to the angle

$$\hat{\alpha}(R) = \frac{2GM}{c^2 R}, \tag{1.1}$$

where G is the universal gravitational constant, whose approximate value is $G = 6.67 \times 10^{-11} \ Nm^{-2}kg^{-2}$, while $c = 299792 \ km\,s^{-1}$ is the speed of light. The quantity given by the previous formula is called the *deflection angle*. Cavendish's notes in which this calculation was described remained unknown until the early 1900s (Will 1988). For this reason, the result is often attributed to a German scientist, Johann von Soldner , who published it in 1802 (von Soldner 1802).

The deflection angle formula allows us to calculate the deflection of a light ray that grazes the surface of the Sun. Considering that the Sun has a radius of $\sim$695700 km and a mass of $\sim$1.989$\times$10^{30} kg, the deflection angle is equal to $\sim$0.875 arc seconds (Fig. 1.1).

Fig. 1.1 Sketch for the deflection of a light ray passing at distance R from a lens of mass M

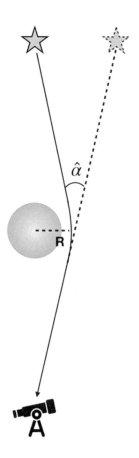

1.2   The Einstein Revolution

As mentioned, in the early 1800s, the wave nature of light was proved. For this reason, the idea that light could feel the effects of gravity was shelved and was no longer discussed for over a century. However, in 1907, Albert Einstein formulated the "Principle of Equivalence" (Einstein 1908), which will be later become one of the cardinal principles of his Theory of General Relativity. We can explain the principle of equivalence with the classic example of Einstein's elevator. Imagine a person in an elevator that falls from the top of a building. The person and the elevator in free fall experience the same acceleration, $\vec{g}$. For this reason, the person feels to be floating in space, precisely as if the elevator was not precipitating at all but was far from any source of gravity. On the contrary, an astronaut in space that started the engines of his spaceship, accelerating to $\vec{g}$, would experience the same feeling as if he was standing on the Earth's surface.

With such reasoning, Einstein concluded that it is not possible to distinguish the effects of gravity from those of acceleration and enunciated the principle of equivalence, according to which the gravitational mass coincides with the inertial mass. Using this principle and assuming that the speed of light is constant, he also postulated that light must feel the effect of gravity. Let us go back to the person closed in the elevator in free fall. If we made a hole on one of the lift's sidewalls and used it to let a ray of light enter the moment the elevator starts to fall, the person inside would see the light propagate at a constant speed from the hole of the entrance to the opposite wall. Not to violate the principle of equivalence, he would observe precisely what he would see if the elevator was floating in space, that is, that light propagates along a straight line with a constant speed. If there were an exit hole on the opposite wall, the light would leave the elevator, obviously after the time needed to travel through it.

Instead, consider an observer outside the elevator: he would observe the light enter the elevator when it begins its free fall from a certain height from the ground and would see it exit the opposite wall after some time, during which the lift falls to a lower altitude. The only way to explain this observation is that gravity is not only responsible for the fall of the elevator but also capable of bending the trajectory of light. The time needed for the light to pass through the lift cabin would be minimal, and the difference between the entry and exit height would be equally small. To carry out this thought experiment by Einstein, we have to use a little imagination.

Einstein published in 1911 the calculation of the deflection angle of a light ray passing at a distance from a mass (Einstein 1911) based on his Theory of Special Relativity. The result was the same as that previously obtained by Cavendish and Soldner. In 1915, however, Einstein formulated the theory of General Relativity, which explains gravity as assemblies of mass and energy curving space–time (Einstein 1916). Einstein repeated the deflection angle calculation within this new framework, which resulted in a factor of two larger than the previously obtained. The deflection of a light ray passing at a distance R from a mass M is, therefore:

$$\hat{\alpha}(R) = \frac{4GM}{c^2 R}. \tag{1.2}$$

In the case of a light ray that grazes the Sun's surface, the deflection angle predicted by the theory of General Relativity is about 1.75 arc seconds.

1.3   How to Prove the Deflection of Light?

Since 1912, Einstein had been looking for observational confirmations of his predictions about the deflection of light. In his correspondence with Sir George Ellery Hale, who was the Director of Mount Wilson Observatory at that time, he asked for advice on how the measure the positions of distant stars around the Sun in daylight (Fig. 1.2). By comparing these positions with those observed at night,

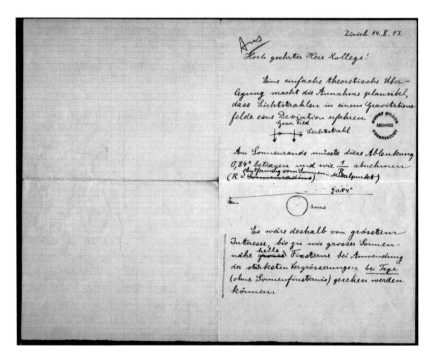

Fig. 1.2 In 1913, Albert Einstein, in a letter to George Ellery Hale, asked for advice on how to measure the deflection of light by the Sun. Note that in the text, Einstein still indicates that the angle of deflection of a light beam grazing the surface of the Sun is 0.84 arc seconds. He has not yet used the equations of the Theory of General Relativity. Credit: Image courtesy of the Observatories of the Carnegie Institution for Science Collection at the Huntington Library, San Marino, California

when the Sun would have been elsewhere, it would have been possible to measure the effect of the mass of the Sun on the trajectory of light. Hale replied that there was no possibility of making these measurements in the light of the Sun, but that he believed the idea of doing them during a total eclipse "very promising".

Several attempts were made to measure the deflection of light during solar eclipses (see, e.g. Coles 1999; Crelinsten 2006; Ellis 2010; Will 2015). The German astronomer Erwin Finlay-Freundlich, urged by Einstein, saw in a solar eclipse of August 1914, observable in Crimea, the right opportunity to make these measurements. He then organized an expedition to Feodosia, which, however, ended badly due to the outbreak of World War I. Finlay-Freundlich was arrested by the Russian army. The American astronomer William Wallace Campbell, director of the Lick Observatory, also attempted the measurement in the Crimea but failed due to bad weather. He was allowed to go home, but his equipment was seized, including special equipment for observing the eclipse.

Campbell himself retried the measurement on the occasion of another solar eclipse observable in Washington State in 1918. Although he was still without his instrumentation, he attempted the measure with more inferior quality equipment, which led him to conclude no deflection of light. When he was about to publish his results and speak out against Einstein's Theory, he learned of the measurements obtained during the two expeditions organized by Sir Arthur Eddington in 1919 (Fig. 1.3).

Fig. 1.3 Digitalized restoration of one of the images of the total solar eclipse in 1919 taken in Sobral. The positions of few stars visible in the obscured sky in Taurus' constellation are indicated with circles. Credit: ESO/Landessternwarte Heidelberg-Königstuhl/F. W. Dyson, A. S. Eddington, & C. Davidson

1.4   The Eddington Expeditions

Einstein presented his Theory of General Relativity to the Prussian Academy of Science in November 1915, when the First World War was underway. This condition made the exchange of scientific publications between England and Germany particularly difficult. This situation did not stop Sir Arthur Eddington, Plumian Professor at Cambridge University, from importing scientific publications from Germany into his country. He did so thanks to the collaboration of the Dutch mathematician, physicist, and astronomer Willem de Sitter. He thus came into possession of Einstein's writings. In 1917 he and his colleague, the Astronomer Royal Frank Dyson, convinced the Royal Astronomical Society to support expeditions to verify the predictions of the Theory of General Relativity. An incredibly tempting occasion occurred on May 29, 1919. A solar eclipse occurred, whose totality covered a vast region that stretched from Africa to South America. As Dyson noted, the exceptional condition was that, during the eclipse, the Sun was transiting in front of the Hyades open star cluster. Potentially, dozens of stars could be observed near the solar disk, obscured by the Moon, enabling the astronomers to take the necessary measures to verify Einstein's theory's predictions. Because of the turbulence in the Earth's atmosphere, the light from distant stars can suffer deflections of similar if not higher amplitude than those caused by the mass of the Sun. However, such deflections are random, and they average out if measured on many sources.

Eddington organized two expeditions, one directly to Principe Island, on the west coast of Africa, in which he participated together with his assistant Edwin Cottingham, and the other to Sobral, in northern Brazil, directed by the astronomer Andrew Crommelin, accompanied by colleague Charles Davidson. Eddington's measurements in Principe were difficult due to bad weather, but they gave a result consistent with Einstein's predictions. His team measured an average deflection of 1.60 ± 0.31 arc seconds for the stars near the solar disk. The measurements made in Sobral gave a different result, equal to 1.98 ± 0.12 arc seconds, but were considered less reliable by Eddington. Recent research has shown that the instruments used in Sobral had indeed undergone alterations due to a sharp change in temperature between the night before the eclipse and the following day (Kennefick 2012).

The results of Eddington's expeditions in support of Einstein's theory were presented in a historic speech by Frank Dyson to the Royal Astronomical Society and Royal Society of London on November 6, 1919. They were later published by Dyson et al. 1920. They had a notable prominence in the press of the time. In Fig. 1.4, we show a New York Times special cable celebrating Einstein's triumph, communicating the news of the measure of the deflection of light ("Lights all askew in the heavens"). The newspaper reported several scholars of the time, who were all in agreement in considering Eddington's measures an important confirmation of Einstein's theories and had different opinions about the latter's impact in people's daily lives. Einstein's theories were, for many scholars of the time, extremely complex and difficult to understand. The New York Times concluded by recalling that Einstein himself had warned his publishers that no more than twelve people

Fig. 1.4 The cable published by the New York Times on November 10, 1919, reporting on the triumph of the Theory of General Relativity of Einstein after the results of Eddington's expeditions were revealed

LIGHTS ALL ASKEW IN THE HEAVENS

Men of Science More or Less Agog Over Results of Eclipse Observations.

EINSTEIN THEORY TRIUMPHS

Stars Not Where They Seemed or Were Calculated to be, but Nobody Need Worry.

A BOOK FOR 12 WISE MEN

No More in All the World Could Comprehend It, Said Einstein When His Daring Publishers Accepted It.

in the world ("A book for 12 wise men") would have been able to understand his theory. Subsequent measurements taken on other eclipses confirmed Eddington's findings.

1.5   Following Intuitions

After his theory was proved right, Einstein himself did not seem particularly convinced that gravitational lensing could have relevant scientific applications. In 1936, however, he was persuaded by an amateur scientist, Rudi Mandl, to explore the consequences of the deflection of light by massive bodies in the universe. Precisely, Mandl, who had grasped the similarity between gravitational lenses and optical lenses, suggested to Einstein that closer stars could amplify distant stars. According to Mandl's idea, when two stars are well aligned, the light from the furthest star can be focused on the observer, as if it was passing through a magnifying glass. Therefore, the distant star should appear brighter as if it was observed with a gravitational telescope. So it was that Einstein published a short article in the journal Science in which he presented the calculations he had made at

Mandl's request (Einstein 1936). Einstein considered the case of two stars perfectly aligned with a terrestrial observer, showing that the distant star would be seen as a ring. The distant star would, therefore, appear much brighter. However, he wrote, "Of course, there is no hope to observe this phenomenon directly," given the tiny size of the ring and the low probability of having two stars so well aligned.

In truth, a little over a decade earlier, a Russian physicist, Orest Chwolson, had anticipated the effects observable when a star acts as a lens on a more distant star (Chwolson 1924). He had discussed the possibility that the farthest source could be seen multiple times and, in the case of perfect alignment, it might appear as a ring. This phenomenon's explanation lies in that light can reach the observer through different trajectories in curved space–time. This particular gravitational lensing regime, in which multiple images and massive distortions of the source's shape can be produced, is called strong lensing. If a ring is created, it is called "Einstein's Ring", although the term "Chwolson Ring" would be more appropriate.

Einstein's pessimism about the possibility of observing strong lensing effects was excessive. As we shall see, some 60 years later, it was realized that there are regions of the sky so dense with stars that it is not so rare to find some of them sufficiently well aligned to produce large amplifications observable even with medium-sized telescopes. Furthermore, Einstein had overlooked the fact that there are far more massive gravitational lenses in the sky than the stars. Since 1924, thanks to the observations conducted by Edwin Hubble, it was known that other galaxies existed besides ours and that these were at great distances from the Earth. Subsequent observations showed that there are also large agglomerations of hundreds and thousands of galaxies held together by gravity in the sky, called galaxy clusters.

In 1937, the American astronomer Fritz Zwicky published a short article entitled "Nebulae as Gravitational Lenses" (Zwicky 1937a). In this article, he explained that the deflection of light from distant sources by massive celestial objects, such as galaxies and clusters of galaxies, would have been of far greater intensity than those produced by single stars. He also wrote that these large natural lenses would allow galaxies to be seen at much greater distances than those achievable with the largest telescopes on Earth, due to the amplification effect already discussed by Einstein. Zwicky was also the first to describe one of the most important modern applications of gravitational lensing: the possibility of using this phenomenon to understand how matter is distributed in the lenses (Zwicky 1937b). Thanks to the studies that he had conducted on the Coma cluster's galaxies' motions, Zwicky had realized in those years that the luminous matter (the stars) was not sufficient to explain the incredible speeds with which the galaxies move in clusters. To explain these large speeds and why the galaxies were not dispersed in space instead of remaining tied, it was necessary to hypothesize that the galaxy cluster contained a large amount of invisible mass, which he called "missing mass", capable of producing a gravitational attraction force far more significant than that attributable to visible matter alone. Today we refer to this invisible matter with the term dark matter. Zwicky sensed that this enormous amount of mass would have had to produce gravitational lensing effects.

In the following decades, not much advancement happened in the field. In the 1960s, various scientists carried out studies that developed the mathematical aspect of gravitational lensing and sought to identify the ideal candidates to observe the effects predicted by the theory.

In the early 1960s, the Dutch astronomer Maarten Schmidt discovered quasars, point objects of enormous brightness, observable even at very great distances from us (Schmidt 1963). Astronomers hypothesized that these sources could be the ideal candidates to suffer the effects of gravitational lensing from nearby galaxies.

Yuri Klimov focused the lensing effects of galaxies on other galaxies, discussing the appearance of multiple images and rings around the lenses (Klimov 1963). Sydney Liebes 1964 reviewed several possible scenarios where lensing effects could be observed, including lensing by stars in the Milky Way on stars in the galactic bulge or the Andromeda galaxy, lensing by and within globular clusters, lensing by non-stellar objects in the Galaxy (asteroids, planets, other floating matter-agglomerates). He even considered the effects of gravitational waves.

The Norwegian astrophysicist Sjur Refsdal 1964 discussed a further consequence of gravitational lensing in addition to multiple images, distortions, and magnification. As we will see in greater detail later on, the light accumulates a time delay when deflected by a gravitational lens. Suppose that a distant source undergoes the effect of gravitational lensing by a galaxy along the line-of-sight. We could not realize the light's delay in a single image because we do not know when the light was emitted. However, as already suggested by Chwolson and Einstein, if the alignment between source and lens was strong, the source could be seen multiple times, and each of its images would have a different time delay. Suppose that the source is a Supernova. Due to gravitational lensing, its explosion would be observed multiple times but with delays between the images.

In the 1970s, many gravitational lensing applications had been conceived, but they remained theoretical speculations, without any observational feedback. The main reason was the lack of adequate tools to observe the events theorized by many scientists. As mentioned above, there must be an excellent alignment between observer, lens, and source for strong lensing events to be detected. Furthermore, for the deflection of light to be significant, the lens and the source must be sufficiently far from the observer. Indicatively, the rule applies that the distance between lens and source must be similar or greater than the distance between observer and lens. These conditions are rarely met, and it is necessary to observe large areas of the sky to discover gravitational lensing events.

Furthermore, gravitational lenses such as galaxies and galaxy clusters are several billion light-years away. Therefore, it is clear that the sources subject to lensing effects must be extraordinarily distant and thus very faint. Consequently, it is necessary to use very sensitive telescopes to be able to observe them, also if one takes into account the lens magnification effect.

1.6   First Observational Discoveries

Technological progress led to the construction of increasingly powerful telescopes capable of collecting light from large areas of the sky. After their invention, Charged-Coupled-Devices (CCD) quickly began to be used in astronomy. These devices allowed to drastically increase the astronomical cameras' sensitivity, quickly replacing the traditional photographic plates.

Finally, the first gravitational lensing event was observed in 1979. A group of astronomers discovered two quasars separated in the sky by 6 arc seconds (Walsh et al. 1979). The two quasars shared several properties, including their spectra and their redshifts ($z = 1.413$). The discovery that the two quasars were chemically similar and were at the same distance made it possible to decipher their nature: they were two multiple images of a single quasar. So, as expected, the first gravitational lensing event involved a quasar. The brightness of the source dominates that of the lens so much so that only the following year, with more in-depth observations, it emerged that the lens consisted of a small group of galaxies at redshift $z = 0.355$ (Young et al. 1980). The system is known by the name Q0957 + 561 and is shown in Fig. 1.5, in a much more recent image than Walsh's original observation. It was obtained with the Hubble Space Telescope. The lens and the source are at distances of approximately 3.7 and 8.7 billion light-years.

The discoveries of other gravitational lenses followed. In 1985 four multiple images of the quasar Q2237 + 0305 (Fig. 1.6) were observed, arranged to form a

Fig. 1.5 A recent observation with Hubble Space Telescope of the twin quasars QSO 0957 + 561, the first example of extra-galactic gravitational lensing. The lens was discovered in 1979. Credit: ESA/Hubble & NASA, reproduced on a non-exclusive basis

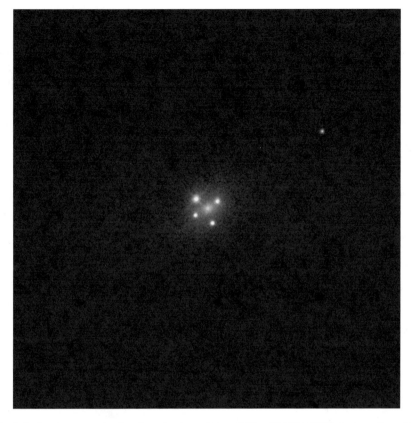

Fig. 1.6 Another example of a galaxy lensing a distant QSO − Q2237 + 0305. Such configuration of multiple images is dubbed "Einstein Cross". Credit: ESA/Hubble & NASA

cross around a lens galaxy (Huchra et al. 1985). This configuration of images was called the "Einstein cross". We can observe it when the lens and source alignment is almost perfect, as in the Einstein ring. We see a cross and not a ring due to how matter, mainly dark, is distributed in the lens. In the following chapters of this book, we will explain how multiple images' geometry depends on the lens's shape. Other double and triple quasars were discovered soon after.

As mentioned, quasars are practically point sources. As such, their multiple images remain point-like and easily separable. On the other hand, if the source that undergoes the lensing effect is a galaxy, a more extended source, multiple images can be deformed by the lens and merge, giving rise to what we call "gravitational arcs."

When Lynds and Petrosian 1986 and Soucail et al. 1987 reported the discovery of arc-shaped images in some galaxy clusters, including that shown in Fig. 1.7 in the cluster Abell 370, the interpretation of these observations seemed doubtful. It was not clear if the arcs were sources belonging to the clusters or were behind them and,

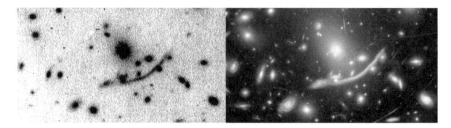

Fig. 1.7 One of the first gravitational arcs ever detected was found in Abell 370. An I-band image from the public Canada-France-Hawaii-Telescope archive is shown in the left panel. It can be compared with a very recent HST observation of the same region of the sky in the right panel

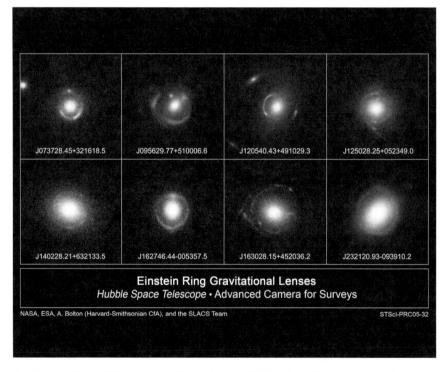

Fig. 1.8 A gallery of Einstein rings observed by the Hubble Space Telescope in the framework of the Sloan Lens ACS (SLACS) program (Bolton et al. 2006). Credits: NASA, ESA, A. Bolton (Harvard-Smithsonian CfA), and the SLACS Team

therefore, distorted by gravitational lensing. However, Soucail et al. 1988 showed that the redshift of the source seen as an arc ($z = 0.724$) was greater than the cluster redshift ($z = 0.37$), thus demonstrating that the one observed in Abell 370 was the first example of a gravitational arc. The Einstein rings shown in Fig. 1.8 are further examples of gravitational lensing effects on extended sources, but in that case, the lenses are galaxies and not galaxy clusters.

1.7   The First Microlensing Observations

As previously mentioned, Einstein in 1936 had concluded that the gravitational lensing effects produced by stars on other more distant stars would have been impossible to observe. He was wrong. His calculations showed that a lens star would produce effects, such as forming multiple images of the source star, on tiny angular scales, so small that not even powerful telescopes could see them. However, he had overlooked some critical aspects of the problem.

There is plenty of stars in our Galaxy, and they are not stationary. On the contrary, they participate in the Galaxy's rotation, with a speed that changes as a function of the distance from the center. Therefore, the stars move relative to each other, and occasionally their relative motion can cause them to remain well aligned with a terrestrial observer for a few dozen days. When this happens, we can observe the gravitational lensing effect experienced by the farthest star. We cannot see the multiple images, which are too close to be distinguished, given the lens's small size. However, we can detect the amplification effect that Rudi Mandl discussed with Einstein. During the alignment, the source star then appears brighter than before the alignment occurs or after the alignment has ceased.

Similar events fall under the gravitational lensing regime known as Microlensing. The term refers to the fact that the effects occur on tiny angular scales, less than one milli-arcsecond.

Bohdan Paczynski 1986 suggested to search for microlensing effects in the halo of the Milky Way by possible dark matter candidates such as Massive Astrophysical Compact Halo Objects (MACHOs). Such search implies monitoring the luminosity of millions of stars in the Magellanic Clouds. The first microlensing events were reported in 1993 by two groups simultaneously (Alcock et al. 1993; Aubourg et al. 1993). Following Paczynski's idea, they had monitored stars in the Large Magellanic Cloud for over a few years, finding three candidates.

An example of a microlensing event is shown in Fig. 1.9. In the boxes on the left, two observations made with a telescope of the Mount Stromlo Observatory, in Australia, of the same region of the sky in the direction of the Large Magellanic Cloud are shown. The first observation dates back to February 1993 while the second to January 1994. Clearly, in the center of the first image, we can see a very bright star that is much weaker in the second.

In the larger image on the right, we can see another observation of the region indicated in the previous images with a green box. In this case, the Hubble Space Telescope was used, much more sensitive and with a much higher spatial resolution than those of the Mount Stromlo telescope. In the center of the image, we can see a reddish star, which is not seen in the images obtained at Mount Stromlo. This star is the gravitational lens that passed in front of the source, another star in the Large Magellanic Cloud. The alignment between the two stars then ceased, and the amplification effect has disappeared, as the 1994 image shows.

Fig. 1.9 A microlensing event occurred between 1993 and 1994 is shown in the left panels. The observations were carried out at the Mt. Stromlo Observatory by looking at a region in the Large Magellanic Cloud. At the center of the upper image, taken in Feb. 1993, we can see a bright source that is no longer visible in the bottom image, dated less than a year later. The lens, a star in our Galaxy's halo, is not detectable in both these two images. It was observed with the Hubble Space Telescope only in 2002, as shown in the right panel (reddish star), which is an enlargement of the green box displayed in the other two images. Credits: NASA, ESA and D. Bennett (University of Notre Dame)

1.8   The Detection of Weak Lensing

Strong lensing causes spectacular effects such as gravitational arcs or multiple images of distant sources that we can observe around lenses such as galaxies or galaxy clusters; it is a rare phenomenon, which requires strong alignment between observer, lens, and source.

Einstein's theory tells us that the light deflection angle is inversely proportional to the distance from the lens, but it is proportional to the lens mass. Therefore, in massive lenses, such as galaxy clusters, whose masses can be of the order of $\sim 10^{15}\ M_\odot$, gravitational lensing effects could be measurable even at great distances from the center.

Indeed, Tyson et al. 1990 observed that distant galaxies were weakly but coherently elongated and tangentially distorted around the clusters Abell 1689 and

CL1409 + 52. These kinds of image distortions manifest the effect of gravitational lensing in the so-called weak regime. The amplitude of the distortion decreases with distance and is now measured around many gravitational lenses.

As we will see, weak lensing effects are beneficial for detecting how matter is distributed over large sky regions, not just around galaxy clusters. The weak lensing signal can be measured over large scales in the universe. The amplitude of the effect, known as *cosmic shear*, is tiny, but it was revealed in the early 2000s (Bacon et al. 2000; Van Waerbeke et al. 2000; Wittman et al. 2000). Today, the cosmic shear is considered one of the most powerful tools for understanding dark energy's nature and plays a central role in numerous cosmological experiments.

References

Alcock, C., Akerlof, C. W., Allsman, R. A., Axelrod, T. S., Bennett, D. P., Chan, S., & Sutherland, W. (1993). Possible gravitational microlensing of a star in the Large Magellanic Cloud. *Nature, 365*(6447), 621–623. https://doi.org/10.1038/365621a0. arXiv: astro-ph/9309052 [astro-ph].

Aubourg, E., Bareyre, P., Bréhin, S., Gros, M., Lachièze-Rey, M., Laurent, B., & Gry, C. (1993). Evidence for gravitational microlensing by dark objects in the Galactic halo. *Nature, 365*(6447), 623–625. https://doi.org/10.1038/365623a0.

Bacon, D. J., Refregier, A. R., & Ellis, R. S. (2000). Detection of weak gravitational lensing by large-scale structure. *MNRAS, 318*(2), 625–640. https://doi.org/10.1046/j.1365-8711.2000. 03851.x. arXiv: astro-ph/0003008 [astro-ph].

Bolton, A. S., Burles, S., Koopmans, L. V. E., Treu, T., & Moustakas, L. A. (2006). The sloan lens ACS survey. I. A large spectroscopically selected sample of massive early-type lens galaxies. *ApJ, 638*(2), 703–724. https://doi.org/10.1086/498884. arXiv: astro-ph/0511453 [astro-ph].

Chwolson, O. (1924). Über eine mögliche Form fiktiver Doppelsterne. *Astronomische Nachrichten, 221*, 329.

Coles, P. (1999). *Einstein and the total eclipse*. Icon.

Crelinsten, J. (2006). *Einstein's jury: The race to test relativity*. Princeton University Press.

de Laplace, P. S. (1796). *Exposition du système du monde*. https://doi.org/10.3931/e-rara-497.

Dyson, F. W., Eddington, A. S., & Davidson, C. (1920). Ix. a determination of the deflection of light by the sun's gravitational field, from observations made at the total eclipse of may 29, 1919. *Philosophical Transactions of the Royal Society of London. Series A, Containing Papers of a Mathematical or Physical Character, 220*(571–581), 291–333.

Einstein, A. (1908). Über das Relativitätsprinzip und die aus demselben gezogenen Folgerun-gen.*Jahrbuch der Radioaktivität und Elektronik, 4*, 411–462.

Einstein, A. (1911). Über den Einfluß der Schwerkraft auf die Ausbreitung des Lichtes. *Annalen der Physik, 340*(10), 898–908. https://doi.org/10.1002/andp.19113401005.

Einstein, A. (1916). Die Grundlage der allgemeinen Relativitätstheorie. *Annalen der Physik, 354*(7), 769–822. https://doi.org/10.1002/andp.19163540702.

Einstein, A. (1936). Lens-Like Action of a Star by the Deviation of Light in the Gravitational Field. *Science, 84*(2188), 506–507. https://doi.org/10.1126/science.84.2188.506.

Ellis, R. S. (2010). Gravitational lensing: A unique probe of dark matter and dark energy. *Philosophical Transactions of the Royal Society A: Mathematical, Physical and Engineering Sciences, 368*(1914), 967–987. https://doi.org/10.1098/rsta.2009.0209.

Fresnel, A. (1819). *Memoire sur la diffraction de la lumiere*.

Huchra, J., Gorenstein, M., Kent, S., Shapiro, I., Smith, G., Horine, E., & Perley, R. (1985). 2237+0305: A new and unusual gravitational lens. *AJ, 90*, 691–696. https://doi.org/10.1086/113777.

Huygens, C. (1690). *Traité de la lumière*. Leiden: Pieter van der Aa.

Kennefick, D. (2012). Not only because of theory: Dyson, Eddington, and the competing myths of the 1919 eclipse expedition. In *Einstein and the changing worldviews of physics* (pp. 201–232). Springer.

Klimov, Y. G. (1963). The Deflection of Light Rays in the Gravitational Fields of Galaxies. *Soviet Physics Doklady, 8,* 119.

Liebes, S. (1964). Gravitational Lenses. *Physical Review, 133*(3B), 835–844. https://doi.org/10. 1103/PhysRev.133.B835.

Lynds, R., & Petrosian, V. (1986). Giant luminous arcs in galaxy clusters. In *Bulletin of the American astronomical society* (Vol. 18, p. 1014).

Michell, J. (1784). On the means of discovering the distance, magnitude, &c. of the fixed stars, in consequence of the diminution of the velocity of their light, in case such a diminution should be found to take place in any of them, and such other data should be procured from observations, as would be farther necessary for that purpose. By the Rev. John Michell, B. D. F. R. S. In a Letter to Henry Cavendish, Esq. F. R. S. and A. S. *Philosophical Transactions of the Royal Society of London Series I, 74,* 35–57.

Newton, I. (1704). *Opticks*. Dover Press.

Paczynski, B. (1986). Gravitational microlensing by the galactic halo. *ApJ, 304,* 1–5. https://doi. org/10.1086/164140.

Refsdal, S. (1964). On the possibility of determining Hubble's parameter and the masses of galaxies from the gravitational lens effect. *MNRAS, 128,* 307. https://doi.org/10.1093/mnras/ 128.4.307.

Schmidt, M. (1963). 3C 273: A star-like object with large red-Shift. *Nature, 197*(4872), 1040. https://doi.org/10.1038/1971040a0.

Soucail, G., Fort, B., Mellier, Y., & Picat, J. P. (1987). A blue ring-like structure in the center of the A 370 cluster of galaxies. *A & A, 172,* L14–L16.

Soucail, G., Mellier, Y., Fort, B., Mathez, G., & Cailloux, M. (1988). The giant arc in a 370-spectroscopic evidence for gravitational lensing from a source at z = 0.724. *Astronomy and Astrophysics, 191,* L19–L21.

Tyson, J. A., Valdes, F., & Wenk, R. A. (1990). Detection of systematic gravitational lens galaxy image alignments: mapping dark matter in galaxy clusters. *ApJL, 349,* L1. https://doi.org/10. 1086/185636.

Van Waerbeke, L., Mellier, Y., Erben, T., Cuilland re, J. C., Bernardeau, F., Maoli, R., & Schneider, P. (2000). Detection of correlated galaxy ellipticities from CFHT data: First evidence for gravitational lensing by large-scale structures. *A & A, 358,* 30–44. arXiv: astro-ph/0002500[astro-ph].

von Soldner, J. (1802). Ueber die Ablenkung eines Lichtstrals von seiner geradlinigen Bewegung, durch die Attraktion eines Weltkörpers, an welchem er nahe vorbei geht. *Astronomisches Jahrbuch für das Jahr 1804*, Astronomisches Jahrbuch für das Jahr 1804.

Walsh, D., Carswell, R. F., & Weymann, R. J. (1979). 0957+561 A, B: Twin quasistellar objects or gravitational lens? *Nature, 279,* 381–384. https://doi.org/10.1038/279381a0.

Will, C. M. (1988). Henry Cavendish, Johann von Soldner, and the deflection of light. *American Journal of Physics, 56*(5), 413–415. https://doi.org/10.1119/1.15622.

Will, C. M. (2015). The 1919 measurement of the deflection of light. *Classical and Quantum Gravity, 32*(12), 124001. https://doi.org/10.1088/0264-9381/32/12/124001.

Wittman, D. M., Tyson, J. A., Kirkman, D., Dell'Antonio, I., & Bernstein, G. (2000). Detection of weak gravitational lensing distortions of distant galaxies by cosmic dark matter at large scales. *Nature, 405*(6783), 143–148. https://doi.org/10.1038/35012001. arXiv: astro-ph/0003014 [astro-ph].

Young, P., Gunn, J. E., Kristian, J., Oke, J. B., & Westphal, J. A. (1980). The double quasar Q0957+561 A, B: A gravitational lens image formed by a galaxy at z=0.39. *ApJ, 241,* 507–520. https://doi.org/10.1086/158365.

Young, T. (1802). The Bakerian Lecture: On the theory of light and colours. *Philosophical Transactions of the Royal Society of London Series I, 92*, 12–48.

Zwicky, F. (1937a). Nebulae as gravitational lenses. *Physical Review, 51*(4), 290–290. https://doi.org/10.1103/PhysRev.51.290.

Zwicky, F. (1937b). On the masses of nebulae and of clusters of nebulae. *ApJ, 86*, 217. https://doi.org/10.1086/143864.

Light Deflection

2

This chapter focuses on the light deflection. We begin with a point mass and derive the deflection angle in both the Newtonian and Relativistic limits. Then, we generalize the result to the cases of ensembles of point masses and extended mass distributions.

2.1   Deflection of a Light Corpuscle

As said at the beginning of the previous chapter, the idea that gravity could bend the light is dated back to the eighteenth century. It was mentioned by Isaac Newton at the end of his book *Optiks*, published in four editions between 1704 and 1730, and explained in the framework of the "Theory of Corpuscular Light" (Newton 1704).

In this framework, the derivation of a photon's deflection angle by a body with mass M is relatively straightforward (Congdon and Keeton 2018). We have to assume that photons are corpuscles with un-specified mass and are therefore capable of experiencing the gravitational pull of the body mentioned above, following Newtonian gravity laws. Besides, we have to assume that they move at a certain initial speed of c.

Newton's law of gravity says that the gravitational force between a corpuscle with mass m and a body with mass M (both assumed to be point-like) is

$$\vec{F} = -m\vec{\nabla}\Phi \, , \tag{2.1}$$

where $\Phi = -GM/r$ is the gravitational potential of the mass M, r is the distance between the bodies, and G is the gravitational constant.

Instead, Newton's second law of motion states that

$$\vec{F} = m\vec{a} \, , \tag{2.2}$$

where a is the acceleration of the corpuscle.

© Springer Nature Switzerland AG 2021   21
M. Meneghetti, *Introduction to Gravitational Lensing*, Lecture Notes
in Physics 956, https://doi.org/10.1007/978-3-030-73582-1_2

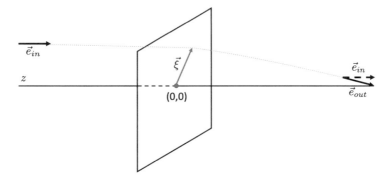

Fig. 2.1 Definition sketch for light deflection

Thus, because of gravity, the corpuscle will be accelerated to

$$\vec{a} = -\vec{\nabla}\Phi \, , \tag{2.3}$$

meaning that it will acquire an extra velocity $\Delta\vec{v}$,

$$\Delta\vec{v} = \int_{t_s}^{t_o} \vec{a}\, dt \tag{2.4}$$

while interacting with the body of mass M between the times t_s and t_o.

Let us assume, as shown in Fig. 2.1, that the corpuscle moves initially parallel to a reference axis z. We may place the body of mass M at the origin of a reference frame in $(\xi, z) = (0, 0)$, where ξ is the coordinate along the axis perpendicular to z. The vector $\vec{\xi}$ also indicates the impact parameter of the corpuscle with respect to the body of mass M. Thus, $r = \sqrt{\xi^2 + z^2}$. We can express z in terms of time t, $z = ct$, so that $dz = cdt$, and assume that the deflection of the corpuscle is small. Then, the time integral can be approximated by the integral along the initial direction of motion of the corpuscle as

$$\Delta\vec{v} \approx \frac{1}{c}\int_{z_s}^{z_o} \vec{a}\, dz = -\frac{1}{c}\int_{z_s}^{z_o} \vec{\nabla}\Phi\, dz \, . \tag{2.5}$$

The gradient of the gravitational potential can be decomposed in the two components along z and ξ:

$$\vec{\nabla}\Phi = \frac{d\Phi}{dz}\vec{e}_z + \frac{d\Phi}{d\xi}\vec{e}_\xi \, . \tag{2.6}$$

Similarly, the velocity change will also have two components along z and ξ. We start by computing

$$\Delta v_{\parallel} = -\frac{1}{c} \int_{z_s}^{z_o} \frac{d\Phi}{dz} dz$$

$$= \frac{1}{c} [\Phi(z_s, \xi) - \Phi(z_o, \xi)] . \qquad (2.7)$$

We can reasonably assume that the photon arrives from far behind the body and travels toward the observer at a long distance. Then, $|z_s| = |z_o| = \infty$. Since $\lim_{|z| \to \infty} \Phi = 0$, the integral above is zero, and the corpuscle does not acquire an additional velocity along the initial direction of motion. The velocity along the z axis continues to be c.

For the component along $\vec{\xi}$, we find

$$\Delta v_{\perp} = -\frac{1}{c} \int_{z_s}^{z_o} \frac{d\Phi}{d\xi} dz$$

$$= -\frac{GM\xi}{c} \int_{z_s}^{z_o} (\xi^2 + z^2)^{-3/2} dz . \qquad (2.8)$$

The integral can be solved by means of the variable change $\tan \theta = z/\xi$, and assuming again that $|z_s| = |z_o| = \infty$. Under these circumstances,

$$\Delta v_{\perp} = -\frac{GM}{c\xi} \int_{-\pi/2}^{\pi/2} \cos \theta d\theta$$

$$= -\frac{2GM}{c\xi} . \qquad (2.9)$$

Thus, while the corpuscle initially moves with velocity

$$\vec{v}_{in} = c\vec{e}_z = c\vec{e}_{in} , \qquad (2.10)$$

after the interaction with the body of mass M, it moves at velocity

$$\vec{v}_{out} = v\vec{e}_{out} = c\vec{e}_{in} - \frac{2GM}{c\xi} \vec{e}_{\xi} . \qquad (2.11)$$

We note that

$$v_{out} = \sqrt{c^2 + \frac{4G^2M^2}{c^2\xi^2}} \approx c . \qquad (2.12)$$

Thus, the deflection of the corpuscle, defined as the difference between the initial and final directions of motion, is

$$\hat{\vec{\alpha}}(\xi) = \vec{e}_{in} - \vec{e}_{out} = \frac{2GM}{c^2\xi}\vec{e}_\xi \ . \tag{2.13}$$

If the photon impact parameter is $\xi = R_\odot$, Eq. 2.13 reduces to

$$\hat{\alpha} = \frac{2GM}{c^2 R_\odot} \approx 0.875" \ , \tag{2.14}$$

when inserting $M = M_\odot = 1.989 \times 10^{30}$ kg and $R_\odot = 6.96 \times 10^8$ m. Thus, using Newtonian gravity and assuming that photons are light corpuscles, we obtain that a photon grazing the surface of the Sun is deflected by 0.875". We will see shortly that this value is just half of what predicted by Einstein in the framework of his Theory of General Relativity.

2.2   Deflection of Light According to General Relativity

2.2.1   Fermat Principle and Light Deflection

Light deflection can be calculated by studying geodesic curves starting from the field equations of general relativity. It turns out that light deflection can equivalently be described by Fermat's principle, as in geometrical optics. This result will be our starting point.

We attempt to treat the deflection of light in a general relativity framework as a refraction problem. We need a refractive index n because Fermat's principle says that light will follow the path which makes extremal the travel time,

$$t_{travel} - \int \frac{n}{c} \mathrm{d}l \ . \tag{2.15}$$

As in geometrical optics, we thus search for the path, $\vec{x}(l)$, for which

$$\delta \int_A^B n(\vec{x}(l)) \mathrm{d}l = 0 \ , \tag{2.16}$$

where the starting point A and the end point B are kept fixed.

Deflection in the Perturbed Minkowski's Space–Time

To find the refractive index, we make a first approximation: we assume that the lens is weak and small compared to the size of the optical system composed of source, lens, and observer. With "weak lens," we mean a lens whose Newtonian gravitational potential Φ is much smaller than c^2, $\Phi/c^2 \ll 1$. Note that this approximation is valid in virtually all cases of astrophysical interest. Consider, for instance, a galaxy cluster: its gravitational potential is $|\Phi| < 10^{-4}c^2 \ll c^2$. Besides, we also assume that the light deflection occurs in a region small enough that we can neglect the expansion of the universe.

Because of the principle of equivalence, we can choose a locally inertial frame where the space–time is flat and described by the Minkowski's metric,

$$\eta_{\mu\nu} = \begin{pmatrix} 1 & 0 & 0 & 0 \\ 0 & -1 & 0 & 0 \\ 0 & 0 & -1 & 0 \\ 0 & 0 & 0 & -1 \end{pmatrix},$$

whose line element is

$$ds^2 = \eta_{\mu\nu}dx^\mu dx^\nu = (dx^0)^2 - (d\vec{x})^2 = c^2dt^2 - (d\vec{x})^2 . \tag{2.17}$$

Now, we consider a weak lens perturbing this metric, such that

$$\eta_{\mu\nu} \to g_{\mu\nu} = \begin{pmatrix} 1 + \frac{2\Phi}{c^2} & 0 & 0 & 0 \\ 0 & -(1 - \frac{2\Phi}{c^2}) & 0 & 0 \\ 0 & 0 & -(1 - \frac{2\Phi}{c^2}) & 0 \\ 0 & 0 & 0 & -(1 - \frac{2\Phi}{c^2}) \end{pmatrix}$$

for which the line element becomes

$$ds^2 = g_{\mu\nu}dx^\mu dx^\nu = \left(1 + \frac{2\Phi}{c^2}\right)c^2dt^2 - \left(1 - \frac{2\Phi}{c^2}\right)(d\vec{x})^2 . \tag{2.18}$$

Example 2.1 (Schwarzschild Metric in the Weak Field Limit). Assuming a spherically symmetric and static potential, the Einstein's field equations can be solved to obtain the *Schwarzschild metric*. The line element is written in spherical coordinates as

$$ds^2 = \left(1 - \frac{2GM}{Rc^2}\right)c^2dt^2 - \left(1 - \frac{2GM}{Rc^2}\right)^{-1}dR^2 - R^2(\sin^2\theta d\phi^2 + d\theta^2) .$$

To obtain a simpler expression, it is convenient to introduce the new radial coordinate r, defined through

$$R = r \left(1 + \frac{GM}{2rc^2}\right)^2$$

and the Cartesian coordinates $x = r \sin\theta \cos\theta$, $y = r \sin\theta \sin\phi$, and $z = r \cos\theta$, so that $dl^2 = dx^2 + dy^2 + dz^2$. After some algebra, the metric can then be written in the form

$$ds^2 = \left(\frac{1 - GM/2rc^2}{1 + GM/2rc^2}\right)^2 c^2 dt^2 - \left(1 + \frac{GM}{2rc^2}\right)^4 (dx^2 + dy^2 + dz^2).$$

In the weak field limit, $\Phi/c^2 = -GM/rc^2 \ll 1$,

$$\left(\frac{1 - GM/2rc^2}{1 + GM/2rc^2}\right)^2 \approx \left(1 - \frac{GM}{2rc^2}\right)^4$$

$$\approx \left(1 - \frac{2GM}{rc^2}\right)$$

$$= \left(1 + \frac{2\Phi}{c^2}\right)$$

and

$$\left(1 + \frac{GM}{2rc^2}\right)^4 \approx \left(1 + 2\frac{GM}{rc^2}\right)$$

$$= \left(1 - \frac{2\Phi}{c^2}\right).$$

Therefore, the Schwarzschild metric in the weak field limit equals

$$ds^2 = \left(1 + \frac{2\Phi}{c^2}\right) c^2 dt^2 - \left(1 - \frac{2\Phi}{c^2}\right) dl^2,$$

thus recovering Eq. 2.18.

Effective Refractive Index

Light propagates at zero eigentime, $ds = 0$, from which we obtain

$$\left(1 + \frac{2\Phi}{c^2}\right) c^2 dt^2 = \left(1 - \frac{2\Phi}{c^2}\right) (d\vec{x})^2. \tag{2.19}$$

The light speed in the gravitational field is thus

$$c' = \frac{|\mathrm{d}\vec{x}|}{\mathrm{d}t} = c\sqrt{\frac{1 + \frac{2\Phi}{c^2}}{1 - \frac{2\Phi}{c^2}}} \approx c\left(1 + \frac{2\Phi}{c^2}\right) , \tag{2.20}$$

where we have used that $\Phi/c^2 \ll 1$ by assumption. The refractive index is thus

$$n = c/c' = \frac{1}{1 + \frac{2\Phi}{c^2}} \approx 1 - \frac{2\Phi}{c^2} . \tag{2.21}$$

With $\Phi \leq 0$, $n \geq 1$, and the light speed c' is smaller than in absence of the gravitational potential.

Deflection Angle
The refractive index n depends on the spatial coordinate $\vec{x}$ and perhaps also on time t. Let $\vec{x}(l)$ be a light path. Then, the light travel time is

$$t_{travel} \propto \int_A^B n[\vec{x}(l)]\mathrm{d}l , \tag{2.22}$$

and the light path follows from

$$\delta \int_A^B n[\vec{x}(l)]\mathrm{d}l = 0 . \tag{2.23}$$

This is a standard variational problem, which leads to the well known Euler equations. In our case we write

$$\mathrm{d}l = \left|\frac{\mathrm{d}\vec{x}}{\mathrm{d}\lambda}\right| \mathrm{d}\lambda , \tag{2.24}$$

with a curve parameter λ which is yet arbitrary, and find

$$\delta \int_{\lambda_A}^{\lambda_B} \mathrm{d}\lambda \, n[\vec{x}(\lambda)]\left|\frac{\mathrm{d}\vec{x}}{\mathrm{d}\lambda}\right| = 0 \tag{2.25}$$

The expression

$$n[\vec{x}(\lambda)]\left|\frac{\mathrm{d}\vec{x}}{\mathrm{d}\lambda}\right| \equiv L(\dot{\vec{x}}, \vec{x}, \lambda) \tag{2.26}$$

takes the role of the Lagrangian (assume that λ is time), with

$$\dot{\vec{x}} \equiv \frac{d\vec{x}}{d\lambda} \tag{2.27}$$

representing a generalized velocity. Finally, we have

$$\left|\frac{d\vec{x}}{d\lambda}\right| = |\dot{\vec{x}}| = (\dot{\vec{x}}^2)^{1/2} \ . \tag{2.28}$$

The Euler equation writes

$$\frac{d}{d\lambda}\frac{\partial L}{\partial \dot{\vec{x}}} - \frac{\partial L}{\partial \vec{x}} = 0 \ . \tag{2.29}$$

Now,

$$\frac{\partial L}{\partial \vec{x}} = |\dot{\vec{x}}|\frac{\partial n}{\partial \vec{x}} = (\vec{\nabla}n)|\dot{\vec{x}}| \ , \ \frac{\partial L}{\partial \dot{\vec{x}}} = n\frac{\dot{\vec{x}}}{|\dot{\vec{x}}|} \ . \tag{2.30}$$

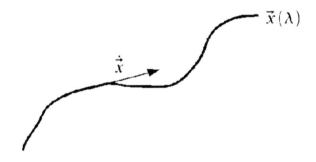

$$\vec{x}(\lambda)$$

$$\dot{\vec{x}}$$

Evidently, $\dot{\vec{x}}$ is a tangent vector to the light path, which we can assume to be normalized by a suitable choice for the curve parameter λ. We thus assume $|\dot{\vec{x}}| = 1$ and write $\vec{e} \equiv \dot{\vec{x}}$ for the unit tangent vector to the light path. Then, we have

$$\frac{d}{d\lambda}(n\vec{e}) - \vec{\nabla}n = 0 \ , \tag{2.31}$$

or

$$n\dot{\vec{e}} + \vec{e} \cdot [(\vec{\nabla}n)\dot{\vec{x}}] = \vec{\nabla}n \ ,$$

$$\Rightarrow n\dot{\vec{e}} = \vec{\nabla}n - \vec{e}(\vec{\nabla}n \cdot \vec{e}) \ . \tag{2.32}$$

The second term on the right hand side is the derivative along the light path, thus the whole right hand side is the gradient of n perpendicular to the light path. Thus

$$\dot{\vec{e}} = \frac{1}{n}\vec{\nabla}_\perp n = \vec{\nabla}_\perp \ln n \ . \tag{2.33}$$

As $n = 1 - 2\Phi/c^2$ and $\Phi/c^2 \ll 1$, $\ln n \approx -2\Phi/c^2$, and

$$\dot{\vec{e}} \approx -\frac{2}{c^2}\vec{\nabla}_\perp \Phi \ . \tag{2.34}$$

The total deflection angle of the light path is now the integral over $-\dot{\vec{e}}$ along the light path,

$$\hat{\vec{\alpha}} = \vec{e}_{in} - \vec{e}_{out} = \frac{2}{c^2}\int_{\lambda_A}^{\lambda_B} \vec{\nabla}_\perp \Phi d\lambda \ , \tag{2.35}$$

or, in other words, the integral over the "pull" of the gravitational potential perpendicular to the light path. Note that $\vec{\nabla}\Phi$ points away from the lens center, so $\hat{\vec{\alpha}}$ points in the same direction.

Born Approximation

As it stands, the equation for $\hat{\vec{\alpha}}$ is not useful, as we would have to integrate over the actual light path. However, since $\Phi/c^2 \ll 1$, we expect the deflection angle to be small. We can then adopt the *Born approximation*, familiar from scattering theory, and approximate the gravitational potential along the deflected trajectory by the potential along the un-deflected trajectory. This approximation allows us to integrate the gradient over the unperturbed light path, i.e. along z (Fig. 2.2).

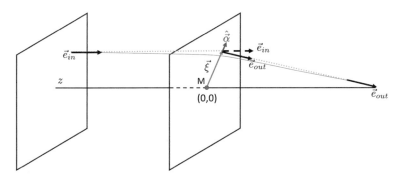

Fig. 2.2 Using Born approximation, we approximate gravitational potential along the deflected trajectory (solid curved line) by the potential along the un-deflected trajectory (dashed line)

Suppose, therefore, that a light ray starts out into $+\vec{e}_z$-direction and passes a lens at $z = 0$, with impact parameter ξ. The deflection angle is then given by

$$\hat{\alpha}(\xi) = \frac{2}{c^2} \int_{-\infty}^{+\infty} \vec{\nabla}_\perp \phi(\xi, z) \mathrm{d}z. \tag{2.36}$$

For point mass, $\Phi = -GM\sqrt{\xi^2 + z^2}$, thus,

$$\hat{\alpha}(\xi) = \frac{4GM}{c^2 \xi} \vec{e}_\xi = \frac{4GM}{c^2 \xi^2} \vec{\xi} \ . \tag{2.37}$$

This equation differs from Eq. 2.13 just by a factor of 2.

2.2.2 Deflection of Light in the Strong Field Limit

For the vast majority of gravitational lenses in the universe, the weak field limit holds. However, compact objects such as neutron stars and black holes can also act as lenses. In these cases, the approximations introduced above break down as photons travel through very strong gravitational fields. In the following, we briefly discuss the deflection angle of a static (i.e. non-rotating) compact lens.

For a general static, stationary, spherically symmetric metric in the form

$$\mathrm{d}s^2 = A(R)\mathrm{d}t^2 - B(R)\mathrm{d}R^2 - C(R)(\mathrm{d}\theta^2 + \sin^2\theta \mathrm{d}\phi^2) \tag{2.38}$$

the analysis of the geodesic equations leads to the following expression for the deflection angle:

$$\hat{\alpha} = -\pi + \frac{2G}{c^2} \int_{R_m}^{\infty} u \sqrt{\frac{B(R)}{C(R)[C(R)/A(R) - u^2]}} \mathrm{d}R \ , \tag{2.39}$$

where u is the impact parameter of the unperturbed photon and R_m is the minimal distance of the deflected photon from the lens (Bozza 2010). It can be shown that

$$u^2 = \frac{C(R_m)}{A(R_m)} \ . \tag{2.40}$$

Note that, in the case of the Schwarzschild metric, $A(R) = 1 - 2GM/Rc^2$, $B(R) = A(R)^{-1}$, and $C(R) = R^2$.

In the weak field limit ($R \geq R_m \gg 2GM/c^2$, i.e. for impact parameters much larger than the lens Schwarzschild radius), Eq. 2.39 reduces to the well know equation

$$\hat{\alpha} = \frac{4GM}{c^2 u} \ . \tag{2.41}$$

The exact solution of Eq. 2.39 in the case of the Schwarzschild metric was calculated by Darwin (1959) to be

$$\hat{\alpha} = -\pi + 4\frac{G}{c^2}\sqrt{R_m/s}\,F(\varphi, m)\,, \qquad (2.42)$$

where $F(\phi, m)$ is the elliptic integral of the first kind,

$$F(\phi, m) = \int_0^\phi \frac{d\varphi}{\sqrt{1 - m\sin^2\varphi}}\,, \qquad (2.43)$$

and

$$s = \sqrt{(R_m - 2M)(R_m + 6M)} \qquad (2.44)$$

$$m = (s - R_m + 6M)/2s \qquad (2.45)$$

$$\varphi = \arcsin\sqrt{2s/(3R_m - 6M + s)}. \qquad (2.46)$$

Figure 2.3 shows how the deflection angle varies as a function of the photon's impact parameter. At large distances, Eq. 2.42 is well approximated by the solution in the weak field limit. For small impact parameters, the solutions in the strong and the weak field limit differ significantly. In particular, the deflection angle in Eq. 2.42 diverges for $u = 3\sqrt{3}GM/c^2$ (or $R_m = 3GM/c^2$). Before reaching that point, the deflection angle exceeds 2π, meaning that the photon loops around the lens before leaving it.

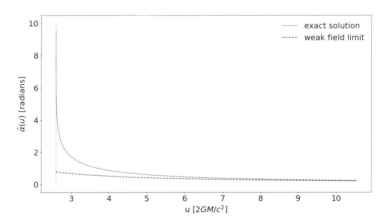

Fig. 2.3 Deflection angle by a compact lens as a function of the photon impact parameter. Shown are the exact solution of the geodesic equations for a Schwarzschild metric (solid line) and the solution in the weak field approximation (dashed line). The dotted vertical line shows the impact parameter, $u = 3\sqrt{3}GM/c^2$, for which the exact solution diverges, indicating that the photon keeps looping around the lens

2.3   Deflection by an Ensemble of Point Masses

The deflection angle in Eq. 2.37 depends linearly on the mass M. This result was obtained by linearizing the equations of general relativity in the weak field limit. Under these circumstances, the superposition principle holds, and we can calculate the deflection angle of an ensemble of lenses as the sum of all contributions by each lens.

Suppose we have a sparse distribution of N point masses on a plane, whose positions and masses are $\vec{\xi}_i$ and M_i, $1 \le i \le N$. The deflection angle of a light ray crossing the plane at $\vec{\xi}$ will be

$$\hat{\vec{\alpha}}(\vec{\xi}) = \sum_i \hat{\vec{\alpha}}_i(\vec{\xi} - \vec{\xi}_i) = \frac{4G}{c^2} \sum_i M_i \frac{\vec{\xi} - \vec{\xi}_i}{|\vec{\xi} - \vec{\xi}_i|^2} \ . \tag{2.47}$$

The formula above is similar to that used to compute the gravitational force between point masses on the plane. While the force depends on the inverse squared distance, the deflection angle scales as ξ^{-1}. In the case of many lenses, the computation of the deflection angle using Eq. 2.47 can become very computationally expensive, as it costs $O(N^2)$. However, as it is usually done to solve numerically N-body problems, algorithms employing meshes or hierarchies (such as the so-called tree algorithms (Barnes and Hut 1986)) can significantly reduce the cost of calculations (e.g. to $O(N \log N)$). For some application of these algorithms in the computation of the deflection angles, we refer the reader to the works of Aubert et al. (2007) and Meneghetti et al. (2010).

2.4   Deflection by an Extended Mass Distribution

We now consider more realistic lens models, i.e. three-dimensional distributions of matter. Even in the lensing by galaxy clusters, the lens's physical size is generally much smaller than the distances between observer, lens, and source. The deflection, therefore, arises along a short section of the light path. For this reason, we can use the *thin screen approximation*: we can approximate the lens by a planar distribution of matter, the lens plane.

Within this approximation, the lensing matter distribution is fully described by its surface density,

$$\Sigma(\vec{\xi}) = \int \rho(\vec{\xi}, z) \, \mathrm{d}z, \tag{2.48}$$

where $\vec{\xi}$ is a two-dimensional vector on the lens plane and ρ is the three-dimensional density.

As long as the thin screen approximation holds, we can obtain the total deflection angle by summing the contributions of all the mass elements $\Sigma(\vec{\xi})\mathrm{d}^2\xi$:

$$\vec{\hat{\alpha}}(\vec{\xi}) = \frac{4G}{c^2} \int \frac{(\vec{\xi} - \vec{\xi}')\Sigma(\vec{\xi}')}{|\vec{\xi} - \vec{\xi}'|^2} \, \mathrm{d}^2\xi' \; . \qquad (2.49)$$

This equation shows that the calculation of the deflection angle is formally a convolution of the surface density $\Sigma(\vec{\xi})$ with the kernel function

$$\vec{K}(\vec{\xi}) \propto \frac{\vec{\xi}}{|\vec{\xi}|^2} \; . \qquad (2.50)$$

This enables the calculation of the deflection angle field in the Fourier space as the product of the Fourier transforms of Σ and K:

$$\tilde{\hat{\alpha}}_i(\vec{k}) \propto \tilde{\Sigma}(\vec{k})\tilde{K}_i(\vec{k}) \; , \qquad (2.51)$$

where $\vec{k}$ is the conjugate variable to $\vec{\xi}$ and the tilde denotes the Fourier Transforms. The subscript $i \in [1, 2]$ indicates the two components along the two axes on the lens plane (remember that $\hat{\alpha}$ is a vector!).

2.5 Python Applications

2.5.1 Light Deflection by a Black-Hole

In our first python application, we write a script to produce Fig. 2.3. A brief python tutorial can be found in Appendix 8.

We need to implement Eqs. 2.40 and 2.42. Then, we compare the resulting deflection angle to that in the weak field limit described by Eq. 2.37.

We start by importing some useful packages:

```
from scipy import special as sy  # need special functions for incomplete \\
# elliptic integrals of the first kind
import numpy as np # efficient vector and matrix operations
import matplotlib.pyplot as plt # a MATLAB-like plotting framework
```

We import the module `special` from the package `scipy` in order to compute the elliptic integral of the first kind in Eq. 2.42. See https://docs.scipy.org/doc/scipy/reference/special.html.

Our goal is to produce a graph. We setup the fonts and the character size, and then we import the `matplotlib` library. We will use this library extensively for visualization purposes in this book. In particular, we will often use the MATLAB-like plotting framework provided by `matplotlib.pyplot`:

```
font = {'family' : 'normal',
        'weight' : 'normal',
        'size'   : 20}

import matplotlib
matplotlib.rc('font', **font)

import matplotlib.pyplot as plt
```

When it is convenient, we will also adopt an object-oriented programming style, using classes to define objects. For example, we can create a class for black hole objects as follows:

```
class point_bh:

    def __init__(self,M):
        self.M=M

    # functions defining the metric.
    def A(self,r):
        return(1.0-2.0*self.M/r)

    def B(self,r):
        return (self.A(r)**(-1))

    def C(self,r):
        return(r**2)

    # compute u from rm
    def u(self,r):
        u=np.sqrt(self.C(r)/self.A(r))
        return(u)

    # functions concurring to the deflection angle calculation
    def ss(self,r):
        return(np.sqrt((r-2.0*self.M)*(r+6.0*self.M)))

    def mm(self,r,s):
        return((s-r+6.0*self.M)/2/s)

    def phif(self,r,s):
        return(np.arcsin(np.sqrt(2.0*s/(3.0*r-6.0*self.M+s)))))

    # the deflection angle
    def defAngle(self,r):
        s=self.ss(r)
        m=self.mm(r,s)
        phi=self.phif(r,s)
        F=sy.ellipkinc(phi, m) # using the ellipkinc function
                               # from scipy.special
        return(-np.pi+4.0*np.sqrt(r/s)*F)
```

A black-hole object is an instance of the class `point_bh`. The only input parameter required to initialize the object (see the `__init__`) is the black-hole mass. The class contains several methods (or functions), which will be used to compute the black-hole deflection angle. For example, it contains the functions $A(R)$, $B(R)$, and $C(R)$. We use them to convert the minimal distance R_m to u. It also contains the functions for computing s, m, φ, which depend on the black-hole mass and the minimal distance R_m. Finally, the function `defAngle` enables to compute the deflection angle using Eq. 2.42. This function uses the method `elipkinc` from `scipy.special` to compute the incomplete elliptic integral of the first kind, $F(\varphi, m)$. Note that φ and m are numpy arrays not scalars, i.e. `elipkinc` returns the integral for several values of (φ, m) with a single call.

Following the same approach, we build another class which deals with point lenses in the weak field limit, i.e. it implements Eq. 2.37:

```
class point_mass:

    def __init__(self,M):
        self.M=M

    # the classical formula
    def defAngle(self,u):
        return(4.0*self.M/u)
```

We can now use the two classes above to build two objects, namely a black-hole lens and a point-mass lens. In both cases, the mass of the lens is fixed to $3M_\odot$. For a mass of this size, the Schwarzschild radius is $R_s \sim 9$km:

```
bh=point_bh(3.0)
pm=point_mass(3.0)
```

We use the `linspace` method from the package `numpy` to initialize an array of minimal distances R_m, which we use to compute $\hat{\alpha}$. We use the function `u(r)` of `point_bh` to convert R_m into an array of impact parameters u:

```
r=np.linspace(3.0/2.0,10,1000)*2.0*bh.M
u=bh.u(r)/2.0/bh.M
```

The deflection angle as a function of u or R_m can be computed in the cases of the exact solution and in the weak field limit using the method `defAngle` applied to bh and pm:

```
a=bh.defAngle(r)
b=pm.defAngle(u*2.0*bh.M)
```

Note that u is in units of the Schwarzschild radius and we assume $G/c^2 = 1$.
For displaying the results, we use the following code:

```
# initialize figure and axes
# (single plot, 15" by 8" in size)
fig,ax=plt.subplots(1,1,figsize=(15,8))
```

```
# plot the exact solution in ax
ax.plot(u,a,'-',label='exact solution')
# plot the solution in the weak field limit
ax.plot(u,b,'--',label='weak field limit',color='red')
# set the labels for the x and the y axes
ax.set_xlabel(r'$u$ $[2GM/c^2]$')
ax.set_ylabel(r'$\hat\alpha(u)$ [radians]')
# add the legend
ax.legend()
```

We also want to show the vertical asymptote at $u_{lim} = 3\sqrt{3}/2$:

```
# plot a vertical dotted line at u=3\sqrt(3)/2
x=[np.min(u),np.min(u)]
y=[0,10]
ax.plot(x,y,':')
```

To conclude, we save the figure in a .png file:

```
# save figure in png format
fig.savefig('bhalpha.png')
```

2.5.2 Light Deflection by an Extended Mass Distribution

In this example, we implement the calculation of the deflection angle field by an extended lens. A two-dimensional map of the lens surface density is provided by the fits file kappa_2.fits. The map was obtained by projecting the mass distribution of a dark matter halo obtained from an N-body simulation (Meneghetti et al. 2017). To be precise, this is the surface density divided by a constant which depends on the lens and source redshifts (we will talk about this constant in the next chapter). This quantity is called *convergence*, κ. In the next chapter, we will show that we can re-write Eq. 2.49 in terms of the convergence as:

$$\vec{\alpha}(\vec{x}) = \frac{1}{\pi} \int \kappa(\vec{x}') \frac{\vec{x} - \vec{x}'}{|\vec{x} - \vec{x}'|^2} d^2 x' \ .$$

As we pointed out in Sect. 2.4, this integral is a convolution. In Fourier Space, it becomes a multiplication:

$$\vec{\tilde{\alpha}}(\vec{k}) = \frac{1}{\pi} \tilde{\kappa}(\vec{k}) \vec{\tilde{K}}(\vec{k}),$$

where $\vec{\tilde{\alpha}}(\vec{k})$, $\tilde{\kappa}(\vec{k})$, and $\vec{\tilde{K}}(\vec{k})$ are the Fourier Transforms of the deflection angle, the convergence and the kernel function,

$$\vec{K}(\vec{x}) = \frac{\vec{x}}{|\vec{x}|^2} \ .$$

We consider a convergence map of $n_{pix} \times n_{pix}$ pixels, i.e. the function to be convolved is sampled at a certain number of positions on a regular grid, where the pixel on the m-th row and n-th column is identified with the couple of indices (m, n). We will use Discrete Fourier Transforms (DFTs) to implement the convolution. For example, the DFT of the convergence map is

$$\tilde{\kappa}_{kl} = \sum_{m=0}^{n_{pix}-1} \sum_{n=0}^{n_{pix}-1} \kappa_{mn} \exp\left\{ -2\pi i \left(\frac{km}{n_{pix}} + \frac{ln}{n_{pix}} \right) \right\}, \qquad (2.52)$$

with $(k, l) \in (1, \ldots, n_{pix})$. A common algorithm to compute DFTs is the so-called Fast Fourier Transform (FFT) algorithm (Cooley and Tukey 1965), which is implemented in many python packages. Here, we use the `numpy.fft` module:

```python
import numpy as np
import numpy.fft as fftengine
```

To deal with lenses described by convergence maps, we define a class called `deflector`. This class, shown here below, contains some methods that will be described in detail later. They allow to

- build the kernel $K(\vec{x})$;
- compute the deflection angle map by convolving the convergence with the kernel;
- perform the so-called zero-padding;
- crop the zero-padded maps.

We initialize a deflector object by reading the fits file containing the lens convergence map. For this purpose, we import the `astropy.io.fits` module.

```python
import astropy.io.fits as pyfits


class deflector(object):

    """
    initialize the deflector using a surface density (convergence) map
    the boolean variable pad indicates whether zero-padding is used
    or not
    """
    def __init__(self,filekappa,pad=False):
        kappa,header=pyfits.getdata(filekappa,header=True)
        self.kappa=kappa
        self.nx=kappa.shape[0]
        self.ny=kappa.shape[1]
        self.pad=pad
        if (pad):
            self.kpad()
        self.kx,self.ky=self.kernel()
```

```
"""
implement the kernel function K
"""

def kernel(self):
    x=np.linspace(-0.5,0.5,self.kappa.shape[0])
    y=np.linspace(-0.5,0.5,self.kappa.shape[1])
    kx,ky=np.meshgrid(x,y)
    norm=(kx**2+ky**2+1e-12)
    kx=kx/norm
    ky=ky/norm
    return(kx,ky)


"""
compute the deflection angle maps by convolving
the surface density with the kernel function
Note that the returned values will be in pixel units
"""

def angles(self):
    # FFT of the surface density and of the two components of the
    # kernel
    kappa_ft = fftengine.rfftn(self.kappa,axes=(0,1))
    kernelx_ft = fftengine.rfftn(self.kx,axes=(0,1),
                                 s=self.kappa.shape)
    kernely_ft = fftengine.rfftn(self.ky,axes=(0,1),
                                 s=self.kappa.shape)
    # perform the convolution in Fourier space and transform the result
    # back in real space. Note that a shift needs to be applied using
    # fftshift
    alphax = 1.0/np.pi*\
             fftengine.fftshift(fftengine.irfftn(kappa_ft*kernelx_ft))
    alphay = 1.0/np.pi*\
             fftengine.fftshift(fftengine.irfftn(kappa_ft*kernely_ft))
    return(alphax,alphay)


"""
returns the surface-density (convergence) of the deflector
"""

def kmap(self):
    return(self.kappa)


"""
performs zero-padding
"""

def kpad(self):
    # add zeros around the original array
    def padwithzeros(vector, pad_width, iaxis, kwargs):
        vector[:pad_width[0]] = 0
        vector[-pad_width[1]:] = 0
        return vector
    # use the pad method from numpy.lib to add zeros (padwithzeros)
    # in a frame with thickness self.kappa.shape[0]
```

```
        self.kappa=np.lib.pad(self.kappa, self.kappa.shape[0],
                              padwithzeros)

    """
    crop the maps to remove zero-padded areas and get back to the
    original region.
    """
    def mapCrop(self,mappa):
        xmin=np.int(0.5*(self.kappa.shape[0]-self.nx))
        ymin=np.int(0.5*(self.kappa.shape[1]-self.ny))
        xmax=xmin+self.nx
        ymax=ymin+self.ny
        mappa=mappa[xmin:xmax,ymin:ymax]
        return(mappa)
```

We begin by building a deflector and using it to compute the deflection angles employing the method `angles`:

```
df=deflector('data/kappa_2.fits')
angx_nopad,angy_nopad=df.angles()
kappa=df.kmap()
```

The function `kmap` returns the convergence map read from the fits file as a numpy array. We visualize this map and the maps of the two components of the deflection angles as in Fig. 2.4 using the following instructions:

```
import matplotlib.pyplot as plt

fig,ax = plt.subplots(1,3,figsize=(16,8))
ax[0].imshow(kappa,origin="lower")
ax[0].set_title('convergence')
ax[1].imshow(angx_nopad,origin="lower")
ax[1].set_title('angle 1')
ax[2].imshow(angy_nopad,origin="lower")
ax[2].set_title('angle 2')
```

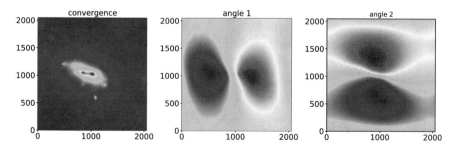

Fig. 2.4 Left panel: the surface density (convergence) map of the lens. Middle and right panels: maps of the two components of the deflection angles

Note that, when creating the instance df of deflector, we did not change the default value of the keyword pad, which is then set to False. Now we explain the usage of this keyword. The computation of the DFTs assumes periodic boundary conditions. In other words, we can imagine that the lens mass distribution is replicated outside the boundaries of the map indefinitely and periodically. Other identical lenses virtually surround the lens, and each of them contributes to the light deflection. Given that the region around the lens considered in this example is relatively small, we expect that the deflection angles will be biased near the borders, where light rays can feel the pull of the mass outside the map. The three panels in Fig. 2.4 show the maps of the convergence and the two components of the deflection angles obtained with this setting.

To mitigate this bias, we can use a method called *zero-padding*. Zero-padding consists of creating a buffer all around the convergence map, where the convergence is set to zero. By doing so, we increase the size of the original map, but we expect to increase the accuracy of the calculations near the borders, because the periodic conditions are better reproduced in this setting. We activate zero-padding by setting the variable pad=True when initializing the deflector. Then, the function kpad does the job:

```
ddf=deflector('data/kappa_2.fits',pad=True)
angx,angy=df.angles()
kappa=df.kmap()
```

In the example shown in Fig. 2.5, we zealously increase the size of the map by a factor of 3 in each dimension. We are not interested in this large, zero-padded area, thus we can get rid of the values outside the footprint of the original convergence map by using the function mapCrop:

```
angx=df.mapCrop(angx)
angy=df.mapCrop(angy)
```

We show the cropped deflection angle maps in the upper panels of Fig. 2.6. For comparison, we also show the maps obtained without zero-padding in the bottom

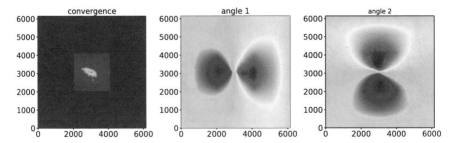

Fig. 2.5 As in Fig. 2.4 but showing the zero-padded convergence map and the two corresponding maps of the deflection angle components

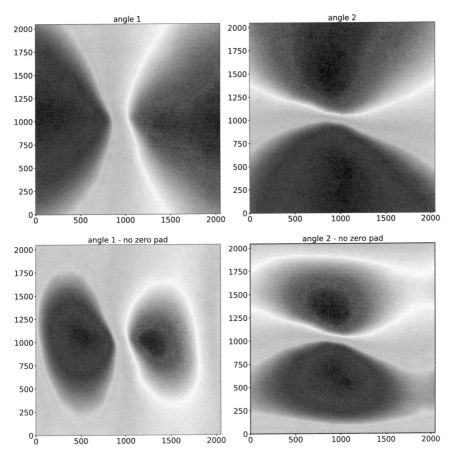

Fig. 2.6 The upper panels show the same two maps displayed in the middle and right panels of Fig. 2.5, cropped to match the original size of the input convergence map. The bottom panels show the maps obtained without padding, for comparison

panels. As expected, we see that the maps differ significantly along the borders. It is imperative to perform zero-padding when using the method outlined in this example to calculate the deflection angles.

References

Aubert, D., Amara, A., & Metcalf, R. B. (2007). Smooth particle lensing. *MNRAS, 376*, 113–124. https://doi.org/10.1111/j.1365-2966.2006.11296.x. eprint: astro-ph/0604360.

Barnes, J., & Hut, P. (1986). A hierarchical O(N log N) force-calculation algorithm. *Nature, 324*, 446–449. https://doi.org/10.1038/324446a0.

Bozza, V. (2010). Gravitational lensing by black holes. *General Relativity and Gravitation, 42*, 2269–2300. https://doi.org/10.1007/s10714-010-0988-2. arXiv: 0911.2187 [gr-qc].

Congdon, A. B., & Keeton, C. (2018). *Principles of gravitational lensing: light deflection as a probe of astrophysics and cosmology*. Springer International Publishing.

Cooley, J. W., & Tukey, J. W. (1965). An algorithm for the machine calculation of complex Fourier series. *Mathematics of computation, 19*(90), 297–301.

Darwin, C. (1959). The gravity field of a particle. *Proceedings of the Royal Society of London Series A, 249*, 180–194. https://doi.org/10.1098/rspa.1959.0015.

Meneghetti, M., Natarajan, P., Coe, D., Contini, E., De Lucia, G., Giocoli, C., & Zitrin, A. (2017). The Frontier Fields lens modelling comparison project. *MNRAS, 472*(3), 3177–3216. https://doi.org/10.1093/mnras/stx2064. arXiv: 1606.04548 [astro-ph.CO].

Meneghetti, M., Rasia, E., Merten, J., Bellagamba, F., Ettori, S., Mazzotta, P., & Marri, S. (2010). Weighing simulated galaxy clusters using lensing and X-ray. *A & A, 514*, A93. https://doi.org/10.1051/0004-6361/200913222. arXiv: 0912.1343.

Newton, I. (1704). *Opticks*. Dover Press.

The General Lens

3

In this chapter, we discuss the consequences of light deflection by masses. First of all, we show that the source apparent and intrinsic positions differ. Light deflection also causes that, in particular circumstances, the source can have multiple images. We discuss how gravitational lensing affects the observed image shapes, distorting them, and changing their apparent size. Finally, we show that gravitational lensing delays the arrival of photons emitted by distant sources at the observer detector. All these consequences of light deflection determine the observable effects of gravitational lensing.

3.1 Lens Equation

Intrinsic and Apparent Source Position
This section seeks a relationship between observed and intrinsic positions of a source in a gravitational lensing event. In the absence of the lens, the light emitted by a distant source reaches an observer, who sees the source at a particular position on the sky, $\vec{\beta}$ (in angular units). This position is the *intrinsic* position of the source. Instead, when the gravitational lens deflects photons, the observer collects them from a different direction, $\vec{\theta}$, which corresponds to the *apparent* (or *observed*) *image* position of the source.

In Fig. 3.1, we sketch a typical gravitational lens system. We place a mass at redshift z_L, corresponding to an angular diameter distance D_L. This lens deflects the light rays coming from a source at redshift z_S (or angular distance D_S). At the bottom of the diagram, an observer collects the photons from the distant source. The angular diameter distance between the lens and the source is D_{LS}.

© Springer Nature Switzerland AG 2021
M. Meneghetti, *Introduction to Gravitational Lensing*, Lecture Notes
in Physics 956, https://doi.org/10.1007/978-3-030-73582-1_3

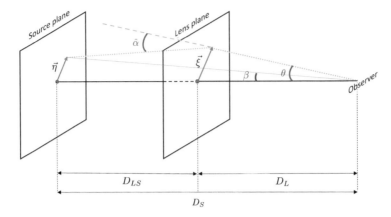

Fig. 3.1   Sketch of a typical gravitational lensing system

Remark 3.1. The angular diameter distance D_A is defined as the ratio of an object's physical transverse size to its angular size (in radians). Therefore, we can use it to convert angular separations in the sky to physical separations on the sources' plane.

This distance does not increase indefinitely with redshift, but it peaks at $z \sim 1$ and then it turns over. Due to the expansion of the universe, the angular diameter distance between z_1 and z_2 (with $z_2 > z_1$) is not the difference between the two individual angular diameter distances:

$$D_A(z_1, z_2) \neq D_A(z_2) - D_A(z_1) \tag{3.1}$$

except for those situations where the universe's expansion can be neglected (i.e. for lenses and sources in our own galaxy). More in-depth discussions can be found in Appendix 9 and in several cosmology books (see e.g. Weinberg 1972).

Thin Screen Approximation

If the physical size of the lens is small compared to the distances D_L, D_{LS}, and D_S, the extension of the lens along the line-of-sight can be neglected in the calculation of the light deflection. We can assume that this occurs on a plane, called the *lens plane*.

Remark 3.2. Given that the source's apparent position, or image position, originates on this plane, the lens plane is often referred to as the *image plane*.

Similarly, we can assume that all photons emitted by the source originate from the same distance D_S, meaning that the source lies on a *source plane*. The approximation of the lens and the source to planar distributions of mass and light is called *thin screen approximation*.

Relating the Intrinsic and Apparent Positions of the Source

We first define an optical axis, indicated in Fig. 3.1 by the dashed line, perpendicular to the lens and source planes and passing through the observer. Then we measure the angular positions on the lens and the source planes relative to this reference direction.

Consider a source at the intrinsic angular position $\vec{\beta}$, which lies on the source plane at a distance $\vec{\eta} = \vec{\beta} D_S$ from the optical axis. The source emits photons (we may now use the term "light rays") that impact the lens plane at $\vec{\xi} = \vec{\theta} D_L$, are deflected by the angle $\hat{\vec{\alpha}}$, and finally reach the observer. The amplitude of the deflection is given by Eq. (2.36).

Due to the deflection, the observer receives the light coming from the source as if it was emitted at the apparent angular position $\vec{\theta}$. We have used vectors to identify the source and image positions on the corresponding planes, either in angular and physical units.

If $\vec{\theta}$, $\vec{\beta}$, and $\hat{\vec{\alpha}}$ are small, the true position of the source and its observed position on the sky are related by a very simple relation, which can be readily obtained from the diagram in Fig. 3.1. This relation is called the *lens equation* and is written as

$$\vec{\theta} D_S = \vec{\beta} D_S + \hat{\vec{\alpha}} D_{LS}, \tag{3.2}$$

where D_{LS} is the angular diameter distance between lens and source.

Defining the reduced deflection angle

$$\vec{\alpha}(\vec{\theta}) \equiv \frac{D_{LS}}{D_S} \hat{\vec{\alpha}}(\vec{\theta}), \tag{3.3}$$

from Eq. (3.2), we obtain

$$\vec{\beta} = \vec{\theta} - \vec{\alpha}(\vec{\theta}). \tag{3.4}$$

It is very common and useful to write Eq. (3.4) in dimensionless form. This can be done by defining a length scale ξ_0 on the lens plane and a corresponding length scale $\eta_0 = \xi_0 D_S / D_L$ on the source plane. Then, we define the dimensionless vectors

$$\vec{x} \equiv \frac{\vec{\xi}}{\xi_0} \; ; \; \vec{y} \equiv \frac{\vec{\eta}}{\eta_0}, \tag{3.5}$$

as well as the scaled deflection angle

$$\vec{\alpha}(\vec{x}) = \frac{D_L D_{LS}}{\xi_0 D_S} \hat{\vec{\alpha}}(\xi_0 \vec{x}).\tag{3.6}$$

Carrying out some substitutions, Eq. (3.4) can finally be written as

$$\vec{y} = \vec{x} - \vec{\alpha}(\vec{x}).\tag{3.7}$$

Solving the Lens Equation
From Eq. 3.4, it is evident that knowing the source intrinsic position and the lens deflection angle field $\vec{\alpha}(\vec{\theta})$, we can find the positions of the image(s) by solving the lens equation for $\vec{\theta}$. As will be discussed later on, we can only achieve this analytically for very simple lens mass distributions. Indeed, the equation is typically highly non-linear. When multiple solutions exist, the source is lensed into *multiple images*.

When observing a lens system, the source's intrinsic position is unknown, while we can measure the position of its images. Then we can recover the intrinsic source position by assuming a model for the mass distribution of the lens, i.e. by solving the lens equation for $\vec{\beta}$. This task is much easier to accomplish because the lens equation is linear in $\vec{\beta}$: for each image, there is a unique solution. Thus, if we can identify multiple images of the same source, and the lens mass model is correct, we should find the lens equation's same solution for all images.

3.2   Lensing Potential

An extended distribution of matter is characterized by its *effective lensing potential*, obtained by projecting the three-dimensional Newtonian potential on the lens plane and by properly re-scaling it:

$$\hat{\Psi}(\vec{\theta}) = \frac{D_{LS}}{D_L D_S} \frac{2}{c^2} \int \Phi(D_L \vec{\theta}, z) dz .\tag{3.8}$$

The lensing potential satisfies two important properties:

1. **the gradient of $\hat{\Psi}$ is the reduced deflection angle**:

$$\vec{\nabla}_{\theta} \hat{\Psi}(\vec{\theta}) = \vec{\alpha}(\vec{\theta}) .\tag{3.9}$$

Indeed, by taking the gradient of the lensing potential we obtain

$$
\vec{\nabla}_{\theta}\hat{\Psi}(\vec{\theta}) = D_{L}\vec{\nabla}_{\perp}\hat{\Psi} = \vec{\nabla}_{\perp}\left(\frac{D_{LS}}{D_{S}}\frac{2}{c^{2}}\int \hat{\Phi}(\vec{\theta}, z)\mathrm{d}z\right)
$$

$$
= \frac{D_{LS}}{D_{S}}\frac{2}{c^{2}}\int \vec{\nabla}_{\perp}\Phi(\vec{\theta}, z)\mathrm{d}z
$$

$$
= \vec{\alpha}(\vec{\theta}). \tag{3.10}
$$

Note that, using the dimensionless notation,

$$
\vec{\nabla}_{x} = \frac{\xi_{0}}{D_{L}}\vec{\nabla}_{\theta} . \tag{3.11}
$$

We can see that

$$
\vec{\nabla}_{x}\hat{\Psi}(\vec{\theta}) = \frac{\xi_{0}}{D_{L}}\vec{\nabla}_{\theta}\hat{\Psi}(\vec{\theta}) = \frac{\xi_{0}}{D_{L}}\vec{\alpha}(\vec{\theta}) . \tag{3.12}
$$

By multiplying both sides of this equation by D_{L}^{2}/ξ_{0}^{2}, we obtain

$$
\frac{D_{L}^{2}}{\xi_{0}^{2}}\vec{\nabla}_{x}\hat{\Psi} = \frac{D_{L}}{\xi_{0}}\vec{\alpha} . \tag{3.13}
$$

This allows us to introduce the dimensionless counterpart of $\hat{\Psi}$:

$$
\Psi = \frac{D_{L}^{2}}{\xi_{0}^{2}}\hat{\Psi}. \tag{3.14}
$$

Substituting Eq. 3.14 into Eq. 3.13, we see that

$$
\vec{\nabla}_{x}\Psi(\vec{x}) = \vec{\alpha}(\vec{x}) . \tag{3.15}
$$

2. **the Laplacian of $\hat{\Psi}$ is twice the *convergence* κ:**

$$
\triangle_{\theta}\hat{\Psi}(\vec{\theta}) = 2\kappa(\vec{\theta}). \tag{3.16}
$$

The *convergence* is defined as a dimensionless surface density

$$\kappa(\vec{\theta}) \equiv \frac{\Sigma(\vec{\theta})}{\Sigma_{\mathrm{cr}}} \text{ with } \Sigma_{\mathrm{cr}} = \frac{c^2}{4\pi G} \frac{D_{\mathrm{S}}}{D_{\mathrm{L}} D_{\mathrm{LS}}}, \tag{3.17}$$

where Σ_{cr} is called the *critical surface density*, a quantity which characterizes the lens system and which is a function of the angular diameter distances of lens and source.

Equation 3.16 is derived from the Poisson equation,

$$\triangle \Phi = 4\pi G \rho . \tag{3.18}$$

The surface-mass density is

$$\Sigma(\vec{\theta}) = \frac{1}{4\pi G} \int_{-\infty}^{+\infty} \triangle \Phi \mathrm{d}z \tag{3.19}$$

and

$$\kappa(\vec{\theta}) = \frac{1}{c^2} \frac{D_{\mathrm{L}} D_{\mathrm{LS}}}{D_{\mathrm{S}}} \int_{-\infty}^{+\infty} \triangle \Phi \mathrm{d}z . \tag{3.20}$$

Let us now introduce a two-dimensional Laplacian

$$\triangle_\theta = \frac{\partial^2}{\partial \theta_1^2} + \frac{\partial^2}{\partial \theta_2^2} = D_{\mathrm{L}}^2 \left(\frac{\partial^2}{\partial \xi_1^2} + \frac{\partial^2}{\partial \xi_2^2} \right) = D_{\mathrm{L}}^2 \left(\triangle - \frac{\partial^2}{\partial z^2} \right) , \tag{3.21}$$

which gives

$$\triangle \Phi = \frac{1}{D_{\mathrm{L}}^2} \triangle_\theta \Phi + \frac{\partial^2 \Phi}{\partial z^2} . \tag{3.22}$$

Inserting Eq. 3.22 into Eq. 3.20, we obtain

$$\kappa(\vec{\theta}) = \frac{1}{c^2} \frac{D_{\mathrm{LS}}}{D_{\mathrm{S}} D_{\mathrm{L}}} \left[\triangle_\theta \int_{-\infty}^{+\infty} \Phi \mathrm{d}z + D_{\mathrm{L}}^2 \int_{-\infty}^{+\infty} \frac{\partial^2 \Phi}{\partial z^2} \mathrm{d}z \right] . \tag{3.23}$$

If the lens is gravitationally bound, $\partial \Phi / \partial z = 0$ at its boundaries and the second term on the right hand side vanishes. From Eqs. 3.8 and 3.14, we find

$$\kappa(\theta) = \frac{1}{2} \triangle_\theta \hat{\Psi} = \frac{1}{2} \frac{\xi_0^2}{D_L^2} \triangle_\theta \Psi \ . \tag{3.24}$$

Since

$$\triangle_\theta = D_L^2 \triangle_\xi = \frac{D_L^2}{\xi_0^2} \triangle_x \ , \tag{3.25}$$

using dimensionless quantities, Eq. 3.24 reads

$$\kappa(\vec{x}) = \frac{1}{2} \triangle_x \Psi(\vec{x}). \tag{3.26}$$

Integrating Eq. (3.16), the effective lensing potential can be written in terms of the convergence as

$$\Psi(\vec{x}) = \frac{1}{\pi} \int_{\mathbf{R}^2} \kappa(\vec{x}') \ln |\vec{x} - \vec{x}'| \mathrm{d}^2 x', \tag{3.27}$$

from which we obtain that the scaled deflection angle is

$$\vec{\alpha}(\vec{x}) = \frac{1}{\pi} \int_{\mathbf{R}^2} \mathrm{d}^2 x' \kappa(\vec{x}') \frac{\vec{x} - \vec{x}'}{|\vec{x} - \vec{x}'|}. \tag{3.28}$$

3.3 First Order Lens Mapping

One of the main consequences of gravitational lensing is image distortion. This distortion is particularly evident when the source has an extended size. For example, background galaxies can appear as very long arcs when lensed by galaxy clusters or other galaxies.

The distortion arises because light bundles are deflected differentially. Ideally, we can determine the shape of the images by solving the lens equation for all the extended source points. In particular, if the source is much smaller than the angular scale on which the lens deflection angle field changes, the relation between source and image positions can locally be linearized.

This situation is sketched in Fig. 3.2. Let us consider a point on the lens (or image) plane at position $\vec{\theta}_0$, where the deflection angle is $\vec{\alpha}_0$. If the deflection angle satisfies the above conditions, at the nearby location $\vec{\theta} = \vec{\theta}_0 + d\vec{\theta}$, the deflection

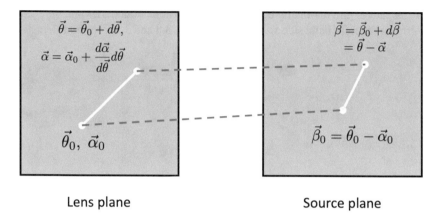

Lens plane   Source plane

Fig. 3.2 Linear mapping between the lens and the source plane, assuming a slowly varying deflection angle

will be

$$\vec{\alpha} \simeq \vec{\alpha}_0 + \frac{d\vec{\alpha}}{d\vec{\theta}} d\vec{\theta} \ . \tag{3.29}$$

Using the lens equation, the points $\vec{\theta}_0$ and $\vec{\theta}$ are mapped on the points $\vec{\beta}_0$ and $\vec{\beta} = \vec{\beta}_0 + d\vec{\beta}$ onto the source plane. Through this mapping, the vector $(\vec{\beta} - \vec{\beta}_0)$ is given by

$$(\vec{\beta} - \vec{\beta}_0) = \left(I - \frac{d\vec{\alpha}}{d\vec{\theta}}\right)(\vec{\theta} - \vec{\theta}_0) \ . \tag{3.30}$$

In other words, the distortion of images can be described by the Jacobian matrix

$$A \equiv \frac{\partial\vec{\beta}}{\partial\vec{\theta}} = \left(\delta_{ij} - \frac{\partial\alpha_i(\vec{\theta})}{\partial\theta_j}\right) = \left(\delta_{ij} - \frac{\partial^2\hat{\Psi}(\vec{\theta})}{\partial\theta_i\,\partial\theta_j}\right), \tag{3.31}$$

where θ_i indicates the i-component of $\vec{\theta}$ on the lens plane.

Equation (3.31) shows that we can write the elements of the Jacobian matrix as combinations of the second derivatives of the lensing potential. For brevity, we will use the shorthand notation

$$\frac{\partial^2\hat{\Psi}(\vec{\theta})}{\partial\theta_i\,\partial\theta_j} \equiv \hat{\Psi}_{ij}. \tag{3.32}$$

We can now split off the isotropic part from the Jacobian, to obtain its trace-less part:

$$\left(A - \frac{1}{2}\mathrm{tr}A \cdot I\right)_{ij} = \delta_{ij} - \hat{\Psi}_{ij} - \frac{1}{2}(1 - \hat{\Psi}_{11} + 1 - \hat{\Psi}_{22})\delta_{ij}$$

$$= -\hat{\Psi}_{ij} + \frac{1}{2}(\hat{\Psi}_{11} + \hat{\Psi}_{22})\delta_{ij}$$

$$= \begin{pmatrix} -\frac{1}{2}(\hat{\Psi}_{11} - \hat{\Psi}_{22}) & -\hat{\Psi}_{12} \\ -\hat{\Psi}_{12} & \frac{1}{2}(\hat{\Psi}_{11} - \hat{\Psi}_{22}) \end{pmatrix} . \qquad (3.33)$$

This allows us to define the *shear* tensor

$$\Gamma = \begin{pmatrix} \gamma_1 & \gamma_2 \\ \gamma_2 & -\gamma_1 \end{pmatrix} , \qquad (3.34)$$

whose components are

$$\gamma_1 = \frac{1}{2}(\hat{\Psi}_{11} - \hat{\Psi}_{22}) \qquad (3.35)$$

$$\gamma_2 = \hat{\Psi}_{12} = \hat{\Psi}_{21} . \qquad (3.36)$$

The shear is manifestly a symmetric tensor. It quantifies the projection of the gravitational tidal field (the gradient of the gravitational force), which describes distortions of background sources.

The eigenvalues of the shear tensor are

$$\pm\sqrt{\gamma_1^2 + \gamma_2^2} = \pm\gamma , \qquad (3.37)$$

where γ is often referred to as the shear modulus. Thus, there exists a rotation $R(\varphi)$ such that the shear tensor (and therefore the Jacobian) can be written in a diagonal form. We remind that tensors transform as

$$A \rightarrow A' = R(\varphi)^T A R(\varphi) , \qquad (3.38)$$

where T indicates the transposed matrix. This shows that the shear components transform under rotations as

$$\gamma_1 \rightarrow \gamma_1' = \gamma_1 \cos(2\varphi) + \gamma_2 \sin(2\varphi)$$

$$\gamma_2 \rightarrow \gamma_2' = -\gamma_1 \sin(2\varphi) + \gamma_2 \cos(2\varphi) . \qquad (3.39)$$

Since the shear components are invariant under rotations by multiples of $\varphi = \pi$, they form a spin-2 tensor. Vectors are invariant by rotations by multiples of $\varphi = 2\pi$, instead. Thus, they have spin 1.

Equations 3.38 and 3.39 give

$$\Gamma' = \begin{pmatrix} \gamma & 0 \\ 0 & -\gamma \end{pmatrix} = R(\varphi)^T \Gamma R(\varphi)$$

$$= R(\varphi)^T \begin{pmatrix} \gamma_1 & \gamma_2 \\ \gamma_2 & -\gamma_1 \end{pmatrix} R(\varphi) , \tag{3.40}$$

and

$$\gamma = \gamma_1 \cos 2\varphi + \gamma_2 \sin 2\varphi$$

$$0 = -\gamma_1 \sin 2\varphi + \gamma_2 \cos 2\varphi . \tag{3.41}$$

From these two last equations, it is easy to see that the components of the shear can be written in terms of the angle φ as

$$\gamma_1 = \gamma \cos 2\varphi$$

$$\gamma_2 = \gamma \sin 2\varphi . \tag{3.42}$$

Thus, we can write the shear tensor as

$$\Gamma = \begin{pmatrix} \gamma_1 & \gamma_2 \\ \gamma_2 & -\gamma_1 \end{pmatrix} = \gamma \begin{pmatrix} \cos 2\varphi & \sin 2\varphi \\ \sin 2\varphi & -\cos 2\varphi \end{pmatrix} . \tag{3.43}$$

Remark 3.3. The factor 2 in front of the angle φ reminds that the shear components are elements of a 2×2 tensor and not of a vector. Often, in the literature, the shear is indicated as a pseudo-vector, $\vec{\gamma} = (\gamma_1, \gamma_2)$, which can be very misleading.

Remark 3.4. The angle φ denotes the direction of the eigenvectors of the shear tensor with eigenvalue γ with respect to the axis θ_1. Thus, any vector $\vec{v}_\gamma$ on the lens plane which is an eigenvector of Γ with eigenvalue γ is mapped onto a parallel vector $\vec{v}_\gamma^s = \gamma \vec{v}_\gamma$ on the source plane. Vice versa, a vector $\vec{v}_\gamma^s$ on the source plane parallel to $\vec{v}_\gamma$ is mapped via the inverse transformation Γ^{-1} onto the vector $\vec{v}_\gamma = \gamma^{-1} \vec{v}_\gamma^s$.

Similarly, a vector $\vec{u}_\gamma$ on the lens plane perpendicular to $\vec{v}_\gamma$ is an eigenvector of Γ with eigenvalue $-\gamma$. It is mapped via the shear tensor onto the vector $\vec{u}_\gamma^s = -\gamma \vec{u}_\gamma$ on the source plane. Obviously, the inverse mapping is given by $\vec{u}_\gamma = -\gamma^{-1} \vec{u}_\gamma^s$.

The remainder of the Jacobian is

$$\left(\frac{1}{2}\mathrm{tr}A \cdot I\right)_{ij} = \left[1 - \frac{1}{2}(\hat{\Psi}_{11} + \hat{\Psi}_{22})\right]\delta_{ij} \tag{3.44}$$

$$= \left(1 - \frac{1}{2}\triangle\hat{\Psi}\right)\delta_{ij} = (1 - \kappa)\delta_{ij} . \tag{3.45}$$

Thus, we can write the Jacobian matrix in terms of the convergence and of the shear as

$$A = \begin{pmatrix} 1 - \kappa - \gamma_1 & -\gamma_2 \\ -\gamma_2 & 1 - \kappa + \gamma_1 \end{pmatrix}$$

$$= (1 - \kappa)\begin{pmatrix} 1 & 0 \\ 0 & 1 \end{pmatrix} - \gamma\begin{pmatrix} \cos 2\varphi & \sin 2\varphi \\ \sin 2\varphi & -\cos 2\varphi \end{pmatrix} . \tag{3.46}$$

The last equation explains the meaning of both convergence and shear. The convergence describes an isotropic transformation, i.e. the images are only re-scaled by a constant factor $1/(1-\kappa)$ in all directions. On the other hand, the shear stretches the source's intrinsic shape along a particular direction, corresponding to the angle φ, shrinking it in the perpendicular direction.

3.3.1 First Order Lensing of a Circular Source

The eigenvalues of the Jacobian matrix are

$$\lambda_t = 1 - \kappa - \gamma \tag{3.47}$$

$$\lambda_r = 1 - \kappa + \gamma . \tag{3.48}$$

Let us consider the reference frame where the Jacobian is diagonal (i.e. rotated by the angle φ, with the θ_1 axis pointing in the same direction of the eigenvectors of Γ with eigenvalue γ). Then,

$$A = \begin{pmatrix} 1 - \kappa - \gamma & 0 \\ 0 & 1 - \kappa + \gamma \end{pmatrix} . \tag{3.49}$$

Consider a circular source, centered in $\vec{\beta}_0$ and with radius r. Then, the points $(\vec{\beta} - \vec{\beta}_0) = (\beta_1 - \beta_{0,1}, \beta_2 - \beta_{0,2})$ on its contour satisfy the equation $(\beta_1 - \beta_{0,1})^2 + (\beta_2 - \beta_{0,2})^2 = r^2$. The source center $\vec{\beta}_0$ corresponds to the center of the image $\vec{\theta}_0$

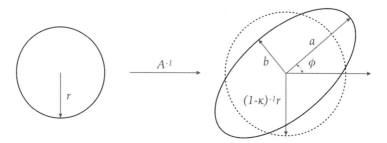

Fig. 3.3 Distortion effects due to convergence and shear on a circular source

via the lens equation. Let us assume $\vec{\beta}_0 = \vec{\theta}_0 = (0, 0)$.[1] Then, the components β_i of $\vec{\beta}$ can be written as

$$\beta_i \simeq \sum_j A_{ij}\theta_j = \sum_j \frac{\partial \beta_i}{\partial \theta_j}\theta_j \, , \tag{3.50}$$

where θ_j are the components of $\vec{\theta}$ and $(i, j) \in (1, 2)$.

Using Eq. 3.49, we obtain that the points on the image contour satisfy the equation

$$r^2 = \beta_1^2 + \beta_2^2 = (1 - \kappa - \gamma)^2\theta_1^2 + (1 - \kappa + \gamma)^2\theta_2^2 \, , \tag{3.51}$$

which is the equation of an ellipse on the lens plane. Thus, a circular source is mapped onto an ellipse, when κ and γ are both non-zero, as shown in Fig. 3.3. This result holds in all cases when the first order lensing approximation applies, i.e. if the source's size is small enough compared to the typical length scale over which the lens deflection field varies significantly.

The semi-major and -minor axes of the ellipse are

$$a = \frac{r}{1 - \kappa - \gamma} \, , \, b = \frac{r}{1 - \kappa + \gamma}. \tag{3.52}$$

The ellipticity is given by

$$\epsilon = \frac{a - b}{a + b} = \frac{\gamma}{1 - \kappa} \, , \tag{3.53}$$

---

[1]This is equivalent to shift the reference frames on the source and on the lens planes such that origins coincide with the source and the image positions.

a result that will be very important when we will discuss the weak lensing regime in Chaps. 6 and 7. Obviously, the ellipse reduces to a circle if $\gamma = 0$. The quantity

$$g = \frac{\gamma}{1 - \kappa} \tag{3.54}$$

is called *reduced shear*.

As said in the previous section, in an arbitrary reference frame, the ellipse will have its axes aligned with the shear tensor's eigenvectors. Note that:

- if $\gamma_1 > 0$ and $\gamma_2=0$, then the major axis of the ellipse will be along the θ_1 axis;
- if $\gamma_1 = 0$ and $\gamma_2>0$, then the major axis of the ellipse will form an angle $\pi/4$ with the θ_1 axis;
- if $\gamma_1 < 0$ and $\gamma_2=0$, then the major axis of the ellipse will be perpendicular to the θ_1 axis;
- if $\gamma_1 = 0$ and $\gamma_2<0$, then the major axis of the ellipse will form an angle $3\pi/4$ with the θ_1 axis.

Figure 3.4 shows the ellipse orientation for different values of the two shear components.

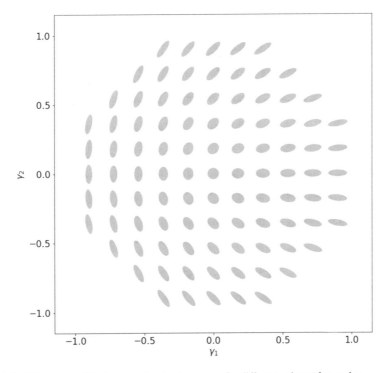

Fig. 3.4 Orientation of the images of a circular source for different values of γ_1 and γ_2

3.4   Magnification

An important consequence of the lensing distortion is magnification. Through the lens equation, the solid angle element $\delta\beta^2$ (or equivalently the surface element $\delta\eta^2$ or δy^2) is mapped onto the solid angle $\delta\theta^2$ (or on the surface element $\delta\xi^2$ or δx^2). Because of the Liouville theorem, in the absence of photons' emission and absorption, the source surface brightness is conserved despite light deflection. Thus, the change of the solid angle under which the source is observed implies that the flux received is magnified (or demagnified).

Given Eq. (3.31), the *magnification* is given by the inverse of the determinant of the Jacobian matrix. For this reason, the matrix $M = A^{-1}$ is called the *magnification tensor*. We therefore define

$$\mu \equiv \det M = \frac{1}{\det A} = \frac{1}{(1-\kappa)^2 - \gamma^2}. \tag{3.55}$$

The eigenvalues of the magnification tensor (or the inverse of the eigenvalues of the Jacobian matrix) measure the amplification in the direction of the eigenvectors of the shear tensor. For an axially symmetric lens, these are tangentially and radially oriented with respect to the lens iso-surface density contours. Thus, the quantities

$$\mu_t = \frac{1}{\lambda_t} = \frac{1}{1-\kappa-\gamma} \tag{3.56}$$

$$\mu_r = \frac{1}{\lambda_r} = \frac{1}{1-\kappa+\gamma} \tag{3.57}$$

are often called the *tangential* and *radial* magnification factors.

The magnification is ideally infinite where $\lambda_t = 0$ and $\lambda_r = 0$. These two conditions define two curves in the lens plane, called the *tangential* and the *radial* *critical lines*.

3.5   Lensing to the Second Order

We extend now the lens equation 3.50 including second order terms and we discuss how a circular source is distorted. Then, the points on its contour have coordinates

$$\beta_i \simeq \sum_j \frac{\partial\beta_i}{\partial\theta_j}\theta_j + \frac{1}{2}\sum_j\sum_k \frac{\partial^2\beta_i}{\partial\theta_j\partial\theta_k}\theta_j\theta_k . \tag{3.58}$$

We can see that the second order term can be described using the tensor

$$D_{ijk} = \frac{\partial^2\beta_i}{\partial\theta_j\partial\theta_k} = \frac{\partial A_{ij}}{\partial\theta_k} . \tag{3.59}$$

Introducing this tensor, Eq. 3.58 reads

$$\beta_i \simeq \sum_j A_{ij}\theta_j + \frac{1}{2}\sum_j \sum_k D_{ijk}\theta_j\theta_k. \tag{3.60}$$

By simple algebra, we can show that

$$D_{ij1} = \begin{pmatrix} -2\gamma_{1,1} - \gamma_{2,2} & -\gamma_{2,1} \\ -\gamma_{2,1} & -\gamma_{2,2} \end{pmatrix}, \tag{3.61}$$

and

$$D_{ij2} = \begin{pmatrix} -\gamma_{2,1} & -\gamma_{2,2} \\ -\gamma_{2,2} & 2\gamma_{1,2} - \gamma_{2,1} \end{pmatrix}. \tag{3.62}$$

Thus, second order lensing effects can be expressed in terms of the derivatives of the shear (or in terms of the third derivatives of the potential).

3.5.1   Complex Notation

It is quite useful to use complex notation to map vectors or pseudo-vectors on the complex plane. In this case, we can also use complex differential operators to write down some relations between the lensing quantities concisely.

In complex notation, any vector or pseudo-vector $\vec{v} = (v_1, v_2)$ is written as

$$v = v_1 + iv_2. \tag{3.63}$$

Similarly we can define the complex deflection angle $\alpha = \alpha_1 + i\alpha_2$ and the complex shear $\gamma = \gamma_1 + i\gamma_2$.

It is also possible to define some complex differential operators, namely

$$\partial = \partial_1 + i\partial_2 \tag{3.64}$$

and

$$\partial^\dagger = \partial_1 - i\partial_2. \tag{3.65}$$

Using this formalism, we can easily see that

$$\partial\hat{\Psi} = \partial_1\hat{\Psi} + i\partial_2\hat{\Psi} = \alpha_1 + i\alpha_2 = \alpha. \tag{3.66}$$

Moreover

$$\partial^\dagger \partial = \partial_1^2 + \partial_2^2 = \triangle \,. \tag{3.67}$$

Thus,

$$\partial^\dagger \partial \hat{\Psi} = \triangle \hat{\Psi} = 2\kappa \,. \tag{3.68}$$

Note that while $\hat{\Psi}$ is a spin-0 scalar field, the application of the ∂ operator gives the deflection angle, i.e. a spin-1 vector field. On the contrary, the $\partial^\dagger$ operator applied to the deflection field provides another spin-0 scalar field (the convergence). Therefore, the ∂ and $\partial^\dagger$ operators are spin raising and lowering operators.

By applying twice the raising operator, we obtain

$$\frac{1}{2}\partial\partial\hat{\Psi} = \frac{1}{2}\partial\alpha = \gamma \ : \tag{3.69}$$

the shear field is indeed a spin-2 tensor field, which is invariant for rotations by angles that are multiples of π.

Note also that

$$\partial^{-1}\partial^\dagger\gamma = \frac{1}{2}\partial^{-1}\partial^\dagger\partial\partial\hat{\Psi} = \partial^\dagger\partial\hat{\Psi} = \kappa. \tag{3.70}$$

We can use the raising and lowering operators to define

$$F = \frac{1}{2}\partial\partial^\dagger\partial\hat{\Psi} = \partial\kappa \tag{3.71}$$

$$G = \frac{1}{2}\partial\partial\partial\hat{\Psi} = \partial\gamma. \tag{3.72}$$

After some math, it can be shown that

$$F = F_1 + iF_2 = (\gamma_{1,1} + \gamma_{2,2}) + i(\gamma_{2,1} - \gamma_{1,2}) \tag{3.73}$$

and

$$G = G_1 + iG_2 = (\gamma_{1,1} - \gamma_{2,2}) + i(\gamma_{2,1} + \gamma_{1,2}) \,. \tag{3.74}$$

The quantities F and G are called *first and second flexion*, respectively (Goldberg and Natarajan 2002). It is easy to show that D_{ijk} can be written in terms of F and G. Thus, they describe second order distortions of the images of lensed sources. Note that F is a spin-1 vector field. Indeed, it is

$$\vec{F} = \vec{\nabla}\kappa \,. \tag{3.75}$$

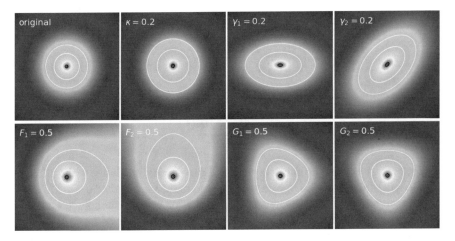

Fig. 3.5 First and second order distortions on the image of a circular source. The unlensed source is shown in the top left panel. The convergence simply changes the size (second-left upper panel). While the shear deforms the image such that it becomes elliptical (third and fourth panels on the upper row), the first and the second flexion introduce curvature and other distortions (panels on the bottom)

Thus, it describes transformations that are invariant under rotations by angles that are multiples of 2π. For this reason, F stretches the images along one particular direction, introducing asymmetries in their shape. On the contrary, G is a spin-3 tensor field. The transformations described by G are invariant under rotations by angles that are multiples of $2\pi/3$. This is manifested in the "triangular" pattern in the image shapes, as shown in Fig. 3.5.

3.6 Time Delay Surface

3.6.1 Gravitational and Geometrical Time Delays

The deflection of light rays causes a delay in the travel-time of light between the source and the observer. This time delay has two components:

$$t = t_{\text{grav}} + t_{\text{geom}}. \tag{3.76}$$

The first one is the *gravitational time delay*, also known as the Shapiro delay (Shapiro 1964). We can derive it by comparing the time required for light to travel through space–time with and without a perturbing gravitational potential, by assuming *same trajectories*.

Let $n = 1 - 2\Phi/c^2$ be the effective refractive index. We have that

$$t_{\text{grav}} = \int \frac{dz}{c'} - \int \frac{dz}{c} = \frac{1}{c} \int (n - 1)dz = -\frac{2}{c^3} \int \Phi dz. \tag{3.77}$$

Using the definition of the lensing potential, this can be written as

$$t_{\text{grav}} = -\frac{D_L D_S}{D_{LS}} \frac{1}{c} \hat{\psi} \;. \tag{3.78}$$

The second term in the time delay is called *geometrical* and is due to the different path length of the deflected light rays compared to the unperturbed ones. This time delay is proportional to the squared angular separation between the source's intrinsic position and the location of its image. This result can be derived from the metric, but it can be estimated also through a simple geometrical construction, shown in Fig. 3.6. The dashed line shows the path of a light-ray emitted by the source S, being deflected by an angle $\hat{\alpha}$, and reaching the observer at O. This light path should be compared to the solid line connecting S and O, which represents the path that the light would follow in absence of the lens. One can trace two circles centered on S and O, which are tangent at the point H along the line $\overline{SO}$. The extra-path of the light in presence of the lens is given by

$$\Delta l \approx \xi \hat{c} \;, \tag{3.79}$$

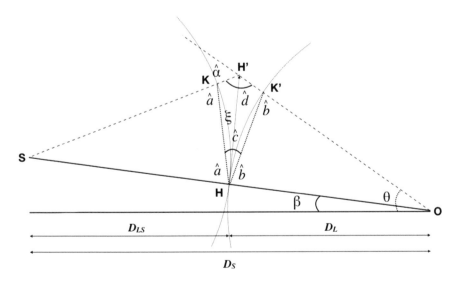

Fig. 3.6 Illustration of the geometrical time delay

using the notation in the figure. On the other hand, since the triangles SHK and OHK' are isosceles, it can be easily shown that the following relations hold

$$\hat{d} = \pi - \hat{\alpha} \, ,$$

$$\hat{a} + \hat{b} + \hat{c} = \pi \, ,$$

$$\hat{a} + \hat{b} = \hat{c} + \hat{d} \, . \tag{3.80}$$

Thus, the angle $\hat{c}$ can be written in terms of the deflection angle $\hat{\alpha}$ as

$$\hat{c} = \frac{\hat{\alpha}}{2} \, . \tag{3.81}$$

Inserting this result in Eq. 3.79, we obtain

$$\Delta l \approx \xi \frac{\hat{\alpha}}{2} = (\vec{\theta} - \vec{\beta}) \frac{D_{\mathrm{L}} D_{\mathrm{S}}}{D_{\mathrm{LS}}} \frac{\hat{\alpha}}{2} = \frac{1}{2} (\vec{\theta} - \vec{\beta})^2 \frac{D_{\mathrm{L}} D_{\mathrm{S}}}{D_{\mathrm{LS}}} \, , \tag{3.82}$$

and the corresponding *geometrical* time delay is

$$t_{\mathrm{geom}} = \frac{\Delta l}{c} \, . \tag{3.83}$$

Both the gravitational and the geometrical time delays occur at the lens position, thus they need to be multiplied by a factor $(1 + z_{\mathrm{L}})$ for accounting for the expansion of the universe. Then, the total time delay introduced by gravitational lensing at the position $\vec{\theta}$ on the lens plane is[2]

$$t(\vec{\theta}) = \frac{(1 + z_{\mathrm{L}})}{c} \frac{D_{\mathrm{L}} D_{\mathrm{S}}}{D_{\mathrm{LS}}} \left[\frac{1}{2} (\vec{\theta} - \vec{\beta})^2 - \hat{\Psi}(\vec{\theta}) \right]$$

$$= \frac{D_{\Delta t}}{c} \tau(\vec{\theta}). \tag{3.84}$$

The quantities

$$D_{\Delta t} = (1 + z_{\mathrm{L}}) \frac{D_{\mathrm{S}} D_{\mathrm{L}}}{D_{\mathrm{LS}}} \tag{3.85}$$

and

$$\tau(\vec{\theta}) = \frac{1}{2} (\vec{\theta} - \vec{\beta})^2 - \hat{\Psi}(\vec{\theta}) \, , \tag{3.86}$$

are often called *time delay distance* and *Fermat potential*, respectively.

---

[2]The dimensionless form of the time delay can be obtained by multiplying and dividing by the factor $(\xi_0/D_{\mathrm{L}})^2$.

3.6.2   Multiple Images and Magnification

Through the effective lensing potential, the lens equation can be written as

$$(\vec{\theta} - \vec{\beta}) - \nabla\hat{\Psi}(\vec{\theta}) = \nabla\left[\frac{1}{2}(\vec{\theta} - \vec{\beta})^2 - \hat{\Psi}(\vec{\theta})\right] = 0. \tag{3.87}$$

Equations (3.84) and (3.87) imply that images satisfy the Fermat Principle, $\nabla t(\vec{\theta}) = 0$. Images therefore are located at the stationary points of the time delay surface given by Eq. (3.84). The elements of the Hessian matrix of this surface are

$$T_{ij} = \frac{\partial^2 t(\vec{\theta})}{\partial\theta_i \partial\theta_j} \propto (\delta_{ij} - \hat{\Psi}_{ij}) = A_{ij}. \tag{3.88}$$

Given that the Hessian matrix of the time delay surface is proportional to the lensing Jacobian and that the magnification $\mu = \det A^{-1}$, it becomes clear that the curvature of the time delay surface at the image position is inversely proportional to the image magnification. In particular, a flat time delay surface implies infinite magnification, while a large curvature means that the magnification is small.

We can also measure the curvature along a specific direction on the time delay surface. This measurement will provide a way to quantify the image distortions. Therefore, the shape of the time delay surface near the stationary points will also provide hints on the images' shape.

We can distinguish between three types of images:

1. type I images arise at the minima of the time delay surface, where the eigenvalues of the Hessian matrix are both positive, hence $\det A > 0$ and $\mathrm{tr} A > 0$. Therefore, they have positive magnifications;
2. type II images arise at the saddle points of the time delay surface, where eigenvalues have opposite signs. Since $\det A < 0$, they have negative magnifications;
3. finally, type III images arise at the maxima of the time delay surface. Here, the eigenvalues are both negative, hence $\det A > 0$ and $\mathrm{tr} A < 0$. These images, therefore, have positive magnification.

Remark 3.5. Note that a negative magnification does not mean that the image is demagnified. The magnification's absolute value accounts for how much larger is the solid angle subtended by the image compared to the unlensed source. Thus, the image is demagnified if $|\mu| < 1$. Instead, the magnification sign is related to the *parity* of the image. The parity determines the orientation of the image relative to the unlensed source.

3.6.3 Examples

Equation 3.84 shows that we obtain the time delay surface by summing the two-dimensional function $\propto (\vec{\theta} - \vec{\beta})^2$, which is a paraboloid with a minimum at the position of the source, and the surface given by $S(\vec{\theta}) = -\hat{\Psi}(\vec{\theta})$. The lensing potential of a centrally concentrated lens has a minimum at the lens center. Thus, because of the negative sign, $S(\vec{\theta})$ has a maximum at the lens center, regardless of the source's position. For simplicity, we choose the reference frame, (θ_1, θ_2) such that the lens center is at $(0, 0)$, and we study how the shape of the time delay surface changes as a function of the position of the source, $\vec{\beta}$.

Axially Symmetric Lenses: One-Dimensional Case
We begin with an axially symmetric lens. Let us forget for the moment that the $t(\vec{\theta})$ is a surface and consider the azimuthal cut of the surface along an arbitrary direction passing through the center of the lens and the position of the source. As an example, we consider a lensing potential whose radial profile is

$$\hat{\Psi}(\theta) \propto \frac{1}{\sqrt{\theta^2 + \theta_c^2}} \, . \tag{3.89}$$

As we will see in Chap. 5, this potential is that of a cored isothermal lens. The core radius θ_c prevents the potential from diverging for $\theta \to 0$.

Figure 3.7 shows the geometrical and the gravitational components of the time delay and their combination for a few positions of the source relative to the lens, given by the vertical dashed lines. Given that β is the parameter defining the shape of the time delay surface, we use the notation $t(\theta) \equiv t(\theta, \beta)$. For $\beta = 0$ (upper left panel), the time delay function $t(\theta, 0)$ has a local maximum at $\theta_0 = 0$, and two minima on both sides of the origin, θ_- and θ_+ (of course, in this one-dimensional example there are no saddle points). Thus, the source at $\beta = 0$ has three images, forming at θ_0, θ_-, and θ_+, with $\theta_- = -\theta_+$.

We shift the source along the positive θ axis, and we notice that the symmetry of the time delay function breaks. In the upper-right panel, the maximum position is on the negative θ axis, $\theta_0 < 0$. One of the two minima, θ_+, moves away from the origin, following the source (i.e. along the positive θ axis), while the other moves toward the maximum. Note also that the difference between the time delays of the images θ_0 and θ_-, $t(\theta_0, \beta) - t(\theta_-, \beta)$, is smaller than in the previous case ($\beta = 0$), and that $t(\theta_+, \beta) < t(\theta_+, 0)$.

We focus now on the curvature of $t(\theta, \beta)$. Clearly, moving the source along the positive θ axis, $t(\theta, \beta)$ flattens off in between θ_- and θ_0. This flattening implies an increasing magnification along the direction connecting the two images. Therefore, these will be stretched toward each other.

Moving the source further away from the lens, we will reach the situation where the two images θ_0 and θ_- will merge. At that point, the function $t(\theta, \beta)$ will only have one minimum, corresponding to the image θ_+. The source will no longer have

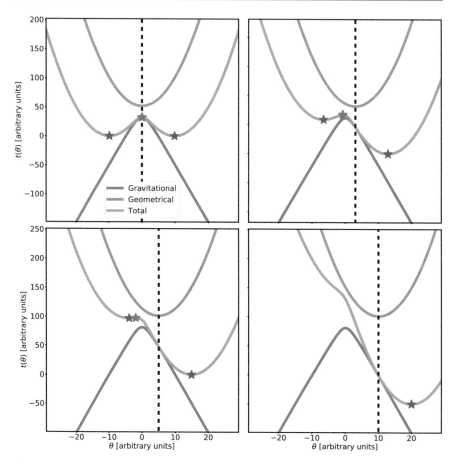

Fig. 3.7 One-dimensional time delay functions for a non-singular isothermal potential. Each panel corresponds to a different position of the source relative to the lens (dashed line). The two components of the time delay are shown separately and then combined. The positions of the images are indicated by stars

multiple images (bottom left panel). As $\beta_- \to \infty$, the image at θ_+ will follow the source at β (bottom right panel).

It is interesting to see how the results are affected by the shape of the lensing potential profile. The examples in Fig. 3.8 show the results obtained by setting the core radius to $\theta_c = 0$ (left four panels) and by using a potential in the form

$$\hat{\Psi} \propto \ln |\theta| . \tag{3.90}$$

This potential is that of the point-mass lens (right panels).

In both cases, the central singularity's presence makes the function $t(\theta, \beta)$ noncontinuously deformable. Consequently, for every choice of β, the central image

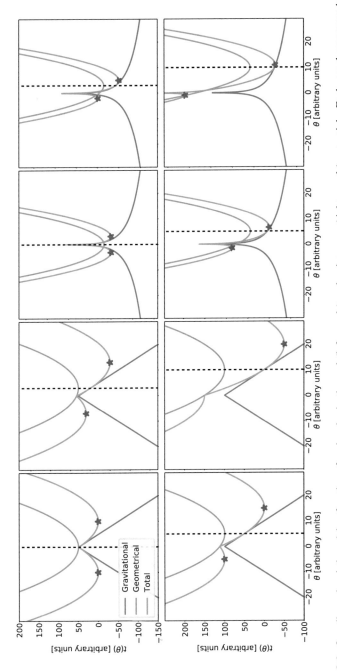

Fig. 3.8 One-dimensional time delay functions for singular isothermal (left panels) and point-mass (right panels) potentials. Each panel corresponds to a different position of the source relative to the lens (dashed line). The two components of the time delay are shown separately and then combined. The positions of the images are indicated by stars

θ_0 is a point of infinite curvature of the time delay surface. For this image, the magnification will be $\mu = 0$. Note that there are always two minima on opposite sides of the lens in the case of the point-mass lens. However, as $\beta \to \infty$, the curvature at $t(\theta_-, \beta)$ becomes increasingly higher, meaning that the image is increasingly demagnified.

Axially Symmetric Lenses: Two-Dimensional Case

The correct representation of the time delay is through a surface, not a one-dimensional function. In Fig. 3.9, we show the two-dimensional analog of Fig. 3.7, where the time delay surfaces correspond to several positions of the source along the θ_1 axis (i.e. $\beta_2 = 0$). We also show the projection of the surfaces on the (θ_1, θ_2) planes and the sections of the surfaces along the θ_1 axis ($\theta_2 = 0$).

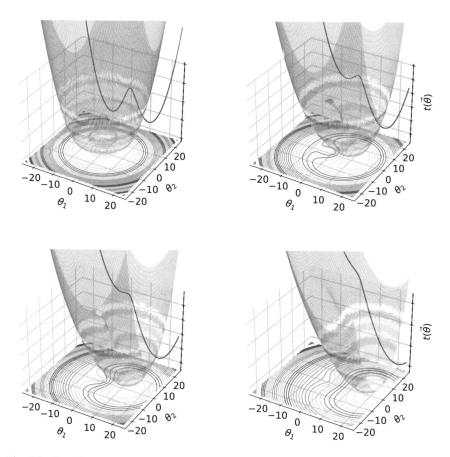

Fig. 3.9 Time delay surfaces for the same lens used in Fig. 3.7. Different panels correspond to different positions of the source relative to the lens. The surfaces are projected onto the plane (θ_1, θ_2) on the bottom of each panel. The blue curves on the vertical plane behind the surfaces show the sections of the surfaces along the θ_1 axis at $\theta_2 = 0$

The upper left panel shows the time delay surface $t(\vec{\theta}, 0)$. In this two-dimensional representation, we see that the minima $\vec{\theta}_-$ and $\vec{\theta}_+$ are part of a ring due to the lens's symmetry properties. We will see later that this ring is called *Einstein ring*. The central image, $\vec{\theta}_0$, still coincides with the center of the lens. Thus a point-source perfectly aligned with this axially symmetric (circular) lens is imaged into a point at the lens center and a ring surrounding the lens.

Along the ring, the curvature of the time delay surface is zero, meaning that the magnification diverges. This condition is met on the lens critical lines. More precisely, the Einstein ring corresponds to the tangential critical line of the lens. The ring is the image of a source at $\vec{\beta} = (0, 0)$. Thus, this point on the source plane is the lens tangential caustic.

Shifting the source relative to the lens breaks the symmetry of the time delay surface. In the upper-right panel of Fig. 3.9, we notice that the image at $\vec{\theta}_-$ is not a minimum, but a saddle point. As the source moves away from the lens, the saddle and the maximum points approach each other. The curvature of $t(\vec{\theta}, \vec{\beta})$ between these two stationary points becomes increasingly smaller in the radial direction. When the two images coincide, the time delay surface is radially flat (bottom left panel): the images $\vec{\theta}_-$ and $\vec{\theta}_0$ merge on another critical line: the radial critical line. The corresponding source position $\vec{\beta} = \vec{\beta}_{rad}$ marks the radial caustic position on the source plane. Due to the lens's symmetry, the radial caustic of an axially symmetric lens is a circle with radius β_{rad}.

For large distances between the lens and the source, only the image $\vec{\theta}_+$ exists, which corresponds to the minimum of the time delay surface.

As discussed earlier, adopting singular potentials, the time delay surface becomes noncontinuously deformable. As a consequence, there are no configurations where the time delay surface can become radially flat. Thus, these lenses do not have a radial critical line. In the case of the singular isothermal lens ($\theta_c = 0$), there is a particular distance of the source from the lens center, β_{cut}, for which $\vec{\theta}_- = 0$. This condition defines a circle on the source plane. This pseudo-caustic is called *cut*. The lens produces two images only if the source lays within the cut. Otherwise, there is only one image.

Elliptical Potentials

While axially symmetric lenses can produce up to three multiple images, depending on the relative position of the source relative to the lens, elliptical lenses behave differently. We can introduce ellipticity in the potentials considered in any of the previous examples by making the substitution

$$|\theta| \to \sqrt{\frac{\theta_1^2}{1 - \epsilon} + \theta_2^2 (1 - \epsilon)}. \tag{3.91}$$

The resulting lens has elliptical iso-potential contours with major axes oriented along the θ_2 axis.

Remark 3.6. Lenses with elliptical potentials are not elliptical lenses. Indeed, their convergence maps do not have elliptical iso-contours. Instead, these contours typically have dumbbell shapes. Introducing large ellipticity in the potential can even lead to nonphysical negative convergence. The lenses with elliptical potentials are *pseudo-elliptical.*

When combined with the paraboloid describing the usual geometrical time delay, the resulting surface can have up to five stationary points, depending on the potential radial profile and the relative positions of lens and source.

The examples displayed in Figs. 3.10 and 3.11 illustrate the case of a lens with cored isothermal potential and ellipticity $\epsilon = 0.4$. The upper panels of Fig. 3.10 show the maps of the lensing potential before and after introducing the ellipticity. The left and the right bottom panels show the lens critical lines and caustics,

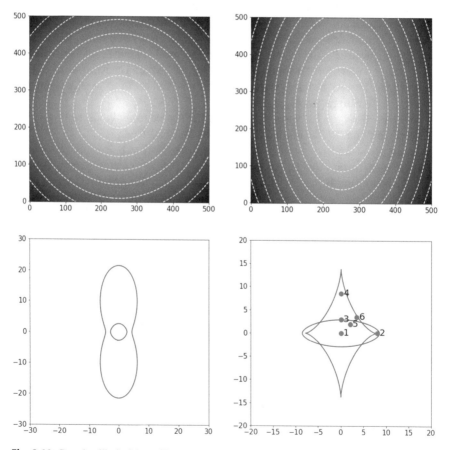

Fig. 3.10 Pseudo-elliptical lens. Upper panels: the lensing potential before and after adding an ellipticity $\epsilon = 0.4$. Bottom panels: critical lines (left) and caustics (right). The blue dots mark the positions of the sources used to generate the time delay surfaces shown in Fig. 3.11

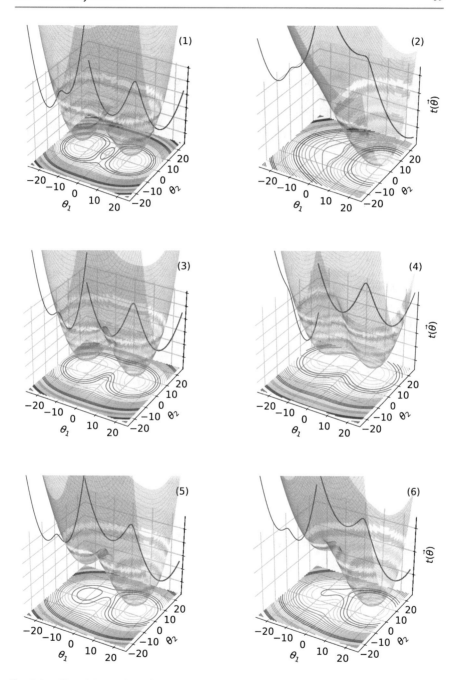

Fig. 3.11 Time delay surfaces for a pseudo-elliptical lens with cored isothermal profile. Different panels correspond to different positions of the source relative to the lens. The surfaces are projected onto the plane (θ_1, θ_2) on the bottom of each panel. The blue curves on the vertical plane behind the surfaces show the sections of the surfaces along the θ_1 axis at $\theta_2 = 0$

respectively. Figure 3.11 displays the time delay surfaces corresponding to the source positions marked by the blue dots in the bottom left panel of Fig. 3.10. In panel (1), the source is at $\vec{\beta} = 0$. Such a source has five multiple images: one maximum at the lens center, two minima along the θ_1 axis, and two saddle points along the θ_2 axis. Both the minima and saddle points are equidistant from the lens center. This image configuration is called *Einstein cross*.

In panel (2), we move the source along the positive θ_1 axis. One of the two minima moves in the same direction, while the maximum moves opposite the source. The saddle points also move opposite to the source approaching the maximum and the minimum. Eventually, for even larger separations between the lens and the source, the saddle points, the maximum, and the minimum merge, producing an image that is both radially and tangentially magnified. When this happens, the source is close to both the radial and the tangential caustics.

Panels (3) and (4) show the case of a source moving along the positive θ_2 axis instead. Two minima and one saddle point move in the same direction. The maximum and the other saddle point approach each other along the negative θ_2 axis. Because the time delay surface is flat between these two images, they are radially magnified. When the source crosses the radial caustic, they merge and disappear, as in panel (3). Moving the source farther, also the two minima and the remaining saddle point merge, forming a very elongated, tangentially magnified image. Such configurations, shown in panel (4), occur when the source is in the caustic cusp's proximity. The most extensive gravitational arcs form in this way.

Finally, in panels (5) and (6), we move the source along the diagonal of the (θ_1, θ_2) plane. As the source departs from the lens center, one of the minima follows the source. The other minimum and one of the saddle points approach each other. The maximum and the other saddle point merge, forming a radially elongated image opposite the source relative to the lens (panel 5). Again, this occurs when the source is on the radial caustic. Moving it farther, as in panel (6), we see that also one of the minima and the remaining saddle point merge, forming a tangentially elongated image when the source is on the tangential caustic.

3.6.4 General Considerations

Here are some other important properties of the continuously deformable time delay surface:

- the height difference at different images of the surface $t(\vec{\theta})$ gives the difference in arrival time between these images. This time delay can be measured if the source is variable and provides one way of potentially measuring the Hubble constant, as we will discuss in Chap. 6;
- in the absence of the lens, the time delay surface is a paraboloid which has a single extremum (a minimum); additional extrema have to come in pairs, thus the total number of images must be odd (as we showed earlier by continuously deforming the time delay surface);

- when two additional images form, they must be a maximum and a saddle point; in between them, the curvature changes from negative to positive. Thus it is zero between them. Remember that $\det A = 0$ is the condition for having a critical point, where the magnification is (formally) infinite. The critical lines thus separate multiple image pairs; these pairs merge and disappear (as discussed above) at the critical lines. In other words, the critical lines separate regions of different image multiplicities.

3.7 Python Applications

3.7.1 Implementing a Ray-Tracing Algorithm

In this example, we implement a simple ray-tracing algorithm. We use the lens equation to propagate a bundle of light rays from the observer position to the source plane through a regular grid covering the lens plane. For each ray passing at the position $\vec{x}^{ij}$, we evaluate the deflection angle $\vec{\alpha}(\vec{x}^{ij})$ and compute the arrival position on the source plane as

$$\vec{y}^{ij} = \vec{x}^{ij} - \vec{\alpha}(\vec{x}^{ij}) \,. \tag{3.92}$$

This example builds on the deflection angles derived in Sect. 2.5.2, for a numerically simulated dark matter halo. In this case, the lens is at redshift $z_L = 0.5$, and the source plane is at $z_S = 9$. The deflection angles are stored in the arrays `angx` and `angy` and the maps contain 512×512 pixels. We start by creating a mesh on the lens plane using the `numpy.meshgrid` method. Suppose coordinates along the x_1 and x_2 axes are represented by the n_{pix}-dimensional arrays $|x_1^i|$ and $|x_2^j|$, with $i, j \in [1, n_{pix}]$ (so that $n_{pix} = 512$ is the number of grid points along one axis on the mesh). We create the mesh as follows:

```
npix=angx.shape[0]
x1=np.linspace(0.0,1.0,npix)*(npix-1) # define x1 coordinates
x2=np.linspace(0.0,1.0,npix)*(npix-1) # define x2 coordinates
x1_,x2_=np.meshgrid(x1,x2) # lens plane mesh
```

This code generates two numpy arrays, `x1_` and `x2_`, with size $n_{pix} \times n_{pix}$. In the first, the values on the $i-$th column will be equal to x_1^i; in the second, the values on the $j-$th row will be equal to x_2^j.

We may now implement the lens equation for the two components along x_1 and x_2:

```
y1=x1_-angx
y2=x2_-angy
```

The resulting arrays `y1` and `y2` have size $n_{pix} \times n_{pix}$. In order to improve the visualization of the results, we down-sample the maps by tracing a lower number of rays through the lens plane. More precisely, we reduce the number of grid points on the lens plane mesh by a factor `ndown=16` along the two axes, x_1 and x_2:

```
ndown=16
x1=np.linspace(0.0,1.0,npix/ndown)*(npix-1) # down-sampled x1,x2 coordinates
x2=np.linspace(0.0,1.0,npix/ndown)*(npix-1) #
x1_,x2_=np.meshgrid(x1,x2) # downsampled grid
```

Using the `map_coordinate` method of `scipy.ndimage`, we interpolate the maps `angx` and `angy` to calculate the deflection angles on the down-sampled ray grid:

```
# now we interpolate the defl. angle maps at (x1_,x2_)
from scipy.ndimage import map_coordinates
# first, we need to reshape x1_ and y1_:
x=np.reshape(x1_,x1_.size)
y=np.reshape(x2_,x2_.size)
# then we interpolate:
angx_=map_coordinates(angx,[[y],[x]],order=1)
angy_=map_coordinates(angy,[[y],[x]],order=1)
# now we reshape the angles back to a mesh
angx_=angx_.reshape((npix/ndown,npix/ndown))
angy_=angy_.reshape((npix/ndown,npix/ndown))
```

Finally, we use the lens equation to calculate the ray positions on the source plane:

```
y1=x1_-angx_
y2=x2_-angy_
```

The result of this calculation is shown in Fig. 3.12. In the left panel, we show the regular ray grid on the lens plane. In the right panel, we show the arrival positions of the light rays on the source plane. We can see that the grid on the source plane is no longer regular, but significantly distorted. The source plane is "crunched," particularly near the center of the lens plane, where many rays converge. This is a manifestation of the lensing magnification. Indeed, a small area on the source plane corresponds to a larger area on the lens plane.

3.7.2 Derivation of the Lensing Potential

Deriving the lensing potential from the lens convergence map requires to solve the Poisson equation in two dimensions (Eqs. 3.26 and 3.27). We can do this numerically by employing Fast Fourier Transform.

The Fourier transform of the Laplace operator is

$$\tilde{\triangle}(\vec{k}) = -4\pi^2 k^2,$$ (3.93)

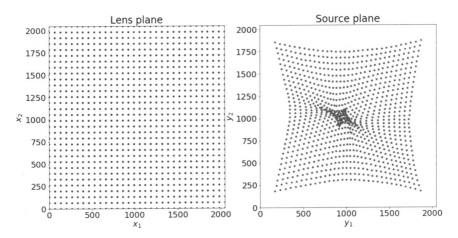

Fig. 3.12 Ray-tracing through a regular grid on the lens plane (left panel). We show the arrival positions of the light rays on the source plane in the right panel. The lens is the same used in Sect. 2.5.2

where $k^2 = k_1^2 + k_2^2$. Therefore, in Fourier space, the Poisson equation reads

$$-4\pi^2 k^2 \tilde{\Psi}(\vec{k}) = 2\tilde{\kappa}(\vec{k}) \, , \qquad (3.94)$$

and the Fourier transform of the lensing potential is

$$\tilde{\Psi}(\vec{k}) = -\frac{\tilde{\kappa}(\vec{k})}{2\pi^2 k^2} \, . \qquad (3.95)$$

As shown in Sect. 2.5.2, the numerical calculation of the Discrete Fourier Transforms (DFTs) can be done using the Fast Fourier Transform algorithms implemented in the `numpy.fft` module. We add the following function to the class `deflector` introduced in Sect. 2.5.2:

```python
def potential(self):
    # define an array of wave numbers (two components k1,k2)
    k = np.array(np.meshgrid(fftengine.fftfreq(self.kappa.shape[0])\
                            ,fftengine.fftfreq(self.kappa.shape[1])))
    #Compute Laplace operator in Fourier space
    kk = k[0]**2 + k[1]**2
    kk[0,0] = 1.0
    #FFT of the convergence
    kappa_ft = fftengine.rfftn(kappa)
    #compute the FT of the potential
    kappa_ft *= - 1.0 / (kk * (2.0*np.pi**2))
    kappa_ft[0,0] = 0.0
    potential=fftengine.irfftn(kappa_ft)
    return potential
```

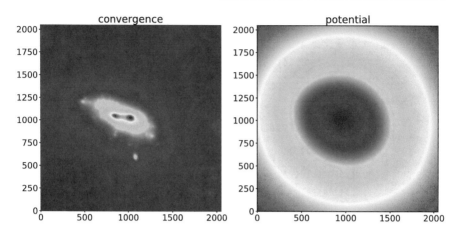

Fig. 3.13 Maps of the convergence and of the lensing potential for the same lens used in Sect. 2.5.2

Then, we use it to compute the lensing potential of the lens in Fig. 2.6:

```
pot=df.potential() # compute the potential
```

It is important to use zero-padding to avoid unwanted boundary effects.

We show the maps of the convergence and the lensing potential for the lens considered in Fig. 3.13. The potential is much smoother than the convergence. This difference reflects the fact that the convergence is obtained from the second derivatives of the potential.

3.7.3 Lensing Maps

Once the potential is known, it is easy to compute maps of many other lenses' properties. For example, the gradient of $\hat{\Psi}$ is the deflection angle. Thus, we can implement a method to compute $\vec{\alpha}$, which is an alternative to that discussed in Sect. 2.5.2.

We use the `numpy.gradient` method to calculate the gradient using finite-differences on the grids. The maps of the deflection angle components α_1 and α_2 are obtained as follows:

```
a2,a1=np.gradient(pot)
```

Note that, because of the axis convention in python, the derivatives of $\hat{\Psi}$ along the second dimension is given first. We do not display the maps, as they are analogous to those shown e.g. in Fig. 2.6.

By computing further gradients of these maps, we obtain the second derivatives of the potential. By combining them, we can compute the convergence (which is

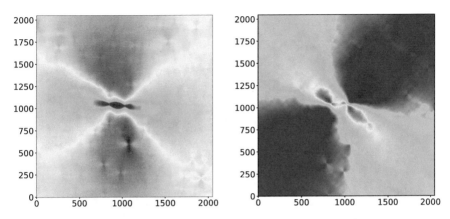

Fig. 3.14 Maps of the shear components for the same lens used in Sect. 2.5.2

already known, as it was the input to derive the potential) and the shear components. The python implementation of Eq. 3.36 is as follows:

```
# First we compute the second derivatives of pot
psi12,psi11=np.gradient(a1)
psi22,psi21=np.gradient(a2)
# Then we combine them to form the first and the second component of
# the shear tensor
gamma1=0.5*(psi11-psi22)
gamma2=psi12
```

In Fig. 3.14, we show the maps both γ_1 and γ_2.

As discussed in Sect. 3.3.1, the shear introduces an an-isotropic distortion of the images. For example, a circular source is mapped onto an elliptical image (in the case of a slowly varying deflection angle). The direction of the axes of the ellipse is given by the angle ϕ in Eq. 3.43, which can be computed using the `arctan2` function:

```
phi=np.arctan2(gamma2,gamma1)/2.0
```

Note that we have to divide by 2 in order to account for the fact that γ is a spin-2 tensor. It is interesting to display the direction into which the shear distorts images and compare it to lens mass distribution. Figure 3.15 shows the direction of the shear using sticks overlaid to the lens convergence. The code to produce the figure is

```
pixel_step=gamma_1.shape[1]/32+1
x,y = np.meshgrid(np.arange(0,gamma_1.shape[1],pixel_step),
                  np.arange(0,gamma_1.shape[0],pixel_step))
fig,ax=plt.subplots(1,1,figsize=(10,10))
ax.imshow(ka,origin='lower',vmax=3)
ax[1].imshow(kappa,origin='lower',vmax=3)
```

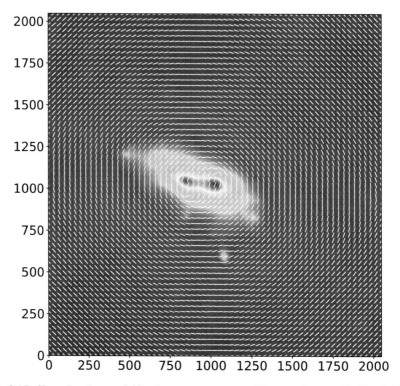

Fig. 3.15 Shear direction overlaid to the convergence map of the same lens used in Sect. 2.5.2

```
# showing only the orientation of the shear. This will create two sticks
# departing from the point where the shear is evaluated (x,y) and directed
# in opposite directions
ax[1].quiver(y,x,np.cos(phi[x,y]),np.sin(phi[x,y]),
        headwidth=0,units="height",scale=x.shape[0],color="white")
ax[1].quiver(y,x,-np.cos(phi[x,y]),-np.sin(phi[x,y]),
        headwidth=0,units="height",scale=x.shape[0],color="white")
```

From the maps of the shear, we can derive the maps of the flexions F and G. Each of these quantities has two components, corresponding to the real and to imaginary parts of the complex quantities in Eqs. 3.73 and 3.74:

```
gamma12,gamma11=np.gradient(gamma_1)
gamma22,gamma21=np.gradient(gamma_2)
F1,F2=gamma11+gamma22,gamma21-gamma12
G1,G2=gamma11-gamma22,gamma21+gamma12
```

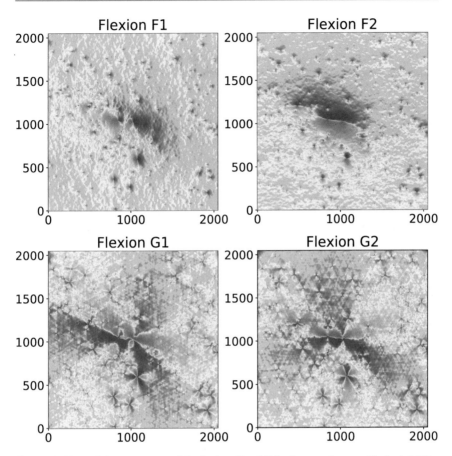

Fig. 3.16 Maps of the components of the flexions F and G for the same lens used in Sect. 2.5.2

We show their maps in Fig. 3.16. These maps show some interesting features:

- the features in the flexion F maps have dipole symmetry, as expected for a spin-1 field;
- the features in the flexion G maps have triangular symmetry, showing the spin-3 nature of this field, which is invariant under rotations by $2\pi/3$ radians;
- in both the cases of the flexion F and G, the signal from the small-scale structures in the convergence map is amplified, because the flexion is obtained via third-order derivatives of the lensing potential.

3.7.4   Critical Lines and Caustics

There are several methods to identify the points belonging to the critical lines. A simple way to visualize them is drawing the zero-level contours in the maps of λ_t and λ_r. For example, to visualize the critical lines of the lens in the previous examples, we can do as follows:

```
from matplotlib.colors import SymLogNorm

gamma=np.sqrt(gamma_1**2+gamma_2**2)
lambdat=1.0-kappa-gamma
lambdar=1.0-kappa+gamma
detA=lambdat*lambdar

fig,ax=plt.subplots(1,3,figsize=(28,8))
ax[0].imshow(lambdat,origin='lower')
ax[0].contour(lambdat,levels=[0.0])
ax[0].set_title('$\lambda_t$',fontsize=25)
ax[1].imshow(lambdar,origin='lower')
ax[1].contour(lambdar,levels=[0.0])
ax[1].set_title('$\lambda_r$',fontsize=25)
ax[2].imshow(detA,origin='lower',norm=SymLogNorm(0.3))
ax[2].contour(detA,levels=[0.0])
ax[2].set_title('$\det A$',fontsize=25)
```

We display the results in Fig. 3.17. The left and the central panels show the maps of λ_t and λ_r, respectively. In the right panel, we show the map of det A. In all panels, we also show the maps' zero-level contours, i.e. the critical lines. In the left and the central panels, the tangential and the radial critical lines are shown separately, while in the right panel, they are displayed simultaneously. Note that these are the critical lines for a specific source redshift. The convergence map used to compute the lensing potential is calculated for $z_{s,norm} = 9$. To obtain the critical lines for

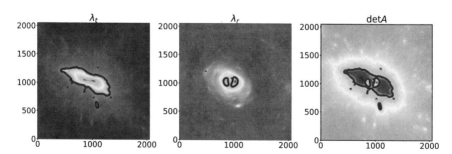

Fig. 3.17 The left and the central panel show the maps of the eigenvalues of the lensing Jacobian with overlaid their zero-level contours, i.e. the critical lines. Instead, the right panel shows the map of det A, i.e. the product of the two previous maps

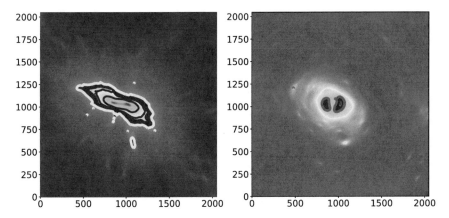

Fig. 3.18 Tangential (left) and radial (right) critical lines of the lens for different source redshift

different source redshift z_s, we need to re-scale both κ and γ by the distance ratio

$$\Xi = \frac{D_{S,norm}}{D_{LS,norm}} \frac{D_{LS}(z_s)}{D_S(z_s)} , \qquad (3.96)$$

where the distances $D_{S,norm}$ and $D_{LS,norm}$ are computed for $z_s = z_{s,norm}$. The code below repeats this operation for 20 equally spaced redshift between z_l and $z_s = 10$. The corresponding critical lines are shown in Fig. 3.18. To compute the distances, we have to assume a cosmological model. We use the `astropy.cosmology` module and import a pre-defined flat ΛCDM cosmological model with density parameter $\Omega_M = 0.3$ and $\Omega_\Lambda = 0.7$. We assume the Hubble parameter to be $H_0 = 70$ km/s/Mpc. The angular diameter distances are computed using the `angular_diameter_distance` and the `angular_diameter_distance_z1z2` methods.

```python
from astropy.cosmology import FlatLambdaCDM
cosmo = FlatLambdaCDM(H0=70, Om0=0.3)

zl=0.5
zs_norm=9.0

zs=np.linspace(zl,10.0,20)
dl=cosmo.angular_diameter_distance(zl)
ds=cosmo.angular_diameter_distance(zs)
dls=[]
for i in range(ds.size):
    dls.append(cosmo.angular_diameter_distance_z1z2(zl,zs[i]).value)


ds_norm=cosmo.angular_diameter_distance(zs_norm)
dls_norm=cosmo.angular_diameter_distance_z1z2(zl,zs_norm)
```

```
fig,ax=plt.subplots(1,2,figsize=(16,8))
ax[0].imshow(lambdat,origin='lower')
ax[1].imshow(lambdar,origin='lower')
for i in range(ds.size):
    kappa_new=kappa*ds_norm.value/dls_norm.value*dls[i]/ds[i].value
    gamma_new=gamma*ds_norm.value/dls_norm.value*dls[i]/ds[i].value
    lambdat_new=(1.0-kappa_new-gamma_new)
    lambdar_new=(1.0-kappa_new+gamma_new)
    ax[0].contour(lambdat_new,levels=[0.0])
    ax[1].contour(lambdar_new,levels=[0.0])

ax[0].contour(lambdat,levels=[0.0],colors="yellow",linewidths=2)
ax[1].contour(lambdar,levels=[0.0],colors="magenta",linewidths=2)
```

The caustics are the "sources" of the critical lines. In other words, if $\vec{\theta}_c$ is a point on the critical lines, then

$$\vec{\beta}_c = \vec{\theta}_c - \vec{\alpha}(\vec{\theta}_c) \qquad (3.97)$$

is the corresponding point on the caustics.

To derive the caustics, we need first to extract the critical points from the contours described above. Then, we measure the deflection angles at the position of the critical points, by interpolating them on the maps of the deflection angles computed earlier. We use again the map_coordinates method from the scipy.ndimage module. In the following, we work in pixel units:

```
fig,ax=plt.subplots(1,2,figsize=(18,8))
# first, we extract the level-0 contours of the map of detA
cs=ax[0].contour(detA,levels=[0.0])

# then, we take the path of each closed contour
contour=cs.collections[0]
p=contour.get_paths()  # p contains the paths of each individual
                       # critical line

sizevs=np.empty(len(p),dtype=int)

from scipy.ndimage import map_coordinates

# if we find any critical line, then we process it
if (sizevs.size > 0):
    for j in range(len(p)):
            # the contours are ensembles of polygons, called paths
        # for each path, we create two vectors containing the x1
        # and x2 coordinates of the vertices
        vs = contour.get_paths()[j].vertices
        sizevs[j]=len(vs)
        x1=[]
        x2=[]
```

```
for i in range(len(vs)):
    xx1,xx2=vs[i]
    x1.append(float(xx1))
    x2.append(float(xx2))
# these are the points we want to map back on the source plane.
# To do that we need to evaluate the deflection angle at their
# positions using scipy.ndimage.interpolate.map_coordinates
# we perform a bi-linear interpolation

a_1=map_coordinates(a1, [[x2],[x1]],order=1)
a_2=map_coordinates(a2, [[x2],[x1]],order=1)

# now we use the lens equation to
# obtain the caustics:
y1=x1-a_1[0]
y2=x2-a_2[0]

# plot the results!
ax[0].plot(x1,x2,'-')
ax[1].plot(y1,y2,'-')
```

```
ax[1].set_xlim([0,2048])
ax[1].set_ylim([0,2048])
```

The left and right panels in Fig. 3.19 show the lens critical lines and the caustics, respectively. We compare the shape of the caustics to the pattern visible in the right panel of Fig. 3.12. Performing ray-tracing, we found that starting from a regular grid of ray positions on the lens plane, we end up with an irregular grid on the source plane, which also covers a smaller area of the sky due to magnification. We can easily see that the arrival positions of light rays on the source plane cluster around the lens's caustics.

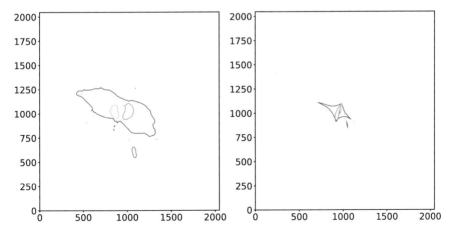

Fig. 3.19 Critical lines (left panel) and caustics (right panel) for $z_s = 9$. We indicate the corresponding pairs of critical lines and caustics with the same colors

3.7.5 Shear and Flexion

In this example, we use Eq. 3.58 to build an application to visualize the lensing distortions due to shear and flexion.

As shown in Sect. 3.5, the elements of the tensor D, D_{ijk}, are expressed as third derivatives of the lensing potential. These in turn can be written in terms of the flexions F and G. After some math, we find that

$$D_{111} = -2\gamma_{11} - \gamma_{22} = -\frac{1}{2}(3F_1 + G_1)$$

$$D_{211} = D_{121} = D_{112} = -\gamma_{21} = -\frac{1}{2}(F_2 + G_2)$$

$$D_{122} = D_{212} = D_{221} = -\gamma_{22} = -\frac{1}{2}(F_1 - G_1)$$

$$D_{222} = 2\gamma_{12} - \gamma_{21} = -\frac{1}{2}(3F_2 - G_2). \tag{3.98}$$

From Eq. 3.60, we find that the two components of $\vec{\beta}$ are

$$\beta_1 = A_{11}\theta_1 + A_{12}\theta_2 + \frac{1}{2}D_{111}\theta_1^2 + D_{121}\theta_1\theta_2 + \frac{1}{2}D_{122}\theta_2^2$$

$$\beta_2 = A_{21}\theta_1 + A_{22}\theta_2 + \frac{1}{2}D_{211}\theta_1^2 + D_{212}\theta_1\theta_2 + \frac{1}{2}D_{222}\theta_2^2. \tag{3.99}$$

We consider a circular source centered at $\vec{\beta}_0 = (0, 0)$. We assign to this source a surface brightness profile. For this exercise, we can choose a generic Sérsic profile (Sérsic 1963) given by

$$I_s(\beta) \propto \exp -b_n[(\beta/r_e)^{1/n} - 1] . \tag{3.100}$$

The shape parameter b_n is given by the approximated formula (Capaccioli et al. 1989)

$$b_n = 1.992n - 0.3271 , \tag{3.101}$$

which is valid for $0.5 < n < 10$. The parameter r_e is the effective radius of the source and n is the Sérsic index.

Since the surface brightness is conserved, we can relate the observed and the intrinsic surface brightnesses, $I(\vec{\theta})$ and $I_s(\vec{\beta})$, using the lens equation:

$$I(\vec{\theta}) = I_s(\vec{\beta}) . \tag{3.102}$$

Thus, we can straightforwardly reconstruct the source image on the lens plane through ray-tracing. First, we cover the lens plane with a grid of $N \times N$ pixels. The pixel centers define the ray positions, $\vec{\theta}_{ij}$. We compute the corresponding positions on the source plane using Eq. 3.99. Finally, we assign the surface brightness $I_s(\vec{\beta}_{ij})$ to the pixel centered on $\vec{\theta}_{ij}$.

Note that in Eq. 3.58 both the image and the source are centered on $(0, 0)$. The image shift $\vec{\theta}_0 = \vec{\beta}_0 + \vec{\alpha}(\vec{\theta}_0) = \vec{\alpha}(\vec{\theta}_0)$ was subtracted for convenience.

We start by implementing a class for Sérsic circular sources. To initialize an instance of this class, we use the parameters n and r_e. These parameters are passed to the initialization function as a dictionary element (`kwargs`). Besides, we also specify the size, `side`, and the number of pixels, N, of the output image:

```python
# import the usual packages: NumPy and matplotlib
import numpy as np
import matplotlib.pyplot as plt

class sersic(object):
    def __init__(self,side,N,**kwargs):
        """
        Initialize the sersic image
        read input parameters from dictionary.
        if not present in kwargs, use some default
        values
        """
        if ('n' in kwargs):
            self.n=kwargs['n']
        else:
            self.n=4

        if ('re' in kwargs):
            self.re=kwargs['re']
        else:
            self.re=50.0

        # number of pixels and field of view:
        self.N=N
        self.side=float(side)

        # define the pixel coordinates on a
        # regular grid using meshgrid
        pc=np.linspace(-side/2.,side/2.,self.N)
        self.x1,self.x2 = np.meshgrid(pc,pc)
         # unlensed image
        self.unlensed = self.brightness(self.x1,self.x2)

    def brightness(self,y1,y2):
        """
        Implement the Sersic brightness profile
        """
        r = np.sqrt(y1**2+y2**2)
        bn = 1.992*self.n - 0.3271
        return (np.exp(-bn*((r/self.re)**(1.0/self.n)-1.0)))
```

The class includes a method, called `brightness`, to calculate the surface brightness at arbitrary positions, using Eq. 3.100. We remind that the source position has coordinates $(0, 0)$. In the initialization function, we call this function to compute the unlensed brightness distribution at coordinates $(x_1.x_2)$ on a regular grid. We store the result in the array `self.unlensed`.

We implement Eq. 3.99 in the function `lens` here below:

```
def lens(self,**kwargs_lens):
    # read the input parameters from kwargs
    if ('kappa' in kwargs_lens):
        self.kappa = kwargs_lens['kappa']
    else:
        self.kappa=0.0

    if ('gamma1' in kwargs_lens):
        self.gamma1 = kwargs_lens['gamma1']
    else:
        self.gamma1=0.0

    if ('gamma2' in kwargs_lens):
        self.gamma2 = kwargs_lens['gamma2']
    else:
        self.gamma2=0.0

    if ('g1' in kwargs_lens):
        self.g1 = kwargs_lens['g1']
    else:
        self.g1=0.0
    if ('g2' in kwargs_lens):
        self.g2 = kwargs_lens['g2']
    else:
        self.g2=0.0

    if ('f1' in kwargs_lens):
        self.f1 = kwargs_lens['f1']
    else:
        self.f1=0.0
    if ('f2' in kwargs_lens):
        self.f2 = kwargs_lens['f2']
    else:
        self.f2=0.0

    # compute elements of the Jacobian matrix
    a11=1.0-self.kappa-self.gamma1
    a22=1.0-self.kappa+self.gamma1
    a12=-self.gamma2
    # compute elements of tensor D
    a111=-0.5*(self.g1+3.0*self.f1)
    a222=-0.5*(3.0*self.f2-self.g2)
    a112=-0.5*(self.f2+self.g2)
```

```
    a221=-0.5*(self.f1-self.g1)

    # lens equation to second order to obtain
    # the source plane coordinates (y1,y2)
    y1 = a11*self.x1 + a12*self.x2 + 0.5*a111*self.x1**2 + \
        a112*self.x1*self.x2 + 0.5*a221*self.x2**2
    y2 = a22*self.x2 + a12*self.x1 + 0.5*a222*self.x2**2 + \
        a221*self.x1*self.x2 + 0.5*a112*self.x1**2

    # compute the surface brightness at positions (y1,y2)
    self.lensed=self.brightness(y1,y2)
```

The lens parameters γ_1, γ_2, F_1, F_2, G_1, and G_2 are stored in the dictionary `kwargs_lens` and passed to the function tt lens. Thus, to produce a lensed image of a circular source with $n = 4$ and $r_e = 4$, including the effects of convergence, shear, and flexion, we can do as follows:

```
kwargs={'n': 4, 're': 4.0, 'q': 1., 'pa': 0.0}
se=sersic(side=5.0,N=250,**kwargs)
kwargs_lens={'kappa': 0.2, 'gamma1': 0.2, 'gamma2': 0.1,
            'f1': 0.3, 'f2': 0.1,'g1': 0.5, 'g2': -0.2}
se.lens(**kwargs_lens)
```

The resulting images have 250×250 pixels and cover a field-of-view of 5×5 arcsec. We can access them as follows:

```
unlensed_image=se.unlensed
lensed_image=se.lensed
```

Using these lines of code, we produced Fig. 3.5.

3.7.6 Full Ray-Tracing Simulation and Time Delay Surface

We can extend the procedure outlined above for lensing a circular source with Sérsic profile to include distortions due to higher-order terms in expanding the deflection field. By design, all these terms are included in a full ray-tracing simulation employing the algorithm discussed in Sect. 3.7.1.

We start by importing some useful packages:

```
# import numpy and matplotlib
import numpy as np
import matplotlib.pyplot as plt

# import map_coordinates from scipy
from scipy.ndimage import map_coordinates

# import fits from astropy
import astropy.io.fits as pyfits
```

```
# import FFT from NumPy
import numpy.fft as fftengine
```

We define a class `gen_lens` for general lenses, which includes functions that can be applied to any kind of lenses. This class plays the role of the *parent* class. Then, we use inheritance to use these functions with classes describing specific lenses. These classes are *child* classes of `gen_lens`. The initialization function of the parent class is empty. It contains only one instruction to set the logic variable `pot_exist=False`, meaning that the parent class itself has no potential defined yet:

```
class gen_lens(object):
    """
    Initialize gen_lens
    """
    # the lens does not have a potential yet
    def __init__(self):
        self.pot_exists=False
```

The functions to derive convergence, shear, and Jacobian determinant (inverse magnification) from the derivatives of the deflection angles are part of the generic lens class, because they can work with any potential:

```
    # convergence
    def convergence(self):
        if (self.pot_exists):
            kappa=0.5*(self.a11+self.a22)
        else:
            print ("The lens potential is not initialized yet")

        return(kappa)

    #shear
    def shear(self):
        if (self.pot_exists):
            g1=0.5*(self.a11-self.a22)
            g2=self.a12
        else:
            print ("The lens potential is not initialized yet")
        return(g1,g2)

    # determinant of the Jacobian matrix
    def detA(self):
        if (self.pot_exists):
            deta=(1.0-self.a11)*(1.0-self.a22)-self.a12*self.a21
        else:
            print ("The lens potential is not initialized yet")
        return(deta)

    # critical lines overlaid to the map of detA, returns a set of
```

```
# contour objects
def crit_lines(self,ax=None,show=True):
    if (ax==None):
        print ("specify the axes to display the critical lines")
    else:
        deta=self.detA()
        #ax.imshow(deta,origin='lower')
        cs=ax.contour(deta,levels=[0.0],colors='white',alpha=0.0)
        if show==False:
            ax.clear()
    return(cs)

# plot of the critical lines in the axes ax
def clines(self,ax=None,color='red',alpha=1.0,lt='-'):
    cs=self.crit_lines(ax=ax,show=False)
    contour=cs.collections[0]
    p=contour.get_paths()
    sizevs=np.empty(len(p),dtype=int)

    no=self.pixel
    # if we found any contour, then we proceed
    if (sizevs.size > 0):
        for j in range(len(p)):
            # for each path, we create two vectors containing
            #the x1 and x2 coordinates of the vertices
            vs = contour.get_paths()[j].vertices
            sizevs[j]=len(vs)
            x1=[]
            x2=[]
            for i in range(len(vs)):
                xx1,xx2=vs[i]
                x1.append(float(xx1))
                x2.append(float(xx2))

            # plot the results!
            ax.plot((np.array(x1)-self.npix/2.)*no,
                    (np.array(x2)-self.npix/2.)*no,lt,color=color,
                    alpha=alpha)

# plot of the caustics in the axes ax
def caustics(self,ax=None,alpha=1.0,color='red',lt='-'):
    cs=self.crit_lines(ax=ax,show=True)
    contour=cs.collections[0]
    p=contour.get_paths() # p contains the paths of each individual
                          # critical line
    sizevs=np.empty(len(p),dtype=int)

    # if we found any contour, then we proceed
    if (sizevs.size > 0):
        for j in range(len(p)):
            # for each path, we create two vectors containing
```

```
# the x1 and x2 coordinates of the vertices
vs = contour.get_paths()[j].vertices
sizevs[j]=len(vs)
x1=[]
x2=[]
for i in range(len(vs)):
    xx1,xx2=vs[i]
    x1.append(float(xx1))
    x2.append(float(xx2))

a_1=map_coordinates(self.a1, [[x2],[x1]],order=1)
a_2=map_coordinates(self.a2, [[x2],[x1]],order=1)

# now we can make the mapping using the lens equation:
no=self.pixel
y1=(x1-a_1[0]-self.npix/2.)*no
y2=(x2-a_2[0]-self.npix/2.)*no

# plot the results!
ax.plot(y1,y2,lt,color=color,alpha=alpha)
```

Note that we discussed all the functions included in the generic lens class in the previous examples.

We now add the functions to compute the time delay surface using Eq. 3.84:

```
# geometrical time delay
def t_geom_surf(self, beta=None):
    x = np.arange(0, self.npix, 1, float)*self.pixel
    y = x[:,np.newaxis]
    if beta is None:
        x0 = y0 = self.npix / 2*self.pixel
    else:
        x0 = beta[0]+self.npix/2*self.pixel
        y0 = beta[1]+self.npix/2*self.pixel

    return 0.5*((x-x0)*(x-x0)+(y-y0)*(y-y0))

# gravitational time delay:
def t_grav_surf(self):
    return -self.pot

# total time delay
def t_delay_surf(self,beta=None):
    t_grav=self.t_grav_surf()
    t_geom=self.t_geom_surf(beta)
    td=(t_grav+t_geom)
    return(t_grav+t_geom)
```

For convenience, we also add a function to display map contour levels:

```python
# display the time delay contours
def show_contours(self,surf0,ax=None,minx=-25,miny=-25,
                  cmap=plt.get_cmap('Paired'),
                  linewidth=1,fontsize=20,nlevels=40,levmax=100,
                  offz=0.0):
    if ax==None:
        print ("specify the axes to display the contours")
    else:
        minx=minx
        maxx=-minx
        miny=miny
        maxy=-miny
        surf=surf0-np.min(surf0)
        levels=np.linspace(np.min(surf),levmax,nlevels)
        ax.contour(surf, cmap=cmap,levels=levels,
                   linewidth=linewidth,
                   extent=[-self.size/2,self.size/2,
                   -self.size/2,self.size/2])
        ax.set_xlim(minx, maxx)
        ax.set_ylim(miny, maxy)
        ax.set_xlabel(r'$\theta_1$',fontsize=fontsize)
        ax.set_ylabel(r'$\theta_2$',fontsize=fontsize)
        ax.set_aspect('equal')
```

Now, we create child classes with specific lensing potentials. The first is the elliptical pseudo-isothermal model with core (PSIEc). Lens models of these kinds will be discussed in detail in Chap. 5.

The PSIEc potential has the form:

$$\hat{\Psi}(\vec{\theta}) = \frac{\text{norm}}{\sqrt{\theta^2 + \theta_c^2}} \,, \tag{3.103}$$

where norm is a normalization factor, $\theta^2 = \theta_1^2/(1 - \text{ell}) + \theta_2^2(1 - \text{ell})$, and θ_c is the core radius. The lens has elliptical iso-potential contours with ellipticity e.

As said, PSIEc is a child class of gen_lens and inherits all its methods. To ensure this inheritance, we indicate the parent class in the class definition:

```python
class PSIEc(gen_lens):
    def __init__(self,co,size=100.0,npix=200,**kwargs):

        # set the cosmological model
        self.co = co

        # core radius
        if ('theta_c' in kwargs):
            self.theta_c=kwargs['theta_c']
        else:
            self.theta_c=0.0
```

```
    # ellipticity
    if ('ell' in kwargs):
        self.ell=kwargs['ell']
    else:
        ell=0.0

    # normalization norm
    if ('norm' in kwargs):
        self.norm=kwargs['norm']
    else:
        self.norm=1.0

    # lens redshift zl
    if ('zl' in kwargs):
        self.zl=kwargs['zl']
    else:
        self.zl=0.5

    # source redshift zs
    if ('zs' in kwargs):
        self.zs=kwargs['zs']
    else:
        self.zs=1.0

    # angular diameter distances
    self.dl = co.angular_diameter_distance(self.zl)
    self.ds = co.angular_diameter_distance(self.zs)
    self.dls = co.angular_diameter_distance_z1z2(self.zl,self.zs)

    # size of the output image
    self.size=size
    # number of pixels
    self.npix=npix

    # pixel scale
    self.pixel=self.size/self.npix

    # calculate the lensing potential
    self.potential()

# lensing potential and its derivatives
def potential(self):
    x = np.arange(0, self.npix, 1, float)
    y = x[:,np.newaxis]
    x0 = y0 = self.npix / 2
    self.pot_exists=True
    #
    self.pot=np.sqrt(((x-x0)*self.pixel)**2/(1-self.ell)
                    +((y-y0)*self.pixel)**2*(1-self.ell)
                    +self.theta_c**2)*self.norm
```

```
self.a2,self.a1=np.gradient(self.pot/self.pixel**2)
self.a12,self.a11=np.gradient(self.a1)
self.a22,self.a21=np.gradient(self.a2)
```

We pass the input parameters to initialize the lens model, θ_c, ell, and norm as a dictionary. The dictionary also includes the lens and the source redshift, zl and zs. Among the input parameters of the initialization function, we also specify the cosmological model, co, and the size and the number of pixels of the output maps, size and npix. We use the function potential to calculate the maps of the lensing potential and its first and second derivatives. The first derivatives are the components of the gradient of the lensing potential, i.e. the deflection angles. The maps of the second derivatives of the potential are attributes of the child class, but we can use them with functions that are part of the parent class. Thus, inheritance enables us to compute converge, shear, magnification and even the time delay surfaces for arbitrary source positions of the PSIEc lens.

Now, we modify the class Sérsic to include ellipticity. We add few more input parameters for the source initialization in the kwargs dictionary. They are the axis ratio, q, the position angle, pa, the coordinates on the source plane, ys1 and ys2, and the source redshift, zs. Also, we introduce an optional argument in the function call, i.e. the general lens gl. If it is not None, then the lensing effects by the lens gl are included via ray-tracing:

```
class sersic(object):

    def __init__(self,size,N,gl=None,**kwargs):

        if ('n' in kwargs):
            self.n=kwargs['n']
        else:
            self.n=4

        if ('re' in kwargs):
            self.re=kwargs['re']
        else:
            self.re=5.0

        if ('q' in kwargs):
            self.q=kwargs['q']
        else:
            self.q=1.0

        if ('pa' in kwargs):
            self.pa=kwargs['pa']
        else:
            self.pa=0.0

        if ('ys1' in kwargs):
            self.ys1=kwargs['ys1']
        else:
```

```
        self.ys1=0.0

if ('ys2' in kwargs):
    self.ys2=kwargs['ys2']
else:
    self.ys2=0.0

if ('zs' in kwargs):
    self.zs=kwargs['zs']
else:
    self.zs=1.0

self.N=N
self.size=float(size)
self.df=gl

# define the pixel coordinates
pc=np.linspace(-self.size/2.0,self.size/2.0,self.N)
self.x1, self.x2 = np.meshgrid(pc,pc)
if self.df != None:
    ds_lens=gl.ds
    dls_lens=gl.dls
    ds=gl.co.angular_diameter_distance(self.zs)
    dls=gl.co.angular_diameter_distance_z1z2(gl.zl,self.zs)
    self.corrf = ds_lens/dls_lens*dls/ds
    if self.zs != gl.zs:
        gl.rescale(self.corrf)
    y1,y2 = self.ray_trace()
else:
    y1,y2 = self.x1,self.x2

self.image=self.brightness(y1,y2)
```

Note that, if the source redshift attributes of the `sersic` and of the `PSIEc` objects are different, then the lens maps are corrected by the factor `self.corrf`, which accounts for the different distance ratios (see Eq. 3.96). The function `rescale` is part of the `genlens` class:

```
# rescale the lensing maps by the factor fcorr
def rescale(self,fcorr):
    if self.pot_exists:
        self.pot=self.pot*fcorr
        self.a1=self.a1*fcorr
        self.a2=self.a2*fcorr
        self.a12=self.a12*fcorr
        self.a11=self.a11*fcorr
        self.a22=self.a22*fcorr
        self.a21=self.a21*fcorr
```

We implement the ray-tracing algorithm in the function `ray_trace`:

```python
def ray_trace(self):
    """
    Ray-tracing through the lens place
    """
    px=self.df.pixel
    x1pix=(self.x1+self.df.size/2.0)/px
    x2pix=(self.x2+self.df.size/2.0)/px

    # compute the deflection angles at the light ray positions
    # on the lens plane. Use the deflection angles of the
    # general lens self.df
    a1 = map_coordinates(self.df.a1,
                         [x2pix,x1pix],order=2)*px
    a2 = map_coordinates(self.df.a2,
                         [x2pix,x1pix],order=2)*px
    # apply the lens equation
    y1=(self.x1-a1) # y1 coordinates on the source plane
    y2=(self.x2-a2) # y2 coordinates on the source plane
    return(y1,y2)
```

The function interpolates the lens deflection angle maps `self.df.a1` and `self.df.a2` on a regular mesh on the lens plane (`self.x1, self.x1`). These coordinates identify the ray positions on the lens plane. Then it uses the lens equation to calculate the ray arrival positions on the source plane, (`y1, y2`).

Finally, we compute the surface brightness at the coordinates (`y1, y2`) using the function `brightness`. Compared to the previous example, we modify this function to allow for positioning the source at arbitrary positions, including the ellipticity of the isophotes, and orienting them according to the input position angle:

```python
def brightness(self,y1,y2):
    # rotate the galaxy by the angle self.pa
    x = np.cos(self.pa)*(y1-self.ys1)+np.sin(self.pa)*(y2-self.ys2)
    y = -np.sin(self.pa)*(y1-self.ys1)+np.cos(self.pa)*(y2-self.ys2)
    # include elliptical isophotes
    r = np.sqrt(((x)/self.q)**2+(y)**2)
    # brightness at distance r
    bn = 1.992*self.n - 0.3271
    brightness = np.exp(-bn*((r/self.re)**(1.0/self.n)-1.0))
    return(brightness)
```

The following code shows how to use the classes above to produce Fig. 3.20.

```python
co = FlatLambdaCDM(H0=70, Om0=0.3)
# lens params
kwargs={'theta_c': 2.0,
        'norm': 10.0,
        'ell': 0.4,
        'zl': 0.5,
        'zs': 1.0}
```

```
el=PSIEc(co,size=80,npix=1000,**kwargs)
# size of the source image
size_stamp=150.0
npix_stamp=1000
xmin,xmax=-el.size/2,el.size/2
ymin,ymax=-el.size/2,el.size/2

fig,ax=plt.subplots(1,2,figsize=(14,8))
# sersic source with no lensing
beta=[0,0]
kwargs={'q': 0.5,
        're': 1.0,
        'pa': np.pi/4.0,
        'n': 1,
        'ys1': beta[0],
        'ys2': beta[1],
        'zs': 1.0}

se_unlensed=sersic(size_stamp,npix_stamp,**kwargs) # same source with lensing by the lens el
se=sersic(size_stamp,npix_stamp,gl=el,**kwargs)

# compute the time delay surface for a source at beta

td=el.t_delay_surf(beta=beta)
# draw caustics (on the left) and critical lines (on the right)
el.caustics(ax=ax[0],lt='--',alpha=1.0)
el.clines(ax=ax[1],lt='--',alpha=1.0)
# show unlensed (on the left) and lensed (on the right) images
ax[0].imshow(se_unlensed.image,origin='lower',
            extent=[-se.size/2,se.size/2,-se.size/2,se.size/2],
            cmap='gray_r')
ax[1].imshow(se.image,origin='lower',
            extent=[-se.size/2,se.size/2,-se.size/2,se.size/2],
            cmap='gray_r')
# show contours of the time delay surface
el.show_contours(td,ax=ax[1],minx=xmin,miny=ymin, nlevels=35,levmax=500,fontsize=20)
```

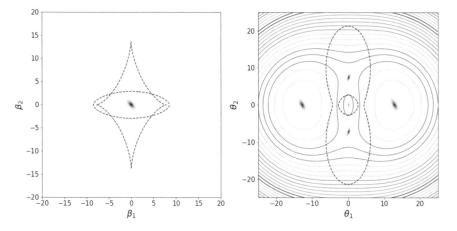

Fig. 3.20 Left panel: Caustics of a PSIEc lens. A Sérsic source is placed at the center of the caustics. Right panel: critical lines (dashed) and lensed images of the source shown in the left panel. The colored solid contours show the levels of equal time delay for the same source

```
x0,x1=-20,20
y0,y1=-20,20
ax[0].set_xlim([x0,x1])
ax[0].set_ylim([y0,y1])
fig.tight_layout()
```

The left panel shows the lens caustics and the unlensed source. The right panel shows the lensed images and some contour levels of the time delay surface. As expected, the multiple images form at the stationary points of the time delay surface.

3.7.7 Lensing by Numerically Simulated Mass Distributions

Our implementation of the general lens class allows us to use its functions with any lens model easily. For example, we consider here the lens analyzed in Sects. 2.5.2 and 3.7.2. The lens mass distribution is in the form of a convergence map stored in a fits file. We modify and extend the class `deflector`, which contains the functions discussed earlier to deal with such kind of mass distributions, as follows:

```python
# child class deflector
class deflector(gen_lens):
    # initialize the deflector using a surface density (convergence) map
    # the boolean variable pad indicates whether zero-padding is used or not

    def __init__(self,co,filekappa,zl=0.5,zs=1.0,
                 pad=False,npix=200,size=100):
        # read input convergence map from fits file
        kappa,header=pyfits.getdata(filekappa,header=True)

        self.co = co
        self.zl = zl
        self.zs = zs

        # angular diameter distances
        self.dl = co.angular_diameter_distance(self.zl)
        self.ds = co.angular_diameter_distance(self.zs)
        self.dls = co.angular_diameter_distance_z1z2(self.zl,self.zs)

        # pixel scale and number of pixels of the input convergence map
        self.pixel_scale=header['CDELT2']*3600.0
        self.kappa=kappa
        self.nx=kappa.shape[0]
        self.ny=kappa.shape[1]

        # pixels and size of the output lensing maps
        # they can be different from the those of the
        # input convergence map
        self.npix=npix
        self.size=size
        self.pixel=float(self.size)/self.npix
        # use zero-padding to compute the lensing potential
        self.pad=pad
```

```
      if (pad):
          self.kpad()
      self.potential()

# performs zero-padding
def kpad(self):
    # add zeros around the original array
    def padwithzeros(vector, pad_width, iaxis, kwargs):
        vector[:pad_width[0]] = 0
        vector[-pad_width[1]:] = 0
        return vector
    # use the pad method from numpy.lib to add zeros (padwithzeros)
    # in a frame with thickness self.kappa.shape[0]
    self.kappa=np.lib.pad(self.kappa, self.kappa.shape[0],
                          padwithzeros)

# calculate the potential by solving the poisson equation
# the output potential map has the same size of the input
# convergence map
def potential_from_kappa(self):
    # define an array of wavenumbers (two components k1,k2)
    k = np.array(np.meshgrid(fftengine.fftfreq(self.kappa.shape[0])\
                        ,fftengine.fftfreq(self.kappa.shape[1])))
    #Compute Laplace operator in Fourier space = -4*pi*k^2
    kk = k[0]**2 + k[1]**2
    kk[0,0] = 1.0
    #FFT of the convergence
    kappa_ft = fftengine.fftn(self.kappa)
    #compute the FT of the potential
    kappa_ft *= - 1.0 / (kk * (2.0*np.pi**2))
    kappa_ft[0,0] = 0.0
    potential=fftengine.ifftn(kappa_ft)
    if self.pad:
        pot=self.mapCrop(potential.real)
    return pot

# interpolate the lensing potential on the grid defined at
# initialization.
# Calculate the first and second-order derivatives.
def potential(self):
    no=self.pixel
    x_ = np.linspace(0,self.npix-1,self.npix)
    y_ = np.linspace(0,self.npix-1,self.npix)
    x,y=np.meshgrid(x_,y_)
    potential=self.potential_from_kappa()
    x0 = y0 = potential.shape[0] / 2*self.pixel_scale-self.size/2.0
    x=(x0+x*no)/self.pixel_scale
    y=(y0+y*no)/self.pixel_scale
    self.pot_exists=True
    pot=map_coordinates(potential,[y,x],order=1)
    self.pot=pot*self.pixel_scale**2/no/no
    self.a2,self.a1=np.gradient(self.pot)
    self.a12,self.a11=np.gradient(self.a1)
```

```
        self.a22,self.a21=np.gradient(self.a2)
        self.pot=pot*self.pixel_scale**2


    # crop the maps to remove zero-padded areas and get back to the original
    # region.
    def mapCrop(self,mappa):
        xmin=int(self.kappa.shape[0]/2-self.nx/2)
        ymin=int(self.kappa.shape[1]/2-self.ny/2)
        xmax=int(xmin+self.nx)
        ymax=int(ymin+self.ny)
        mappa=mappa[xmin:xmax,ymin:ymax]
        return(mappa)
```

This class and the class PSIEc from the previous example have several similarities. For example, they are both derived from the genlens class and contain a function, called potential, to compute the lensing potential on a grid of arbitrary size and resolution. When we call this function, we calculate also the first and the second order derivatives of the potential map and register them as class attributes. Then, we can use the methods defined in the genlens class to compute the lens properties.

Let us consider the following example:

```
size=200.0
npix=500
# create instance of the deflector class
df=deflector(co,'data/kappa_2.fits',
             zl=0.5, zs=9.0, pad=True,
             npix=npix,size=size)

# set the source position and Sersic parameters:
beta=[-30,8]

kwargs={'q': 0.5,
        're': 1.0,
        'pa': np.pi/4.0,
        'n': 1,
        'ys1': beta[0],
        'ys2': beta[1],
        'zs': 9.0}

#
xmin,xmax=-df.size/2,df.size/2
ymin,ymax=-df.size/2,df.size/2

fig,ax=plt.subplots(1,2,figsize=(14,8))

se_unlensed=sersic(size_stamp,npix_stamp,**kwargs)
se=sersic(size_stamp,npix_stamp,gl=df,**kwargs)
td=df.t_delay_surf(beta=beta)
df.caustics(ax=ax[0],lt='--',alpha=1.0)
```

```
df.clines(ax=ax[1],lt='--',alpha=1.0)
ax[0].imshow(se_unlensed.image,origin='lower',
            extent=[-se_unlensed.size/2,se_unlensed.size/2,
                    -se_unlensed.size/2,se_unlensed.size/2],
            cmap='gray_r')
ax[1].imshow(se.image,origin='lower',
            extent=[-se.size/2,se.size/2,-se.size/2,se.size/2],
            cmap='gray_r')
df.show_contours(td,ax=ax[1],minx=xmin,miny=ymin,nlevels=25,
                    levmax=1600,fontsize=20)
x0,x1=-40,10
y0,y1=-25,25
ax[0].set_xlim([x0,x1])
ax[0].set_ylim([y0,y1])
x0,x1=-80,50
y0,y1=-65,65
ax[1].set_xlim([x0,x1])
ax[1].set_ylim([y0,y1])
fig.tight_layout()
```

The code above produces Fig. 3.21. If we place the source near the cusp of the tangential caustic, it produces three images, which are distorted and merge into a tangential arc. We also display some contour levels, showing that the arc shape reflects the curvature of the time delay surface.

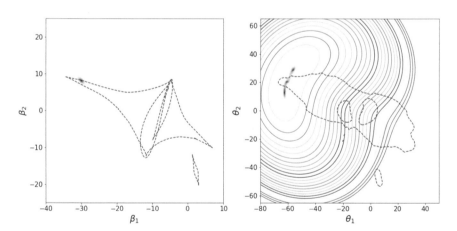

Fig. 3.21 Left panel: Caustics of a numerically simulated lens. A Sérsic source is placed near the cusp of the caustics. Right panel: critical lines (dashed lines) and lensed images of the source shown in the left panel. We also show, using different colors, the contour levels of the time delay surface

References

Capaccioli, M., della Valle, M., Rosino, L., & D'Onofrio, M. (1989). Properties of the nova population in M31. *AJ, 97,* 1622–1633. https://doi.org/10.1086/115104.

Goldberg, D. M., & Natarajan, P. (2002). The galaxy octopole moment as a probe of weak-lensing shear fields. *ApJ, 564*(1), 65–72. https://doi.org/10.1086/324202. arXiv: astro-ph/0107187 [astro-ph].

Sérsic, J. L. (1963). Influence of the atmospheric and instrumental dispersion on the brightness distribution in a galaxy. *Boletin de la Asociacion Argentina de Astronomia La Plata Argentina, 6,* 41.

Shapiro, I. I. (1964). Fourth test of general relativity. *Physical ReviewLetters, 13*(26), 789–791. https://doi.org/10.1103/PhysRevLett.13.789.

Weinberg, S. (1972). *Gravitation and cosmology: principles and applications of the general theory of relativity.*

Part II
Applications

Microlenses

4

We dedicate this chapter to microlensing, i.e. the lensing effects of small-sized lenses in the universe. A broad range of masses belongs to this class of lenses: planets, stars, star clusters, and more generally compact objects in our or other galaxies. Given the lenses' small size, microlenses are (to first order) assimilated to point masses or ensembles of point masses. Microlensing effects are mostly detectable and searched within our galaxy by monitoring vast amounts of stars in the bulge of the Milky Way or toward the Magellanic Clouds. Nevertheless, microlensing effects are also relevant in extra-galactic gravitational lensing. Indeed, small mass lenses in galaxies other than ours produce small scale perturbations to the lensing signal of their hosts, which can be revealed in some situations. In this chapter, however, we will focus on microlenses in the Milky Way.

4.1 The Point-Mass Lens

4.1.1 Deflection Angle and Lensing Potential

In Chap. 2, we already derived the formula for the deflection angle of a point mass. By choosing the lens position as the center of the reference frame (i.e. by counting the angles β and θ starting from the lens position), the deflection angle turns out to be

$$\hat{\vec{\alpha}}(\vec{\xi}) = \frac{4GM}{c^2} \frac{\vec{\xi}}{|\vec{\xi}|^2} = \frac{4GM}{c^2 D_{\mathrm{L}}} \frac{\vec{\theta}}{|\vec{\theta}|^2} = \hat{\vec{\alpha}}(\vec{\theta}) \,, \tag{4.1}$$

© Springer Nature Switzerland AG 2021
M. Meneghetti, *Introduction to Gravitational Lensing*, Lecture Notes
in Physics 956, https://doi.org/10.1007/978-3-030-73582-1_4

where, as usual, we have used the relation between the physical length ξ, the angle θ, and the angular diameter distance D_L, $\xi = D_L\theta$. Given that

$$\vec{\alpha}(\vec{\theta}) = \frac{D_{LS}}{D_S}\hat{\vec{\alpha}}(\vec{\theta}) = \vec{\nabla}\hat{\Psi}(\vec{\theta}) \qquad (4.2)$$

and that

$$\nabla \ln |\vec{x}| = \frac{\vec{x}}{|\vec{x}|^2} \ , \qquad (4.3)$$

we can see that the lensing potential of the point-mass lens is given by

$$\hat{\Psi}(\vec{\theta}) = \frac{4GM}{c^2}\frac{D_{LS}}{D_L D_S} \ln |\vec{\theta}| \ , \qquad (4.4)$$

as we anticipated in Sect. 3.6.3.

4.1.2 Lens Equation

The vector $\hat{\vec{\alpha}}$ points away from the lens. Therefore, we may omit the vector sign in many of the following equations. Then

$$\hat{\alpha}(\theta) = \frac{4GM}{c^2 D_L \theta} \ . \qquad (4.5)$$

The lens equation reads

$$\beta = \theta - \frac{4GM}{c^2 D_L \theta}\frac{D_{LS}}{D_S} \ . \qquad (4.6)$$

This equation is quadratic in θ, i.e. for a given position of the source β, there always exist two images whose positions can be determined by solving the lens equation.

4.1.3 Multiple Images

Equation 4.6 can be written in a more concise way by introducing the *Einstein radius*,

$$\theta_E \equiv \sqrt{\frac{4GM}{c^2}\frac{D_{LS}}{D_L D_S}} \ . \qquad (4.7)$$

The importance of this quantity will be clear shortly.

By inserting Eq. 4.7 into Eq. 4.6, we obtain

$$\beta = \theta - \frac{\theta_E^2}{\theta} \ . \tag{4.8}$$

Dividing by θ_E and setting $y = \beta/\theta_E$ and $x = \theta/\theta_E$, i.e. by expressing all angles in units of the Einstein radius, we obtain the lens equation

$$y = x - \frac{1}{x} \ . \tag{4.9}$$

Multiplication with x leads to

$$x^2 - xy - 1 = 0 \ , \tag{4.10}$$

which has two solutions:

$$x_\pm = \frac{1}{2}\left[y \pm \sqrt{y^2 + 4} \right] \ . \tag{4.11}$$

The right panel of Fig. 4.1 shows a sequence of sources at different angular distances from the lens (indicated by a red star). Each source is shown using a different color, so that its images can be easily recognized in the left panel. For convenience the sources have been placed on the axis $y_2 = 0$.

Each source has two images, both on the axis $x_2 = 0$, but one at $x_+ > 0$ and one at $-1 < x_- < 0$. Thus, they are on opposite sides with respect to the lens, and the image at x_- is always within a circle of radius $x = 1$. Such circle coincides with the image of the source at $y = 0$, $x_\pm = \pm 1$; that is, a source exactly behind the point lens has a ring-shaped image with radius θ_E, also called *Einstein ring*. The size of the Einstein radius is

$$\theta_E \approx (10^{-3})'' \left(\frac{M}{M_\odot} \right)^{1/2} \left(\frac{D}{10\text{kpc}} \right)^{-1/2} \ ,$$

$$\approx 1'' \left(\frac{M}{10^{12}M_\odot} \right)^{1/2} \left(\frac{D}{\text{Gpc}} \right)^{-1/2} \ , \tag{4.12}$$

where

$$D \equiv \frac{D_{\text{L}} D_{\text{S}}}{D_{\text{LS}}} \tag{4.13}$$

is the *effective lensing distance*.

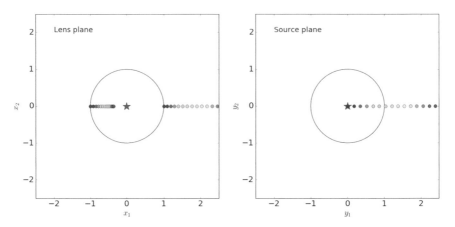

Fig. 4.1 Solutions of the lens equation for a point-mass lens. In both panels, the lens is given by the star at the center of the figure. The Einstein ring is shown in black. In the right panel, the positions of several source are indicated by colored circles. The corresponding images, as obtained by Eq. 4.11, are shown in the left panel

As $\beta \rightarrow \infty$, we see that $\theta_- = x_-\theta_E \rightarrow 0$, while $\theta_+ = x_+\theta_E \rightarrow \beta$: when the angular separation between the lens and the source becomes large, the source is unlensed. Formally, there is still an image at $\theta_- = 0$, but as we have seen when discussing the properties of the time delay surface, this central image has zero magnification.

4.1.4 Critical Lines, Caustics, and Magnification

Given that the lens is axially-symmetric, the Jacobian determinant is

$$\det A(x) = \frac{y}{x}\frac{\mathrm{d}y}{\mathrm{d}x} \; . \tag{4.14}$$

From Eqs. 4.14 and 4.6, the eigenvalues of the Jacobian matrix are

$$\lambda_t(x) = \frac{y}{x} = \left(1 - \frac{1}{x^2}\right)$$

$$\lambda_r(x) = \frac{\mathrm{d}y}{\mathrm{d}x} = \left(1 + \frac{1}{x^2}\right) \; . \tag{4.15}$$

Obviously, the second eigenvalue is never zero. Therefore, the point-mass lens only has one critical line, namely a circle with equation $x^2 = 1$. This is the equation of the Einstein ring, which thus coincides with the tangential critical line.

Using the lens equation, as seen above, this line is mapped onto the tangential caustic, which is a point at $\beta = 0$ ($y = 0$).

The magnification is the inverse of the Jacobian determinant, thus

$$\mu(x) = \left[1 - \frac{1}{x^4}\right]^{-1} . \tag{4.16}$$

4.1.5 Source Magnification

From Eq. 4.11, we can see that, at the image positions,

$$\frac{x}{y} = \frac{1}{2}\left(1 \pm \frac{\sqrt{y^2 + 4}}{y}\right)$$

$$\frac{dx}{dy} = \frac{1}{2}\left(1 \pm \frac{y}{\sqrt{y^2 + 4}}\right) . \tag{4.17}$$

Thus, the image magnification can be written as a function of the source position as

$$
\begin{aligned}
\mu_{\pm}(y) &= \frac{1}{4}\left(1 \pm \frac{\sqrt{y^2 + 4}}{y}\right)\left(1 \pm \frac{y}{\sqrt{y^2 + 4}}\right) \\
&= \frac{1}{4}\left(1 \pm \frac{\sqrt{y^2 + 4}}{y} \pm \frac{y}{\sqrt{y^2 + 4}} + 1\right) \\
&= \frac{1}{4}\left(2 \pm \frac{2y^2 + 4}{y\sqrt{y^2 + 4}}\right) = \frac{1}{2}\left(1 \pm \frac{y^2 + 2}{y\sqrt{y^2 + 4}}\right) .
\end{aligned}
\tag{4.18}
$$

Note that, for $y > 0$, $\mu_-(y) < 0$, and $\mu_+(y) > 0$, showing that the parity of the two images is different.[1]

The total source magnification is

$$\mu(y) = \mu_+(y) + |\mu_-(y)| = \frac{y^2 + 2}{y\sqrt{y^2 + 4}} , \tag{4.19}$$

---

[1] For $y > 0$, $x_+ > 0$, and $x_- < 0$. Thus, $\mu_t = x/y$ is positive at x_+ and negative at x_-. Given that $\mu_r = dx/dy > 0$, the magnifications of the two images, $\mu_\pm = \mu_t(x_\pm)\mu_r(x_\pm)$ have the same signs of $x_\pm$.

while the sum of the *signed* magnifications is $\mu = 1$. By means of a power series expansion of the function above, we see that $\mu \propto 1 + 2/y^4$ for $y \to \infty$, i.e. the magnification drops quickly as the source moves away from the lens.

The magnification ratio of the two images is

$$
\left| \frac{\mu_+}{\mu_-} \right| = \frac{1 + \frac{y^2+2}{y\sqrt{y^2+4}}}{\frac{y^2+2}{y\sqrt{y^2+4}} - 1}
$$

$$
= \frac{y^2 + 2 + y\sqrt{y^2 + 4}}{y^2 + 2 - y\sqrt{y^2 + 4}} . \tag{4.20}
$$

Given that

$$
\frac{1}{2} \left(y + \sqrt{y^2 + 4} \right)^2 = y^2 + 2 + y\sqrt{y^2 + 4} \tag{4.21}
$$

and

$$
\frac{1}{2} \left(y - \sqrt{y^2 + 4} \right)^2 = y^2 + 2 - y\sqrt{y^2 + 4} , \tag{4.22}
$$

we find that

$$
\left| \frac{\mu_+}{\mu_-} \right| = \left(\frac{y + \sqrt{y^2 + 4}}{y - \sqrt{y^2 + 4}} \right)^2
$$

$$
= \left(\frac{x_+}{x_-} \right)^2 . \tag{4.23}
$$

We can see that $\lim_{y\to\infty} \mu_- = 0$ and that $\lim_{y\to\infty} \mu_+ = 1$. Moreover, a Laurent series expansion shows that for large y,

$$
\left| \frac{\mu_+}{\mu_-} \right| \propto y^4 , \tag{4.24}
$$

i.e. the image at x_+ dominates the magnification budget pretty quickly as the source is moved away from the lens.

4.1.6 Microlensing Cross Section

A source at $y = 1$ has two images at

$$x_\pm = \frac{1 \pm \sqrt{5}}{2} \, , \qquad (4.25)$$

and their magnifications are

$$\mu_\pm = \left[1 - \left(\frac{2}{1 \pm \sqrt{5}} \right)^4 \right]^{-1} . \qquad (4.26)$$

Thus, the total source magnification is $\mu = |\mu_+| + |\mu_-| = 1.17 + 0.17 = 1.34$. In terms of magnitudes, this corresponds to $\Delta m = -2.5 \log \mu \sim 0.3$ only. The image at x_+ contributes for $\sim 87\%$ of the total magnification. As seen above, for $y > 1$ the magnification drops quickly meaning that the only chance to detect microlensing events via magnification effects is by finding sources well aligned with the lenses, i.e. within their Einstein rings. For this reason, the area of the Einstein ring is generally assumed to be the cross-section for microlensing,

$$\sigma_{micro} = \pi \theta_E^2 \, . \qquad (4.27)$$

This is the solid angle within which a source has to be placed in order to produce a detectable microlensing signal.

4.2 Microlensing Light-Curve

The typical lens's Einstein radius gives the order of magnitude of the image separation in microlensing events. For one solar mass star within our galaxy, this is of the order of $\sim 10^{-3}$ arcseconds, thus undetectable with the current instrumentation.

However, we can detect microlenses in the Milky Way or its surrounding by exploiting the motion of lenses relative to sources due to the (differential) rotation of our galaxy. If the source and the lens are in relative motion, i.e. if the distance between the lens and the source, y, is a function of time, then Eq. 4.19 shows that the magnification is a function of time as well, $\mu \equiv \mu(t)$. Therefore, a source with intrinsic flux f_s will appear to have a flux $f(t) = \mu(t) f_s$, while being lensed. The curve describing the variation of the source flux as a function of time during the microlensing event is called the *microlensing light-curve*.

We assume that a straight line can approximate the path of the source relative to the lens, as shown by the blue dashed line in the diagram in Fig. 4.2. The source (indicated by the blue dot) reaches the minimum dimensionless distance y_0 from the lens at time t_0. y_0 is the dimensionless *impact parameter* of the source. Assuming

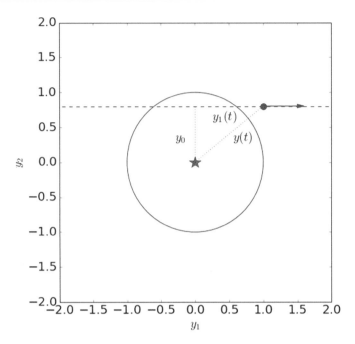

Fig. 4.2 Illustration for the lens position and source trajectory. The dimensionless impact parameter is y_0. $y_1(t)$ is the dimensionless distance of the source from the point of minimum distance from the lens. Finally, $y(t)$ is the dimensionless distance of the source from the lens

that the source moves with transverse velocity v relative to the lens, we can write the dimensionless distance of the source from the point of minimum distance from the lens as

$$y_1(t) = \frac{v(t - t_0)}{D_L \theta_E} , \qquad (4.28)$$

where D_L indicates, as usual, the angular diameter distance between the observer and the lens.

If the source moves at velocity v, it will take a time

$$t_E = \frac{D_L \theta_E}{v} = \frac{\theta_E}{\mu_{rel}} \qquad (4.29)$$

to cross the Einstein radius of the lens. In the equation above, we have introduced the source's proper motion relative to the lens, $\mu_{rel} = v/D_L$. Since, as we discussed in the previous section, the magnification significantly deviates from unity only for sources with $|y| \lesssim 1$, we can assume that the *Einstein radius crossing time*, t_E, is the timescale of the microlensing event. If we use the definition of Einstein radius given in Eq. 4.7, we see that

$$t_E \approx 19 \text{ days} \sqrt{4\frac{D_L}{D_S}\left(1 - \frac{D_L}{D_S}\right)} \left(\frac{D_S}{8\text{kpc}}\right)^{1/2} \left(\frac{M}{0.3M_\odot}\right)^{1/2} \left(\frac{v}{200\text{km/s}}\right)^{-1}.$$

$$(4.30)$$

To write this equation, we have used the approximation $D_{LS} = D_S - D_L$, which is valid only for non-cosmological distances, and thus applies in our galaxy.

Inserting Eq. 4.29 into Eq. 4.28, we obtain

$$y_1(t) = \frac{(t - t_0)}{t_E}.$$

$$(4.31)$$

Thus,

$$y(t) = \sqrt{y_0^2 + y_1^2(t)} = \sqrt{y_0^2 + \frac{(t - t_0)^2}{t_E^2}}.$$

$$(4.32)$$

Combining Eqs. 4.32 and 4.19, we obtain the microlensing light-curve:

$$\mu(t) = \frac{y(t)^2 + 2}{y(t)\sqrt{y(t)^2 + 4}}.$$

$$(4.33)$$

Some examples of light-curves corresponding to different values of the impact parameter y_0 are shown in Fig. 4.3.

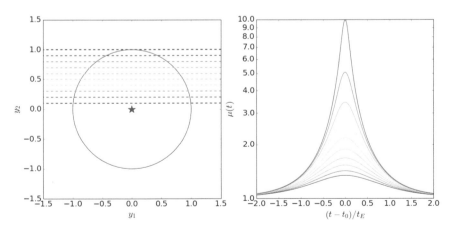

Fig. 4.3 Left panel: source trajectories corresponding to different values of the impact parameter y_0, varying from 0.1 (red) to 1 (purple). Right panel: color-coded light-curves corresponding to the source trajectories displayed in the left panel

4.2.1   Light-Curve Fitting

A peak at t_0 characterizes the shape of the *standard* microlensing light-curve. The value of y_0 determines the peak height: the closest the source trajectory passes to the lens, the highest is the peak. The Einstein crossing time t_E, which in turn depends on the lens mass, the relative transverse velocity, the lens, and the source distances, determines the peak width. Therefore from an observed light-curve well fitted by the standard model, one cannot infer the distances, the velocity, and the lens mass uniquely. This degeneracy is the so-called *microlensing degeneracy*.

Besides, to fit an observed light-curve, we need the baseline flux, f_0. Generally, a *blending parameter*, b_s, which describes the fraction of light contributed by the lensed source, is also included. In the presence of crowded fields, the measured flux is indeed the sum of the fluxes from the source, the lens, and the other unrelated stars within the seeing disk. Since these may be wavelength-dependent, the blending parameter is generally different in different filters. Note that blending biases the magnification estimate low.

While the standard light-curve model works well in many cases, there are situations where one or more of the standard model's assumptions break down. In these cases, it is possible to derive extra constraints that partially lift the microlensing degeneracy. Non-standard light-curves occur, for example, when the source or the lens is not point-like (finite source size and lens size effects) or when the motion of the source relative to the lens is not linear.

If the lens has a finite size, one can always find a time interval when the lens obscures the inner (and the outer, depending on the lens size) lensed image in the early rising and final declining stages of the light-curve. Thus, the impact of these effects is to dim the wings of the light-curve.

The finite source size effect is particularly important in the high-magnification limits (Gould 1994, Lee et al. 2009, Witt & Mao 1994). In this case, the light-curve is broadened near the peak, as the result of the fact that different parts of the source experience different magnifications. Assuming that the surface brightness of the source is uniform, the magnification near the peak of the light-curve can be approximated by the following formula (Gould 1994):

$$\mu'(y) \simeq \mu(y) \frac{4y}{\pi \rho} E(\vartheta_{max}, y/\rho) , \qquad (4.34)$$

where $E(\vartheta, \varphi)$ is the Elliptical integral of the second kind and ϑ_{max} is defined as

$$\vartheta_{\max} = \begin{cases} \frac{\pi}{2} & y \leq \rho \\ \arcsin(\rho/y) & y > \rho \end{cases} . \qquad (4.35)$$

Fitting the light-curve allows measuring $\rho = \theta_\star/\theta_E$, i.e. the source size, $\theta_\star$, in units of the Einstein radius. Thus, if we can independently measure $\theta_\star$, the finite source effect allows us to measure the Einstein radius's size. For example, we can derive the source size from empirical relations between the surface brightness

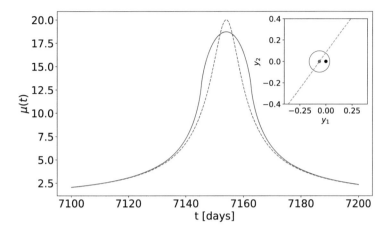

Fig. 4.4 Source's finite size effects on the microlensing light-curve. We consider two sources moving on the trajectory indicated by the dashed gray line in the smaller frame. The red source is point-like, while the blue one has a radius of $0.1\theta_E$, where θ_E is the Einstein radius of the lens indicated by the black dot at $\vec{y} = 0$. The corresponding microlensing light-curves are given by the dashed red and solid blue curves in the larger frame. The second light-curve is wider and smoother near the peak

and the colors. Kervella et al. (2004) proposed the following relation between the angular diameter for A0-M2 dwarf stars or A0-K0 sub-giants, the $V - K$ color, and K-band magnitude:

$$\log 2\theta_\star = 0.0755(V - K) + 0.5170 - 0.2K \ . \tag{4.36}$$

Measuring θ_E and t_E then allows to measure $\mu_{rel} = \theta_E/t_E$.

Note that, given the large difference between the sizes of θ_E ($\sim$1 mas) and $\theta_\star$ ($\sim$0.5 μas), we can measure the finite source size effect only in very high-magnification events (when the source overlaps with the point-like caustic). We illustrate this case in Fig. 4.4.

4.3   Microlensing Parallax

For a point-mass lens in the Milky Way, using the fact that $D_{LS} = D_S - D_L$, the Einstein radius can be written as

$$\theta_E = \sqrt{\frac{4GM}{c^2}\frac{D_{LS}}{D_L D_S}} = \sqrt{\frac{4GM}{c^2}\left(\frac{1}{D_L} - \frac{1}{D_S}\right)} = \sqrt{\frac{4GM}{c^2}\pi_{rel}} \ . \tag{4.37}$$

In the above equation, $\pi_{rel} = 1/D_L - 1/D_S$ is the *relative parallax* of the lens and the source. From Eq. 4.37, we can also see that

$$\theta_E = \frac{4GM}{c^2}\frac{\pi_{rel}}{\theta_E} = \frac{4GM}{c^2}\pi_E \,, \qquad (4.38)$$

where we have introduced the *microlensing parallax*, $\pi_E = \pi_{rel}/\theta_E$. By setting

$$k = \frac{4G}{c^2} = 8.14 \text{ mas } M_\odot^{-1} \,, \qquad (4.39)$$

we find that

$$M = \frac{\theta_E}{k\pi_E} \,, \qquad (4.40)$$

which shows that measuring the microlensing parallax, π_E, and the Einstein radius, θ_E, allows to break the microlensing degeneracy and measure the lens mass.

The diagram in Fig. 4.5 also shows that the microlensing parallax is the inverse of the projection of the physical Einstein radius, $r_E = \theta_E D_L$, onto the observer plane, $\tilde{r}_E$. Indeed,

$$\tilde{r}_E = D_L\hat{\alpha}(\theta_E) = \frac{r_E}{\theta_E}\hat{\alpha}(\theta_E) = \frac{kM}{\theta_E} = \frac{1}{\pi_E} \,. \qquad (4.41)$$

Measuring the microlensing parallax is possible in two situations. First, the observer moves while the microlensing event is ongoing. Second, the observer uses two telescopes to observe the event from different locations simultaneously. There are three types of microlensing parallax:

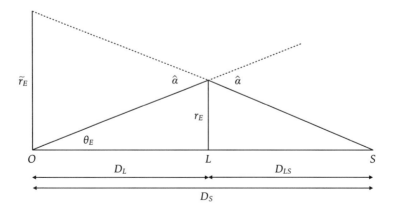

Fig. 4.5 Relationship between projected Einstein radius, $\tilde{r}_E$ and angular Einstein radius θ_E

1. orbital or annual parallax: the observer on the earth participates in the orbital motion of the earth around the sun;
2. satellite parallax: the observer uses a ground-based facility and a space telescope to observe the same event;
3. terrestrial parallax: the observer uses a network of ground-based facilities to observe the same event.

Because of the different line-of-sight to the source, the observer will see a variation of the lens position relative to the sources. Since the magnification in Eq. 4.33 depends on the absolute separation of lens and source in units of the Einstein radius, the source light-curve will be affected.

4.3.1   Orbital Parallax

The orbital motion of the earth around the sun implies that the line-of-sight to the source changes in the course of the microlensing event for a ground-based observer. In the frame where the observer and the source are fixed, the lens moves both because of its intrinsic motion relative to the source (which we can assume to be on a linear trajectory) and because of the earth's orbital motion. While the first motion produces the standard light-curve, the second distorts it. Alcock et al. (1995) reported the first detection of orbital microlensing parallax. We show the light-curves of this microlensing event in the B and R−bands in Fig. 4.6. The data-

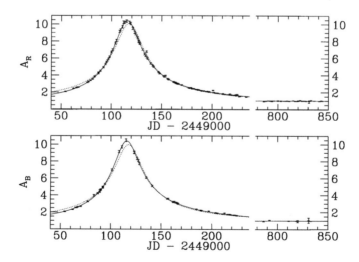

Fig. 4.6 First detection of the orbital parallax effect. The two panels show the event's light-curves in the B and the R-bands. The data are shown with their $\pm 1\sigma$ error bars. The dotted curve is the best-fit standard light-curve model (i.e. assuming a constant relative velocity). The solid line shows the best-fit model accounting for the earth's orbital motion. Figure from Alcock et al. (1995). Reproduced by permission of the AAS

points do not follow a standard microlensing light-curve (dotted line). Instead, they are fitted well by a model that accounts for the orbital parallax effect (solid curve).

Remark 4.1. An important aspect of the microlensing parallax is that it is a vector rather than a scalar. Indeed, the observed parallax effects on the light-curve depend on how the apparent separation between lens and source changes due to the observer's motion. For example, we may decompose this motion into two components, one along and one perpendicular to the lens velocity. Because of the first component, the lens accelerates toward the source. The effect introduces an asymmetric distortion of the light-curve's shape (relative to t_0). Instead, the second component changes the impact parameter y_0, bringing the lens trajectory closer or farther from the source. Consequently, the distortion of the light-curve will be symmetric. To describe both these effects, we need two parallax components.

Since the timescale of the observer baseline variation is ~ 1 year, we can measure the orbital parallax effect in events that are not too short (~ 100 days). Besides, the effect is more easily observable in spring and fall. In these seasons, the earth's motion around the sun is mostly perpendicular to the direction of the galactic bulge, where typical sources in microlensing events are located.

4.3.2 Satellite Parallax

The satellite parallax effect is measurable, combining observations from the ground and a space observatory. At the moment, the instrument which offers the best opportunity to measure this effect is the *Spitzer* space telescope. This telescope is one of the large Space Observatories of NASA, observing at infrared wavelengths from an earth trailing orbit. *Spitzer* happens to be at a distance of ~ 1 AU from the earth. When projected on the observer plane, the typical Einstein radius size, $\tilde{r}_E$, in galactic microlensing events is $\sim$ few-10 AU. Thus, *Spitzer* offers the right baseline to detect parallax effects in most microlensing events involving from M-dwarfs to $\sim$ solar mass lenses.

We show an example in Fig. 4.7, which refers to the first detection of satellite parallax for a single star (Yee et al. 2015). The event, dubbed OGLE-2014-BLG-0939, was observed by the OGLE collaboration from the ground and then followed up by *Spitzer*. Both light-curves are well-represented by standard microlensing curves (blue), but they have substantially different maximum magnifications and times of maximum, whose differences yield a measurement of the "microlens parallax" vector, $\vec{\pi}_E$.

4.3.3 Terrestrial Parallax

Terrestrial parallax effects are detectable in short microlensing events observed from two or more observatories on the earth. Similar to the case of satellite parallax, using

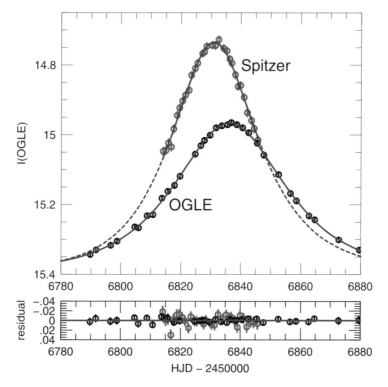

Fig. 4.7 Satellite parallax effect in the event OGLE-2014-BLG-0939. The figure shows the two light-curves measured in the *I*-band by the OGLE collaboration (black) and from *Spitzer* (red). Both light-curves are well described by standard light-curve models (blue). The residuals are shown in the bottom panel. Figure from Yee et al. (2015). Reproduced by permission of the AAS

multiple telescopes on the ground to watch the same microlensing event allows measuring the parallax by comparing the light-curves measured from each location.

Figure 4.8 refers to the event OGLE-2007-BLG-224 (Gould et al. 2009). Three light-curves were measured from three locations in Chile (red), Canary Island (blue), and South Africa (green). The differences between the light-curves are due to terrestrial parallax and allow to measure it.

Note that the shape of the light-curves near the peak is rounded due to finite source effects. Thus, fitting the light-curve and accounting for the source's finite size allows measuring the Einstein radius. By measuring the microlensing parallax, we can determine both the mass and the distance of the lens.

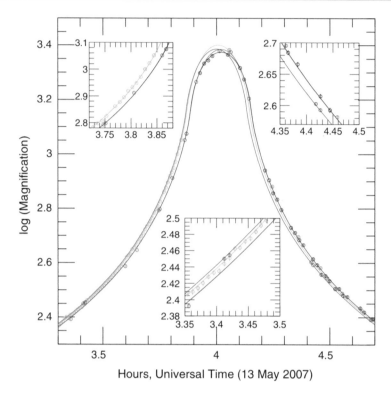

Fig. 4.8 Terrestrial parallax effect in the event OGLE-2007-BLG-224. The blue, red, and green light-curves refer to the measurements conducted from Chile, Canary Island, and South Africa, respectively. Figure from Gould et al. (2009). Reproduced by permission of the AAS

4.4   Astrometric Microlensing

During the microlensing event, the positions and the relative magnifications of the images vary as a function of time. Consequently, the images' light centroid moves along a trajectory, which we can calculate easily. This effect is called *astrometric microlensing* (Dominik & Sahu 1998, Hog et al. 1995, Miyamoto & Yoshii 1995, Proft et al. 2011) and it was recently detected in few microlensing events (Dong et al. 2019, Sahu et al. 2017, Zurlo et al. 2018).

First of all, we consider the motion of the images forming outside and inside the Einstein ring in response to the source's motion. Thus, as done earlier for photometric microlensing, we consider the frame where the lens is fixed, and the source is moving. The vector giving the source position as a function of time, $\vec{y}(t)$, has two components, namely $y_{\parallel} = (t - t_0)/t_E$ and $y_{\perp} = y_0$, parallel and perpendicular to the direction of motion of the source, respectively. The two images always lay on the line passing through the lens and the source. Their distances from the lens in units of the Einstein radius are given in Eq. 4.11. For the images external and internal to the Einstein ring, we have that

$$x_{\pm,\parallel} = \frac{1}{2}(1 \pm Q)y_{\parallel}$$

$$x_{\pm,\perp} = \frac{1}{2}(1 \pm Q)y_{\perp} \, , \tag{4.42}$$

where

$$Q = \frac{\sqrt{y^2 + 4}}{y} \, . \tag{4.43}$$

We show the path of the two images of a source moving on a trajectory with $y_0 = 0.2$ in Fig. 4.9.

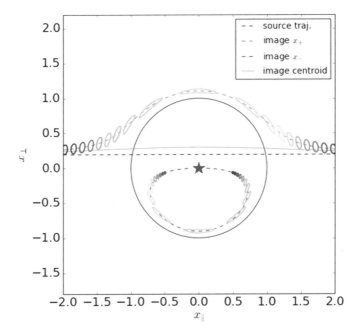

Fig. 4.9 Illustration of a microlensing event. The dashed line gives the source trajectory (corresponding to an impact parameter $y_0 = 0.2$). The red star indicates the lens position in (0,0). The green and the red dashed lines show the trajectories of the images external and internal to the Einstein ring (black circle), respectively. For better visualizing the magnification of the images, we assign to the source a circular shape. As the source moves from left to right, the two images' color changes from blue to red. The external image size is always bigger than the size on the internal image, showing that the former will generally provide a larger fraction of the flux. Consequently, the light centroid will follow a path (solid cyan line), which will differ from the source's path, being pulled toward the external image

The corresponding magnifications are given in Eq. 4.18. The light centroid of the images can be computed as

$$\vec{x}_c = \frac{\vec{x}_+\mu_+ + \vec{x}_-|\mu_-|}{\mu_+ + |\mu_-|} \, . \tag{4.44}$$

As discussed in Sect. 4.1.5, the magnified flux received from the two images is strongly unbalanced in favor of the image external to the Einstein radius for most of the time. Therefore, the light centroid is pulled toward the external image.

In practice, however, we observe microlensing events by monitoring the source rather than the lens. In other words, the source position is fixed (in the galaxy bulge, for example). Therefore, we will not measure the centroid shift relative to the lens, as calculated above. Rather, we may observe a centroid moving relative to the unlensed source position during the microlensing event. This shift is given by

$$\delta\vec{x}_c = \vec{x}_c - \vec{y} \, . \tag{4.45}$$

As shown in Fig. 4.10, this has the characteristic shape of an ellipse. The ellipse axis ratio and size depend on the source's impact parameter, y_0, as discussed below.

Since the lens, the source, and the images are aligned on the plane of the sky, $\vec{x}_c$ and $\vec{y}$ in Eq. 4.45 are aligned too. Thus, using Eqs. 4.11 and 4.18, we can compute the amplitude of the shift as

$$\delta x_c = \frac{\frac{1}{4}\left[(y + \sqrt{y^2+4})\left(1 + \frac{y^2+2}{y\sqrt{y^2+4}}\right) - (y - \sqrt{y^2+4})\left(1 - \frac{y^2+2}{y\sqrt{y^2+4}}\right)\right]}{\frac{y^2+2}{y\sqrt{y^2+4}}} - y$$

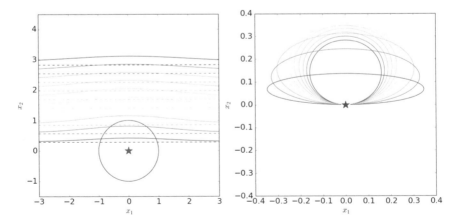

Fig. 4.10 Left panel: Light centroid trajectories (solid lines) for sources with decreasing impact parameter (from blue to red). The unlensed source trajectories are given by the dashed lines. Right panel: the corresponding light centroid paths compared to the paths of the unlensed sources

$$\frac{\frac{1}{4}\left(y + \sqrt{y^2+4} + \frac{y^2+2}{\sqrt{y^2+4}} + \frac{y^2+2}{y} - y + \sqrt{y^2+4} + \frac{y^2+2}{\sqrt{y^2+4}} - \frac{y^2+2}{y}\right)}{\frac{y^2+2}{y\sqrt{y^2+4}}} - y$$

$$= \frac{y}{y^2+2} . \qquad (4.46)$$

Given the sign of the result, $\delta \vec{x}_c$ points in the same direction of $\vec{y}$. An interesting property is that, for $y \gg \sqrt{2}$,

$$\delta x_c \approx \frac{1}{y} , \qquad (4.47)$$

meaning that the amplitude of the astrometric microlensing effect decreases with the distance of the source from the lens much slower than the photometric microlensing effect. In addition,

$$\frac{d(\delta x_c)}{dy} = \frac{2 - y^2}{(y^2+2)^2} , \qquad (4.48)$$

which shows that the shift has a maximum amplitude for $y = \sqrt{2}$, where $\delta x_c = \delta x_{c,max} = (2\sqrt{2})^{-1} \approx 0.354$. Assuming $\theta_E \approx 1$ mas, then $\delta\theta_c = \delta x_c \theta_E \approx 1/3$ mas.

We can now decompose the shift into the components parallel and perpendicular to the source trajectory:

$$\delta x_{c,\parallel} = \frac{y_\parallel}{y^2+2} = \frac{(t-t_0)/t_E}{[(t-t_0)/t_E]^2 + y_0^2 + 2} \qquad (4.49)$$

$$\delta x_{c,\perp} = \frac{y_\perp}{y^2+2} = \frac{y_0}{[(t-t_0)/t_E]^2 + y_0^2 + 2} . \qquad (4.50)$$

The functions $\delta x_{c,\parallel}(t)$ and $\delta x_{c,\perp}(t)$ are shown in Fig. 4.11.

We can see that $\delta x_{c,\parallel}$ is negative for $t < t_0$ and positive otherwise. We define $p \equiv (t-t_0)/t_E$. By taking the derivative with respect to p, we see that

$$\frac{d(\delta x_{c,\parallel})}{dp} = \frac{y_0^2 + 2 - p^2}{(p^2 + y_0^2 + 2)^2} . \qquad (4.51)$$

Thus, the function is extremal at the times t_m such that $(t_m - t_0)/t_E = \pm\sqrt{y_0^2 + 2}$, where

$$\delta x_{c,\parallel,min}, \delta x_{c,\parallel,max} = \pm\frac{1}{2\sqrt{y_0^2 + 2}} . \qquad (4.52)$$

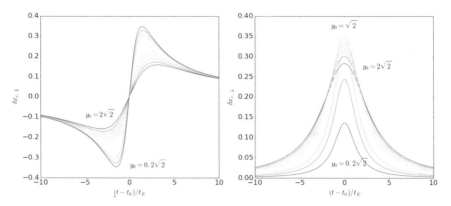

Fig. 4.11 Components of the light centroid shift as a function of time. The left and the right panels show the shift components parallel and perpendicular to the trajectory of the source, respectively. Different colors are used to illustrate the results for different impact parameters y_0

For $y_0 \ll 1$, the minimum and the maximum occur at $(t_m - t_0)/t_E \approx \pm\sqrt{2}$ and $\delta x_{c,\parallel,min/max} \approx \pm \delta x_{c,max}$.

On the other hand, $\delta x_{c,\perp}$ has a maximum at $(t - t_0)/t_E = 0$, i.e. at $t = t_0$, where it reaches the value

$$\delta x_{c,\perp,max} = \frac{y_0}{y_0^2 + 2} \ . \tag{4.53}$$

Since $\delta x_{c,\parallel}(t = t_0) = 0$, the shift is only perpendicular to the motion of the source at this time. For $y_0 = \sqrt{2}$, $\delta x_{c,\perp,max}$ reaches its maximum amplitude $\delta x_{c,max}$.

If Fig. 4.12, we show how the total amplitude of the centroid shift varies as a function of time for a variety of values of y_0. We can see that

$$\frac{d(\delta x_c)}{dp} = p \frac{2 - y_0^2 - p^2}{\sqrt{y_0^2 + p^2}(y_0^2 + p^2 + 2)^2} \ . \tag{4.54}$$

For $y_0 < \sqrt{2}$, $\delta x_c(t)$ has two maxima at t_{max} such that $(t_{max} - t_0)/t_E = \pm\sqrt{2 - y_0^2}$ and one minimum at $t_{min} = t_0$. On the contrary, for $y_0 > \sqrt{2}$, only one maximum exists at $t_{max} = t_0$. These results can be interpreted as follows. For a small impact parameter y_0, the centroid shift is mainly parallel to the motion of the lens relative to the source, thus two maxima exist, which correspond to the minimum and to the maximum of $\delta x_{c,\parallel}$ shown in Fig. 4.11. On the contrary, for large impact parameters, the dominant component of the shift is the one perpendicular to the direction of motion of the lens relative to the source, which peaks at $t = t_0$.

When combined, these motions generate the elliptical paths shown in the right panel of Fig. 4.10. The ellipses are centered in $(0, y_0)$. Their major-axes and the

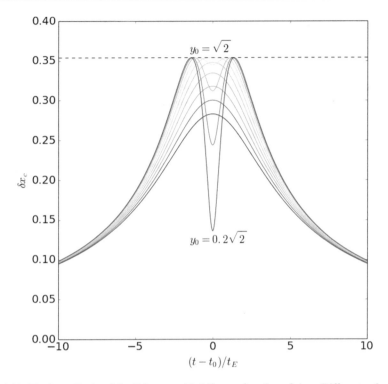

Fig. 4.12 Total amplitude of the light centroid shift as a function of time. Different colors are used to illustrate the results for different impact parameters y_0

semi-minor axes are oriented along the $\delta_{\parallel}$ and $\delta_{\perp}$ directions, respectively. As it results from Eqs. 4.52 and 4.53, their sizes are

$$a = \frac{1}{2} \frac{1}{\sqrt{y_0^2 + 2}} \tag{4.55}$$

$$b = \frac{1}{2} \frac{y_0}{y_0^2 + 2} . \tag{4.56}$$

For $y_0 \to \infty$, the ellipse degenerates to a circle with radius $r \approx 1/2y_0$ (thus, with size tending to zero). For $y_0 \to 0$, the ellipse degenerates to straight line, as $b \to 0$ and $a \to 1/2$.

Measuring the size of the ellipse defining the centroid path allows an independent measurement of the Einstein radius. Therefore, the combination of astrometric and photometric microlensing effects enables us to break the microlensing degeneracy.

4.5   Photometric Microlensing: Optical Depth and Event Rates

4.5.1   Optical Depth

The optical depth to some distance D_S is the probability that a source at that distance gives rise to a detectable microlensing event. As discussed earlier, we can assume that the lens cross section (in steradians) coincides with the solid angle enclosed by the Einstein ring, $\pi \theta_E^2$. Therefore, the optical depth can be computed as the sum of the cross sections of all lenses up to distance D_S, divided by the area of the sky. Let us assume that the number density of lenses varies as a function of the lens distance as $n(D_L)$. Then, the number of lenses within the solid angle Ω at distances between D_L to $D_L + dD_L$ is

$$dN_L = \Omega D_L^2 n(D_L) dD_L .$$
(4.57)

Thus, the optical depth is

$$\tau(D_S) = \frac{1}{\Omega} \int_0^{D_S} [\Omega D_L^2 n(D_L)](\pi \theta_E^2) dD_L .$$
(4.58)

If all lenses have the same mass M, then $n(D_L) = \rho(D_L)/M$, where $\rho(D_L)$ is the mass density. Note that, since $\theta_E \propto M^{1/2}$, the optical depth depends on the total mass density but not on the lens mass. This result can be generalized also to the case where the lens has a mass distribution (a.k.a. the *lens mass function*), if the spatial distribution of the lenses is independent on the mass. Indeed, we can write

$$n(D_L) = \int n(D_L, M) dM = \int \rho_M(D_L) M^{-1} dM .$$
(4.59)

Under this assumption, and using Eq. 4.7, we obtain

$$\tau(D_S) = \frac{4\pi G}{c^2} \int_0^{D_S} \rho(D_L) D_L^2 \frac{D_{LS}}{D_L D_S} dD_L$$

$$= \frac{4\pi G}{c^2} \int_0^{D_S} \rho(D_L) D_L \frac{D_S - D_L}{D_S} dD_L$$

$$= \frac{4\pi G}{c^2} \int_0^{D_S} \rho(D_L) \frac{D_L}{D_S} \left(1 - \frac{D_L}{D_S}\right) D_S dD_L .$$
(4.60)

Setting $x = D_L/D_S$, $dx = dD_L/D_S$, the optical depth is thus

$$\tau(D_S) = \frac{4\pi G}{c^2} D_S^2 \int_0^1 \rho(x) x(1 - x) dx .$$
(4.61)

By taking the derivative with respect to x, we see that

$$\frac{d\tau}{dx} \propto \rho(x)x(1-x) \, . \tag{4.62}$$

The function $f(x) = x(1-x)$, which weights the contribution of lenses to the optical depth, has a maximum at $x = 0.5$, i.e. the lenses at about halfway between the observer and the sources contribute the most to the optical depth. However, one must consider $\rho(x)$ to establish where most of the microlensing signal comes.

Note that, since sources are not all at a distance D_S, we should calculate the total optical depth by integrating $\tau(D_S)$ over the distribution of D_S.

We consider a very crude model of the galaxy, assuming that it is a spherical, self-gravitating system of lenses of mass M. Then, with $\rho(x) = \rho_0 =$const, we obtain

$$\tau(D_S) = \frac{4\pi G}{c^2}\rho_0 D_S^2 \int_0^1 x(1-x)dx = \frac{2}{3}\frac{\pi G}{c^2}D_S^2 \rho_0 \, . \tag{4.63}$$

Let us consider the case of microlensing in the Milky Way. The sphere centered on the center of the galaxy and with radius D_S contains a mass $M_{gal} = \frac{4}{3}\pi D_S^3 \rho_0$, thus

$$\tau(D_S) = \frac{GM_{gal}}{2c^2 D_S} = \frac{V_{circ}^2}{2c^2} \, , \tag{4.64}$$

where $V_{circ} \approx 220$ km/s is the circular velocity. Assuming that the sources are all at a galactocentric distance, the optical depth for microlensing is

$$\tau \approx 2.6 \times 10^{-7} \, . \tag{4.65}$$

Thus, we need to monitor many millions of stars to have a realistic yield of microlenses, meaning that observational campaigns have to target regions where stars' numerical density is very high. Several microlensing experiments have been carried out since the 1990s, which have targeted the galactic bulge and the Magellanic Clouds.

Remark 4.2. The calculations reported above assume that the mass density of lenses is constant. However, this is an over-simplification: in the case of microlensing toward the galactic bulge, we should account for the far more complex structure of our Galaxy, which includes several components (bulge, bar, disk, halo), each of which has its mass density. The true optical depth toward the galactic center is ∼3–10 times larger due to the flattened nature of the galactic disk and a bar's presence.

Optical Depth of an Exponential Disk

We report the calculations for computing the optical depth to the galactic center for lenses in the galactic disk (i.e. having an exponential density profile). This exercise was proposed in the excellent review paper by Mao (2008).

The mass density in the exponential disk is described, with respect to an observer near the sun, by the function

$$\rho(R) = \rho_0 \exp\left(-(R - R_0)/R_D\right), \tag{4.66}$$

where ρ_0 is the mass density in the solar neighborhood, R is the distance of the lens from the galactic center, R_0 is the distance of the sun from the galactic center, and R_D is the disk scale (i.e. a parameter which defines how quickly the density falls as a function of radius). Using the standard notation, we have that $R = D_{LS}$ and $R_0 = D_S$, thus

$$\rho(D_L) = \rho_0 \exp\left(D_L/R_D\right). \tag{4.67}$$

Scaling the distances by D_S, the density can be written as

$$\rho(x) = \rho_0 \exp x/x', \tag{4.68}$$

where $x' = R_D/D_S$. This can be inserted into Eq. 4.61 to obtain

$$\tau(D_S) = \frac{4\pi G}{c^2} \rho_0 D_S^2 \int_0^1 \exp\left(x/x'\right) x(1-x) dx. \tag{4.69}$$

Solving the last integral, we find that

$$\tau(D_S) = \frac{4\pi G}{c^2} \rho_0 D_S^2 x'^2 [2x' - 1 + \exp\left(1/x'\right)(2x' - 1)]. \tag{4.70}$$

Assuming $D_S = 8$ kpc, $R_D = 3$ kpc, $\rho_0 = 0.1 M_\odot \text{pc}^{-3}$, we obtain

$$\tau \approx 2.9 \times 10^{-6}. \tag{4.71}$$

4.5.2 Event Rate

The optical depth gives the probability that a source is undergoing a microlensing event at any given time. We are interested in knowing the rate of microlensing events we may observe while monitoring a certain number of sources for a specific time.

To calculate the event rate it is more natural to imagine that sources form a static background in front of which lenses move at some transverse velocity v. For simplicity, we can assume that this velocity is the same for all the lenses (although, in a realistic case, both the lenses and the sources have some velocity distributions).

The lens cross section has diameter $2r_E$, where r_E is the physical size of the Einstein radius on the lens plane, $r_E = D_L \theta_E$. In order to compute the probability to observe a microlensing event in a given time dt, we have to consider that the lens, while moving in front of the sources at velocity v, swipes a certain area. The area swept in the time dt is

$$dA = 2r_E v dt = 2r_E^2 \frac{dt}{t_E} . \tag{4.72}$$

Multiplying by the number of lenses in the solid angle Ω between D_L and $D_L + d D_L$, and then dividing by Ω, we obtain the probability that a source undergoes a new microlensing event in the time dt

$$d\tau = \frac{1}{\Omega} \int_0^{D_S} n(D_L) \Omega d A d D_L = 2 \int_0^{D_S} n(D_L) r_E^2 \frac{dt}{t_E} d D_L . \tag{4.73}$$

If we monitor $N_\star$ sources during the time dt, we obtain the expected number of microlensing events observed by multiplying the probability that one source undergoes a microlensing event by the number of stars monitored. Finally, by dividing by the time dt, we obtain the *event rate*:

$$\Gamma = \frac{d(N_\star \tau)}{dt} = \frac{2N_\star}{\pi} \int_0^{D_S} n(D_L) \frac{\pi r_E^2}{t_E} d D_L . \tag{4.74}$$

Assuming that all the Einstein crossing times are identical, we obtain

$$\Gamma = \frac{2N_\star}{\pi t_E} \tau . \tag{4.75}$$

Therefore, if $t_E \approx 19$ days,

$$\Gamma \approx 1200 \text{yr}^{-1} \frac{N_\star}{10^8} \frac{\tau}{10^{-6}} \left(\frac{t_E}{19 \text{days}} \right)^{-1} , \tag{4.76}$$

meaning that by monitoring $\sim 10^8$ stars, we would expect to observe ~ 1200 microlensing events per year. For comparison, the OGLE-IV collaboration, by monitoring 2×10^8 stars in the galactic bulge, detected ~ 1500–2000 event candidates/year between 2011 and 2017.

Note that while the optical depth does not depend on the mass, the event rate is mass-dependent because of $\Gamma \propto t_E^{-1} \propto M^{-1/2}$. This property is crucial because it means that we can use the distribution of event timescales to probe the kinematics of the Milky Way and the stellar population in the Galaxy.

4.6 Results from MACHO Searches

Since 1991, several collaborations between groups of astronomers around the globe were born to monitor the densest regions of stars inside and nearby our Galaxy: the bulge and the Small and the Large Magellanic Clouds (SMC and LMC). The primary motivation for these observational campaigns was to search for *Massive Astrophysical Compact Halo Objects* (MACHOs), i.e. very faint or invisible compact objects such as black holes, neutron stars, white and brown dwarfs. These compact objects were among the *baryonic* candidates for dark matter. As suggested by Paczynski (1986), compact objects in the halo of the Galaxy would produce a microlensing signal in addition to that produced by known stellar populations in the Galaxy. As discussed earlier, the optical depth for microlensing is minimal, and the event timescales can vary between a fraction of days and hundreds of days. Thus, to have the chance to detect some events, hundreds of millions of stars have to be monitored with a sufficiently short cadence. This requires to build networks of telescopes dedicated to these observations.

The LMC and SMC are sites that host sources that might be microlensed by MACHOs. The galactic bulge is less interesting for searching MACHOs, but searching for microlensing events in this direction can serve to probe the Galaxy structure. Lately, MACHOs searches have also been extended toward the galaxy M31 (Andromeda).

We give a list of collaborations involved in microlensing searches in Table 4.1. One of them (the Optical Gravitational Lensing Experiment, OGLE) is still in operation (OGLE IV). Some of the most interesting results found by these groups can be summarized as follows:

Table 4.1 List of some of the collaborations which have undertaken searches for microlensing events from dark matter candidates

Group	Target	Operation time	Reference
DUO (Disk unseen objects)	Bulge	1994–1997	Alard et al. (1995)
EROS (Experience pour la Recherche d'Objets sombres)	SMC, LMC	1990–2003	http://eros.in2p3.fr
MACHO	Bulge, LMC	1992–2003	http://wwwmacho.anu.edu.au
MOA (Microlensing Observations in Astrophysics	1995–2013	Bulge, LMC, SMC	http://www.phys.canterbury.ac.nz/moa/
OGLE (Optical Gravitational Lensing Experiment)	1992–*now*	Bulge, LMC, SMC	http://ogle.astrouw.edu.pl
POINT-AGAPE (Andromeda Galaxy and Amplified Pixels Experiment)	1999–2006	M31	Paulin-Henriksson et al. (2003)

- the relatively high rate of detection favored a barred model of the Galaxy;
- toward the Magellanic Clouds, no short events (timescales from a few hours up to 20 days) have been seen by any group. This places strong limits on Jupiters in the dark halo: specifically, compact objects in the mass range 10^{-6}–0.05 solar masses contribute less than 10% of the dark matter around our Galaxy. This is a significant result, as these objects were previously thought to be the most plausible form of baryonic dark matter, and (for masses below 0.01 solar masses) they would have been virtually impossible to detect directly;
- the detections of microlensing events toward the bulge are most likely caused by known stellar populations. Black holes can contribute to 2% of the total mass of the halo. Sumi et al. (2011) reported, however, an excess of short events, which indicates the presence of free-floating planets in the disk of the Milky Way.

4.7   Multiple Point Masses

4.7.1   Generalities

Deflection Angle

The deflection angle of an ensemble of N point masses was given in Eq. 2.47. Even for such lens, the proper choice of an angular scale allows to write the deflection angle in a convenient form. Generalizing the case of a single point mass, we can define an equivalent Einstein radius for a mass equal to the sum of the point masses, $M_{tot} = \sum_{i=1}^{N} M_i$. The reduced deflection angle can be written as

$$\vec{\alpha}(\vec{\theta}) = \sum_{i=1}^{N} \frac{D_{\mathrm{LS}}}{D_{\mathrm{L}} D_{\mathrm{S}}} \frac{4GM_i}{c^2} \frac{(\vec{\theta} - \vec{\theta}_i)}{|\vec{\theta} - \vec{\theta}_i|^2} \frac{M_{tot}}{M_{tot}} = \sum_{i=1}^{N} m_i \frac{\theta_E^2}{|\vec{\theta} - \vec{\theta}_i|^2} (\vec{\theta} - \vec{\theta}_i) , \qquad (4.77)$$

where we have set $m_i = M_i / M_{tot}$. By further dividing by θ_E, we obtain

$$\vec{\alpha}(\vec{x}) = \sum_{i=1}^{N} \frac{m_i}{|\vec{x} - \vec{x}_i|^2} (\vec{x} - \vec{x}_i) , \qquad (4.78)$$

where $x = \theta / \theta_E$.

Lens Equation

The lens equation in the dimensionless form then reads

$$\vec{y} = \vec{x} - \sum_{i=1}^{N} \frac{m_i}{|\vec{x} - \vec{x}_i|^2} (\vec{x} - \vec{x}_i) . \qquad (4.79)$$

Witt (1990) showed that it is convenient to use the complex notation instead of the vectorial form to write this lens equation. Using this notation, $z = x_1 + i x_2$ and

$z_s = y_1 + iy_2$ are the positions on the lens and on the source planes. The complex deflection angle is $\alpha(z) = \alpha_1(z) + i\alpha_2(z)$ which can then be written as

$$\alpha(z) = \sum_{i=1}^{N} m_i \frac{(z - z_i)}{(z - z_i)(z^* - z_i^*)} = \sum_{i=1}^{N} \frac{m_i}{z^* - z_i^*} , \qquad (4.80)$$

where the $*$ symbol denotes the complex conjugate. The lens equation is

$$z_s = z - \sum_{i=1}^{N} \frac{m_i}{z^* - z_i^*} . \qquad (4.81)$$

By taking the complex conjugate of both sides, we can then solve for z^*:

$$z^* = z_s^* + \sum_{i=1}^{N} \frac{m_i}{z - z_i} . \qquad (4.82)$$

This can be inserted into Eq. 4.81 to obtain a complex polynomial equation of degree $N^2 + 1$. Thus, the lens equation formally has up to $N^2 + 1$ solutions, some of which may, however, be spurious. Rhie (2001; 2003) probed that, in fact, a lens composed of $N > 3$ point masses produces a maximum of $5(N - 1)$ images.

Critical Lines

To find the critical lines, we first need to compute the determinant of the lensing Jacobian. In Sect. 3.3, this was found to be

$$\det A = \frac{\partial y_1}{\partial x_1} \frac{\partial y_2}{\partial x_2} - \left(\frac{\partial y_1}{\partial x_2} \right)^2 . \qquad (4.83)$$

Using the complex differential operators, we see that

$$\frac{\partial z_s}{\partial z} = \frac{1}{2} \left(\frac{\partial}{\partial x_1} - i \frac{\partial}{\partial x_2} \right) (y_1 + iy_2) = \frac{1}{2} \left(\frac{\partial y_1}{\partial x_1} + \frac{\partial y_2}{\partial x_2} \right) + \frac{i}{2} \left(\frac{\partial y_2}{\partial x_1} - \frac{\partial y_1}{\partial x_2} \right)$$
$$\qquad (4.84)$$

$$\frac{\partial z_s}{\partial z^*} = \frac{1}{2} \left(\frac{\partial}{\partial x_1} + i \frac{\partial}{\partial x_2} \right) (y_1 + iy_2) = \frac{1}{2} \left(\frac{\partial y_1}{\partial x_1} - \frac{\partial y_2}{\partial x_2} \right) + \frac{i}{2} \left(\frac{\partial y_2}{\partial x_1} + \frac{\partial y_1}{\partial x_2} \right) . \qquad (4.85)$$

The imaginary part of Eq. 4.84 is zero, because $\partial y_1 / \partial x_2 = \partial y_2 / \partial x_1$. Thus,

$$\left(\frac{\partial z_s}{\partial z} \right)^2 = \frac{1}{4} \left[\left(\frac{\partial y_1}{\partial x_1} \right)^2 + \left(\frac{\partial y_1}{\partial x_2} \right)^2 + 2 \frac{\partial y_1}{\partial x_1} \frac{\partial y_2}{\partial x_2} \right] \qquad (4.86)$$

and

$$\left(\frac{\partial z_s}{\partial z^*}\right)\left(\frac{\partial z_s}{\partial z^*}\right)^* = \frac{1}{4}\left[\left(\frac{\partial y_1}{\partial x_1}\right)^2 + \left(\frac{\partial y_1}{\partial x_2}\right)^2 - 2\frac{\partial y_1}{\partial x_1}\frac{\partial y_2}{\partial x_2}\right] + \left(\frac{\partial y_1}{\partial x_2}\right)^2. \quad (4.87)$$

By taking the difference of Eqs. 4.86 and 4.87, we obtain that

$$\left(\frac{\partial z_s}{\partial z}\right)^2 - \left(\frac{\partial z_s}{\partial z^*}\right)\left(\frac{\partial z_s}{\partial z^*}\right)^* = \frac{\partial y_1}{\partial x_1}\frac{\partial y_2}{\partial x_2} - \left(\frac{\partial y_1}{\partial x_2}\right)^2 = \det A. \quad (4.88)$$

Using the lens equation in the form given in Eq. 4.81, we finally see that

$$\frac{\partial z_s}{\partial z} = 1 \quad (4.89)$$

and

$$\frac{\partial z_s}{\partial z^*} = \sum_{i=1}^{N} \frac{m_i}{(z^* - z_i^*)^2}. \quad (4.90)$$

Thus,

$$\det A = 1 - \left|\sum_{i=1}^{N} \frac{m_i}{(z^* - z_i^*)^2}\right|^2. \quad (4.91)$$

It follows that the critical lines are defined by

$$\left|\sum_{i=1}^{N} \frac{m_i}{(z^* - z_i^*)^2}\right|^2 = 1. \quad (4.92)$$

The sum in the above equation must be satisfied on the unit circle. The complex solutions of this equation can be found by solving

$$\sum_{i=1}^{N} \frac{m_i}{(z^* - z_i^*)^2} = e^{i\phi} \quad (4.93)$$

for each $\phi \in [0, 2\pi)$. The above equation is a complex polynomial of order $2N$ with respect to z. Thus, for each value of ϕ, there are $2N$ or less critical points. By varying ϕ continuously, the solutions will trace out $2N$ (or less) critical lines. Critical lines corresponding to different solutions may join smoothly (see e.g. Witt 1990). Note that for $N = 1$, $m_1 = 1$ and, by taking $z_1 = 0$, we obtain that the critical line is the Einstein ring ($|z| = 1$). In the case of multiple point masses, however, the critical lines are much more complicated, as we will see in the next section.

As usual, we can map the critical lines onto the source plane through the lens equation to obtain the caustics.

4.7.2 Binary Lenses

Lens Equation

The binary lens is a particular case of multiple point-mass lenses, where $N = 2$. In this case, the lens equation is

$$z_s = z - \frac{m_1}{z^* - z_1^*} - \frac{m_2}{z^* - z_2^*} \,. \qquad (4.94)$$

Since the choice of the reference frame is arbitrary, we may opt for selecting the real axis to pass through the two lenses, as shown in Fig. 4.29. We may further set $z_2 = -z_1$.

As discussed in the previous section, the lens equation for the binary lens can be reduced to a complex polynomial equation of degree 5:

$$c_0 + c_1 z + c_2 z^2 + c_3 z^3 + c_4 z^4 + c_5 z^5 = 0 \,, \qquad (4.95)$$

where

$$c_0 = z_1^2 [4(\Delta m)^2 z_s + 4m\,\Delta m z_1 + 4\Delta m z_s z_s^* z_1 + 2m z_s^* z_1^2 + z_s z_s^{*2} z_1^2 - 2\Delta m z_1^3 - z_s z_1^4]$$

$$c_1 = -8m\,\Delta m z_s z_1 - 4(\Delta m)^2 z_1^2 - 4m^2 z_1^2 - 4m z_s z_s^* z_1^2 - 4\Delta m z_s^* z_1^3 - z_s^{*2} z_1^4 + z_1^6$$

$$c_2 = 4m^2 z_s + 4m\,\Delta m z_1 - 4\Delta m z_s z_s^* z_1 - 2z_s z_s^{*2} z_1^2 + 4\Delta m z_1^3 + 2z_s z_1^4$$

$$c_3 = 4m z_s z_s^* + 4\Delta m z_s^* z_1 + 2z_s^{*2} z_1^2 - 2z_1^4$$

$$c_4 = -2m z_s^* + z_s z_s^{*2} - 2\Delta m z_1 - z_s z_1^2$$

$$c_5 = z_1^2 - z_s^{*2} \,. \qquad (4.96)$$

In the above equations we have introduced $\Delta m = (m_1 - m_2)/2$ and $m = (m_1 + m_2)/2$ (Witt & Mao 1995).

Critical Lines and Caustics

It can be shown that Eq. 4.95 has 3 or 5 images depending on the distance between the two point masses and on their mass ratio. This can be better understood by looking at the structure of the critical lines and caustics. The critical lines can be found as explained above by solving the Eq. 4.93. In the case of the binary lens, this assumes the form

$$\frac{m_1}{(z^* - z_1^*)^2} + \frac{m_2}{(z^* - z_2^*)^2} = \frac{m_1}{(z^* - z_1^*)^2} + \frac{m_2}{(z^* + z_1^*)^2} = e^{i\phi} \tag{4.97}$$

for $\phi \in [0, 2\pi)$. By getting rid of the fractions, the equation can be turned into

$$z^4 - z^2 (2z_1^{*2} + e^{i\phi}) - z z_1^{*2} (m_1 - m_2) e^{i\phi} + z_1^{*2} (z_1^{*2} - e^{i\phi}) = 0 . \tag{4.98}$$

The left side is a fourth-degree polynomial. Thus, for each ϕ there are up to four solutions of this equation.

In the python application in Sect. 4.9.5, we show how to find the solutions of the above equation and to derive the critical lines and caustics of the binary lens. Depending on the ratio of the two masses, $q = m_1/m_2 = M_1/M_2$, and the separation between the two lenses in units of the equivalent Einstein radius, $d = |z_1 - z_2|$, the resulting caustics can be one, two, or three.

We distinguish between wide, intermediate, and close systems based on the topology of the caustics, as shown in Fig. 4.13. More precisely:

- in wide systems, there exists two separate extended caustics, which correspond to the point-like caustics associated with the individual lenses. The caustics' shape resembles an astroid with four cusps, resulting from each lens's reciprocal perturbation. Indeed, the presence of two masses breaks the symmetry of the point-mass lens. The wide topologies are typical of binary systems with large separations between the two lenses;
- in intermediate systems, there exists a single caustic, characterized by six cusps. This caustic is the result of the merging of the two individual astroid-like caustics in wide systems, which happens when the lenses are brought closer to each other (or when the masses of the individual lenses are increased, thus making the equivalent Einstein radius bigger, and reducing d);
- finally, in close systems, there are three caustics. Two have triangular-like shapes, and one is an astroid-like caustic with four cusps. The triangular shape caustics are at the same distance from the axis connecting the two lenses. In binary lenses with $q \sim 1$, they are located nearly on an axis passing through the central caustic.

The transitions between these topologies occur when two critical lines merge at one point (Mollerach & Roulet 2002). This happens where not only $\det A = 0$ but also $\partial \det A / \partial z^* = 0$ (the gradient of the Jacobian determinant is zero). These are saddle points of the *surface* $\det A(\vec{\theta})$. In particular, it can be shown that the transition between wide and intermediate regimes occurs for a separation between the lenses (in units of the equivalent Einstein radius) of

$$d_{WI} = (m_1^{1/3} + m_2^{1/3})^{3/2} . \tag{4.99}$$

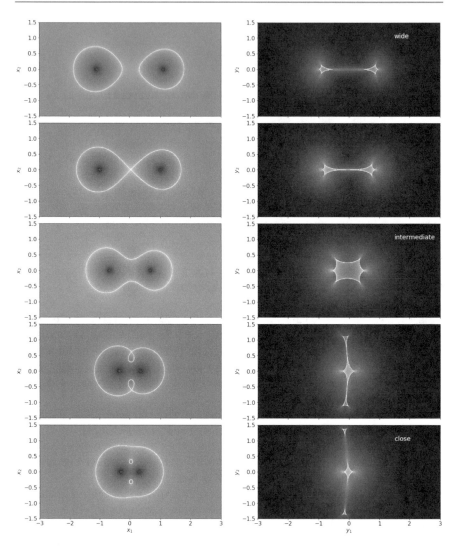

Fig. 4.13 Critical lines (left panels) and caustics (right panels) of a binary lens system for different values of the separation between the point masses, d. From the top to the bottom, we show examples of wide, intermediate, and close topologies. The results refer to the case of a lens with $M_2 = 1M_\odot$ and $q = 0.8$. The critical lines and the caustics are overlaid to the maps of the magnification on the lens and on the source planes, respectively

On the other hand, the transition between intermediate and close regimes occurs when

$$d_{IC} = (m_1^{1/3} + m_2^{1/3})^{-3/4} \ . \tag{4.100}$$

A detailed discussion about caustic topologies in binary lenses is presented in Erdl and Schneider (1993).

If q significantly differs from unity, i.e. one mass component is dominant over the other, the shape of the caustics changes as shown in Fig. 4.16. In particular:

- in wide lenses, the caustic of the primary (i.e. most massive) component is smaller than the caustic of the secondary. Besides, it is highly asymmetric and elongated toward the secondary. The two caustics are shifted toward the primary lens.
- in intermediate lenses, the caustic is thinner on the side of the primary and fatter on the side of the secondary;
- in close lenses, the triangular caustics are located behind the primary lens, opposite to the secondary lens. They are still equidistant from the axis passing through the two point masses.

Multiple Images

Solving Eq. 4.95 leads to finding the multiple images of a source at z_s. We obtain these solutions using numerical methods, as illustrated in Sect. 4.9.6. Here, we give some brief and qualitative statements about the occurrence of multiple images.

- a source outside the caustics has only three images, meaning that two lens equation solutions are spurious. Two of these images form inside the critical lines and, for large distances of the source from the binary lens, their positions are very close to the point masses. They are analogs of the image with negative parity in the single point-mass lens case (see Eq. 4.11). The image outside the critical lines corresponds to a local minimum of the time delay surface and has positive parity;
- when the source is on the caustic, two additional images appear on the critical line (thus they are formally indistinguishable and have infinite magnification);
- when the source is inside the caustics, five images exist.

Remark 4.3. Note that, due to the singularity of the two point-masses' lensing potential, there are always at least three images of a single source.

The above statements can be verified by looking at Fig. 4.14. In the left panel, we show the critical lines of a binary lens. The case chosen corresponds to values of $q = 1$ and $d = 1$, respectively. We show the corresponding caustics in the right panel, where we also display the trajectory of a source in motion relative to the binary lens. Similarly to the case of a single point lens, we can describe this trajectory using few parameters, as shown in Fig. 4.29, namely the minimal distance of the source from the center of the reference frame, y_0, the time t_0 when it reaches this distance, and the inclination angle θ_S relative to the real axis. More details are given in Sect. 4.9.5. We use a color sequence to show the source position as a function of time. We show the images, also color-coded, in the left panel.

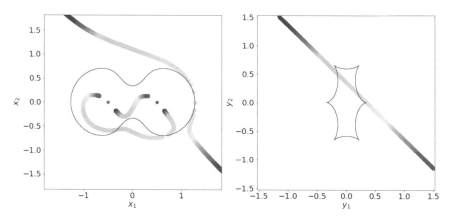

Fig. 4.14 Images of a source moving behind a binary lens. In the right panel, we show the lens caustics and the trajectory of the source. We color-code the source position as a function of time. We show the corresponding images in the left panel, together with the lens critical line. The binary lens consists of two stars of equal mass at a distance of $d = 1$. Thus the caustic is resonant

The source trajectory crosses the lens caustics at two points. Let us call the corresponding crossing times t_1 (light blue) and t_2 (green). Before t_1 and after t_2, there are three images of the background source. Between t_1 and t_2, the images are five. In particular, two images appear at the time t_1 on the lens critical line. While the source moves across the caustic, one of the images moves outside the critical line, while the other follows a trajectory that brings it near the star of mass m_1. The outer image approaches the critical line again at t_2 when it merges with one of the inner images. The two images disappear once the source has crossed the caustic again.

Image Magnifications and Light-Curves
As in the case of microlensing by single lenses, the multiple images are not spatially resolved. Thus, the microlensing events involving binary lenses also manifest through variations of the source luminosity. However, the source light-curves are significantly more complex than those of the single point-mass lenses.[2]

The shape of the light-curve reflects how the sum of the magnifications of the images changes as a function of time:

$$\mu(t) = \sum_{j=1}^{N_{ima}} |\mu_j(t)| . \tag{4.101}$$

---

[2]Even in the case of binary lenses, microlensing events produce astrometric signatures.

The magnification of image j is obtained by inserting the image position $z_j(t)$ in Eq. 4.91:

$$\mu_j(t) = \left[1 - \left|\sum_{i=1}^{N} \frac{m_i}{z_j^*(t) - z_i^*}\right|^2\right]^{-1}. \qquad (4.102)$$

In Sect. 4.9.7, we discuss how to compute the light-curve of a source in the background of a binary lens. The shape of the light-curve reflects the pattern of the magnification along the trajectory of the source. In the left panel of Fig. 4.13, we overlay the caustics to the source plane magnification maps (the maps in the left panels show the image-plane magnification instead). We compute the total source magnification at each position using Eq. 4.102. Some features of the magnification maps are important to remind:

- lobes of high magnification surround the cusps of the caustics. Thus, when they pass near the cusps, sources experience high magnifications, which appear as bumps in their light-curves;
- the folds of the caustics mark sharp transitions in magnification from outside to inside the caustics. The magnification rises suddenly when a source crosses the caustics fold, declines more gently while the source is inside the caustic, before rising again when the source approaches another fold. Then, after crossing the fold, the magnification drops. Because of the magnification pattern, caustic crossings appear in the light-curves as sharp transitions characterized by asymmetric profiles. Witt and Mao (1995) showed that, while the source is inside the caustic, the total magnification cannot be smaller than 3;
- extended regions of high magnification are present along the direction connecting the caustics of wide and close lenses. Therefore, the passage of the source in between caustics also produces a bump in the light-curve.

We show in Fig. 4.15 the light-curve corresponding to the example in Fig. 4.14, regarding a source moving past a six-cusp caustic in an intermediate lens. As discussed, the signatures of the passage of a source across the caustic are two very sharp spikes in the light-curve. In between them, the light-curve has a characteristic "U"-shape. The passage of the source near the lens's cusp produces another single bump in the light-curve.

Finite source size effects affect the sharpness of these transitions in the light-curves. Indeed, if the source's size is not negligible, the light-curve appears smoothed, while the event lasts longer (because the source takes more time to cross the fold of the caustic). Because of the more extended size of the caustics, we can detect finite source size effects in binary lenses more easily than in single lenses (where we can see them only if the impact parameter y_0 is very small). As discussed earlier, such effects are essential to constrain the caustics' size and derive θ_E, helping to break the model degeneracies.

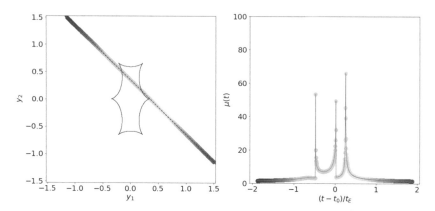

Fig. 4.15 Light-curve of the source whose multiple images are shown in Fig. 4.14

4.8   Planetary Microlensing

A system made of a planet orbiting a star is a particular binary lens where the star dominates the mass budget. For Jupiter-like planet orbiting a solar mass star $q \sim 10^{-3}$. Conversely, for an Earth-like planet $q \sim 3 \times 10^{-6}$. Given the small mass ratio, the light-curve is very similar to the standard microlensing light-curve by a single star most of the time. The presence of a secondary lens (the planet) produces localized perturbations to the magnification pattern, which manifest through short-time variations of the standard light-curve. The features produced by the planet have shapes that strongly depend on the source's trajectory relative to the lens. They depend on whether the source passes near the perturbed caustic of the star or the planetary caustics.

4.8.1   Perturbations of the Central Caustic

As discussed earlier, the caustic shape is essential. Also, in the case of planetary microlensing, three types of caustic (and critical line) topologies are possible: wide, intermediate (or resonant), and close.

Figure 4.16 illustrates how the caustics and the critical lines of a binary lens change when we vary the mass ratio of q while keeping d fixed. The examples shown in the three columns refer to the three topologies mentioned above. First, we focus on the central caustic, i.e. the caustic of the primary lens, in the cases of wide and close topologies (left and right columns). Four cusps and four folds characterize the caustic shape. Because of the small value of q, the caustic is very asymmetric and elongated in the planet's direction, with one cusp pointing toward it. Three additional cusps are located on the back of the caustic (relative to the planet).

In both cases of wide and close topologies, the caustic becomes smaller as q decreases. Also, the angular distance between the star and its caustic becomes

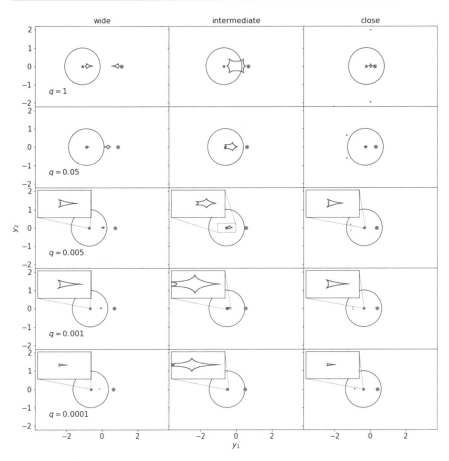

Fig. 4.16 Caustics of a binary lens for different values of the mass ratio q. The left, central, and right panels refer to wide, intermediate, and close topologies. In each panel, the two mass components are indicated with a red star and with a circle. In the cases with $q \ll 1$, the green circle represents a planet-like mass. The black circle in each panel gives the equivalent Einstein radius. The insets show zooms over the central caustics

smaller. This result is not surprising: as long as the two masses are well separated, putting most of the mass into one of the two lenses makes the overall system very similar to a single point-mass lens, which is only weakly perturbed by the secondary. Similarly, bringing the star and the planet close to each other, the central caustic shrinks (close topology), as the lens also becomes more and more similar to a single point lens.

The insets in Fig. 4.16 show that the central caustic of wide and close binary lenses are degenerate for $q \ll 1$. In particular, close star-planet pairs with separation d_c have central caustics identical to those of wide systems with separation $d_w = d_c^{-1}$ (as in the examples shown in the left and in the right panels of Fig. 4.16). This degeneracy is called "wide-close degeneracy").

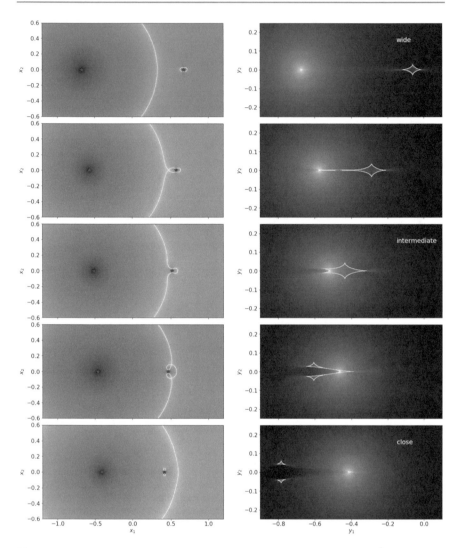

Fig. 4.17 As in Fig. 4.13, but for the case of a lens with $M_2 = 1M_\odot$ and $q = 10^{-3}$

Figure 4.17 is analogous to Fig. 4.13 but refers to the case of a star-planet system where the primary lens has mass $M_2 = 1M_\odot$ and $q = 10^{-3}$. The magnification maps in the right panels show some interesting features that are important to interpret the light-curves' shape. They show that the central caustic back is a region of de-magnification in close and wide systems. Dips in the light-curves are, therefore, strong indications of the presence of a planet.

Figure 4.18 shows some examples of the effects of the central caustic perturbations by planets on the source light-curve. The dotted line gives the source trajectory in each of the left panels. The central panels zoom over the central caustic. The

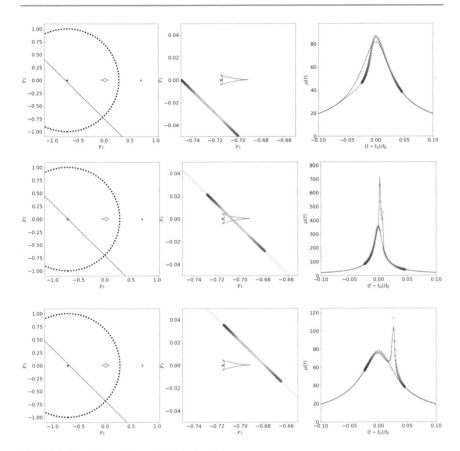

Fig. 4.18 Central caustic perturbations by planets

source trajectory is indicated here by a color sequence, which allows the source position's read-off when a perturbation of the light-curve appears in the right panels. From the upper to the lower panels, the source trajectory passes behind, across, or in front of the central caustic (relative to the planet's position given by the blue dot in the left panels). The standard light-curve, i.e. the light-curve we would measure in the absence of the planet, is given by the right panels' dashed black lines.

When the source passes behind the central caustic (upper panels), we notice a negative deviation of its light-curve from the standard light-curve of a point lens, followed by a positive variation when the source passes close to one of the back cusps.

If the source crosses the caustic (middle panels), two peaks appear in the light-curve when the source enters and exits the central caustic. Finally, if the source passes near the pronounced cusp between the star and the planet, where the magnification is incredibly strong, a single, sharp peak in the light-curve shows up (bottom panels).

It is worth noticing that all these perturbations are near the standard light-curve peak, i.e. they occur in a high-magnification regime. This fact is particularly relevant for event detectability. In some sense, events like these are predictable because we can discover the primary microlensing event earlier than the planet perturbation appears. If the search strategy consists of detecting the primary event and triggering a follow-up (as in the first generation of microlensing surveys), the cadence at which the light-curve is monitored can be adjusted to make the detection of such planet perturbations possible. Besides, photometric measurements in such events can be more accurate (Gaudi 2012).

4.8.2   Perturbations of the Planetary Caustic

Another way to detect a planet around a star is by using signatures of the planetary caustics. We can estimate these caustics' distance from the central caustic (which is roughly coincident with the host star's projected position) from the lens equation. If the planet is at a distance d from the host star, then the distance between the planetary and the central caustic is $s \sim |d - d^{-1}|$.

In wide topologies, the planetary caustic is an astroid-like caustic with four cusps and four folds. Han (2006) showed that the size of this caustic scales as $\sim q^{1/2} d^{-2}$. Signatures of this caustic in the light-curve are single- or double-peaks occurring when the source passes close to a cusp or across the folds. We show some examples of these perturbations in the upper panels of Fig. 4.19. The perturbations can be detected near the primary light-curve's peak if the source trajectory is nearly perpendicular to the binary lens axis, in the wings otherwise. In all cases, planetary caustic events occur in low- to medium-magnification regimes as the source passes at a relatively large distance from the star.

In close topologies, there are two planetary caustics with three cusps and three folds. They are located on the opposite side of the star relative to the planet (see the bottom panels in Fig. 4.19). The two caustics' positions are symmetric with respect to the axis passing through the star and the planet. Han (2006) showed that the caustic size scale as $\sim q^{1/2} d^3$. The separation between the two caustics is $\sim 2q^{1/2}(d^{-2} - 1)^{1/2}$, and their distance from the star is $\sim d^{-1} - d$. The signatures of these caustics in the source light-curve can be single- or double-peaks corresponding to the cases where the source passes near the cusps or across the caustics. Besides, the region in between the two triangular caustics is characterized by a relatively low magnification, as shown in Fig. 4.17. Thus, we see a dip in the primary microlensing event light-curve when the source passes near these planetary caustics and crosses the axis of the binary lens.

Remark 4.4. Negative deviations from the standard light-curve are signatures of planets.

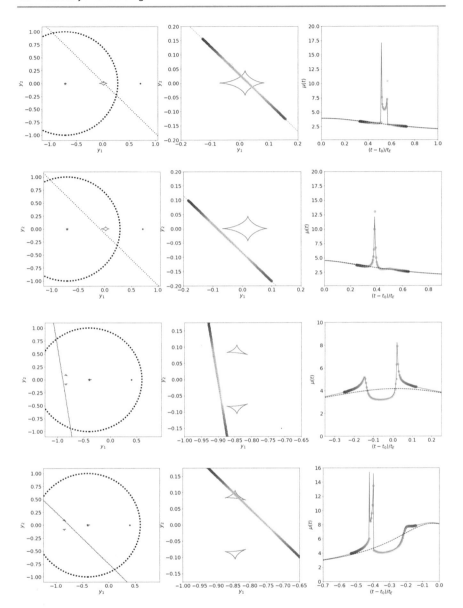

Fig. 4.19 Planetary caustic perturbations of the light-curves

4.8.3   Perturbations of the Resonant Caustic

Intermediate (or resonant) caustics are possible only for a narrow range of separations between the star and the planet. Let us consider the transition between wide and intermediate topologies (Eq. 4.99), which can be written in terms of the mass of the primary lens m_2 and of q as

$$d_{WI} = (1 + q^{1/3})^{3/2} m_2^{1/2} .$$ (4.103)

We recall that d is the angular separation between the two masses in units of the equivalent Einstein radius, θ_E, which is in turn related to the Einstein radius of the primary by

$$\theta_E' = m_2^{1/2} \theta_E .$$ (4.104)

Thus, d_{WI} can be re-written in units of the Einstein radius of the primary star as

$$d_{WI}' = d_{WI}/m_2^{1/2} = (1 + q^{1/3})^{3/2} \sim 1 + 3/2 q^{1/3} .$$ (4.105)

For very small values of q, we obtain that

$$d_{WI}' \sim d_{WI} \sim 1 + 3/2 q^{1/3} .$$ (4.106)

Similarly, from Eq. 4.100), we obtain that

$$d_{IC}' \sim d_{IC} \sim 1 - 3/4 q^{1/3} .$$ (4.107)

Therefore, the range of distances between the star and the planet for which the caustic is resonant is

$$d_{WI}' - d_{IC}' \sim \frac{9}{4} q^{1/3} .$$ (4.108)

This range is very slim for small values of q. In particular, both d_{WI}' and d_{IC}' are close to one for $q \ll 1$. Thus, in the case of resonant caustics, the planet is as far from the star as one Einstein radius. This is shown in the examples shown in the central column of Fig. 4.16.

Remark 4.5. For a given q, even small changes of d have dramatic effects on the resonant caustic shape. This is shown in Fig. 4.20 for a star-planet lens with $q = 10^{-3}$. Extremely different caustic shapes are obtained for star-planet pairs with d varying in the range $d \in [0.95, 1.1]$. Consider a source that enters the resonant caustic at the time t. Due to the relatively large extension of the caustic, the time $\Delta t = t' - t$ needed to the source to exit the caustic can be long enough that, in the meanwhile, d has changed because of the orbital motion of the planet. Consequently, the overall shape of the resonant caustic has changed too. Modeling the variation of the caustic shape as a function of time in resonant caustic crossing events allows to set constraints on the orbital motion of the planet (see, e.g. Gaudi et al. 2008)

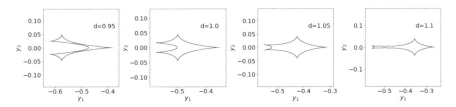

Fig. 4.20 Shape of the resonant caustic of a star-planet lens with $q = 10^{-3}$ for several values of d

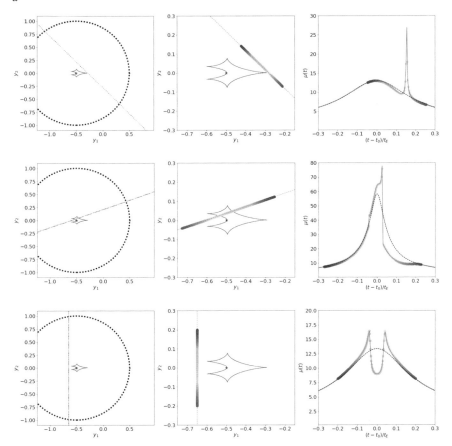

Fig. 4.21 Resonant caustic perturbations of the light-curves

We show some examples of light-curve perturbations by a resonant caustic in Fig. 4.21. The source's passage in front, across, or behind the caustic corresponds to either positive or negative deviations from the standard light-curve. These perturbations are located near the standard light-curve's peak, and, because the caustic is close to the star, these events occur in intermediate- to high-magnification regimes.

An essential property of the resonant caustic is that it is generally weak, as shown in Fig. 4.17. As the source approaches the caustic folds, the magnification rises sharply (see the middle panels of Fig. 4.21). However, such short magnification boosts can be washed-out by finite source size effects, resulting in being difficult to detect.

4.8.4   Perturbations of the Inner and Outer Images

As discussed above, planetary microlensing events cause deviations from the standard light-curve of the primary microlensing event produced by the star. Here, we show that they result from the perturbation of either the inner or the outer images in the primary microlensing events.

Some examples are shown in Fig. 4.22. The upper panels refer to a wide caustic topology, as shown in the left panel. We consider the positions of the source at

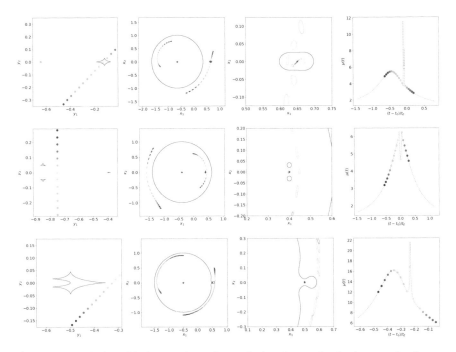

Fig. 4.22 Perturbation of the inner and outer images in the primary microlensing event by planets. We assume that the primary star has mass $M_2 = 1 M_\odot$ and that $q = 10^{-3}$. Each row of panels refers to a different caustic topology (wide, close, and resonant, from the upper to the lower panels). We show the caustics and the source positions at different times in the first panel on the left in each row. The time is color-coded from blue to red. The second panel shows the critical lines and the positions of source images. The positions of the star and the planet are given by the red and the blue stars. Even in this case, the colors help associate the images to the left panel's corresponding source positions. In the third panel, we zoom over the planetary critical lines to show how the planet's images are affected. The sources are assumed to have circular shapes, with radius $r = 0.01\theta_E$. The fourth panel shows the source light-curve during the microlensing event

different times along its trajectory, indicated with different colors. The trajectory is such that the source crosses the planetary caustic (see green color). We show the source images and the critical lines in the second panel. When the source is far from the planet, it has two images, one inside and one outside the Einstein ring of the host star. In the third panel, we zoom over the planetary critical line. For illustration purposes, we have assumed that the source has a circular shape and that its radius is $r = 0.01\theta_E$. Thus, the images are extended, and we can appreciate their distortion and magnification. When the source is inside the planetary caustic, the outer image is affected by the planet. Specifically, it splits into additional multiple images. Therefore, the anomaly in the source light-curve shown in the right panel is the consequence of the perturbation by the planet of the outer image in the primary microlensing event.

Similarly, the middle panels show the planetary perturbation of the inner image in a close topology case. In this case, the inner image is de-magnified when it is close to the planetary critical lines. Consequently, we observe a dip in the source light-curve.

The bottom panels illustrate the case of a source passing near the cusp of a resonant caustic. In this case, the planet magnifies the outer image in the primary event.

Remark 4.6. The examples shown above illustrate that microlensing is sensitive to planets near the host star's Einstein radius. For solar mass stars in between us and the bulge, the Einstein radius's size is $\sim$2–4 AU, which is thus the typical orbital size of planets that can be detected via microlensing.

4.8.5 Analysis of the Light-Curve in a Planetary Caustic Crossing Event

The connection between anomalies in the light-curve and image locations allows to derive meaningful information about the star and the planet from the light-curves (Gaudi & Gould 1997). For example, let us consider the case illustrated in the upper panels of Fig. 4.22. We can use the standard light-curve model given in Eq. 4.33 for describing the primary event, which allows us to obtain t_E, t_0, and y_0.

Remark 4.7. Note that y_0 denotes the minimal distance of the source from the star here, as in the case of single point lenses. It is not the minimal distance from the mid-point between the star and the planet, which was chosen as the origin of the reference frame for binary lenses (see Fig. 4.29). Similarly, t_0 is the time when the distance between the star and the source is y_0. Let us re-define the minimal distance of the source from the mid-point between the star and the planet as y_0', and the corresponding source passage time as t_0'. From the geometrical construction in Fig. 4.29, we can derive the following relations between t_0' and y_0' and the parameters

y_0 and t_0 of the primary event:

$$t_0' = t_0 + t_E \cos\theta_S \Re(z_2) , \tag{4.109}$$

$$y_0' = y_0 + \sin\theta_S \Re(z_2) . \tag{4.110}$$

The time of the planetary anomaly, t_p, can be used to find the planet's distance from the host star. Indeed, the anomaly in the light-curve marks the passage of the source on the planetary caustic. The distance of the planetary caustic from the central caustic (and from the host star) is thus

$$y_p(t_p) = \sqrt{y_0^2 + \left(\frac{t_p - t_0}{t_E}\right)^2} . \tag{4.111}$$

Using the lens equation, the distance of the planet from the host star is thus

$$d = \frac{y_p \pm \sqrt{y_p^2 + 4}}{2} . \tag{4.112}$$

The sign to choose in the above equation depends on the caustic topology. In the case considered here, we choose the plus sign because the planet perturbs the outer image.

From the light-curve, we can also measure the length of the planetary perturbation, which is approximately equal to the Einstein crossing time of the planet, $t_{E,p}$, As shown in Eq. 4.30, t_E and $t_{E,p}$ are proportional to square root of the masses of host star and of the planet, respectively. Therefore, the mass ratio q can be estimated as

$$q \approx \left(\frac{t_{E,p}}{t_E}\right)^2 . \tag{4.113}$$

Of course, a more detailed model of the system is necessary to measure the lens parameters accurately. Specific software has been developed to perform fast and accurate numerical calculations of light-curves to be fitted to the data (see, e.g. Bozza 2010, Bozza et al. 2018). This software also includes features which were not addressed in detail here, such as orbital motions. Note that the microlensing degeneracy discussed in the case of single point masses persists. Finite source size and microlensing parallax effects can be used to break the degeneracy (Udalski et al. 2005). In rare cases, HST imaging or adaptive optics on large telescopes allows resolving the host star (Batista et al. 2015, Bennett et al. 2006; 2015).

Modeling planetary events remains, however, a challenging job. As discussed earlier, a fundamental degeneracy exists between close and wide systems. Besides, similar features in the light-curves can be produced by planetary microlensing and close/wide binary lenses (Han & Gaudi 2008). Although rare, binary sources lensed by single lenses can produce features similar to those produced by a planetary caustic perturbation (Jung et al. 2017a;b).

4.8.6 Planetary Microlensing Detections

The search for planets using microlensing is a relatively young field of research. Mao and Paczynski (1991) and Gould and Loeb (1992) were the first to point out that planets could be found in this way. To date, ~90 planets have been discovered using microlensing.[3] Compared to other methods to find planets, like, e.g. the radial velocity shifts, or the transits methods, which led to discovering thousands of planets so far, this number may seem tiny. Nevertheless, microlensing is unique in many ways, and for this reason, complementary to the methods mentioned above.

First of all, the processes that lead to a planet's formation are yet not fully understood (Armitage 2010, e.g.). Planets form from gaseous proto-planetary disks, but several models exist that explain their growth. For example, in the accretion model, terrestrial planets form by accretion onto a solid core. If the core is sufficiently massive ($\sim 8 M_\oplus$), subsequent accretion of gas leads to gas giant planets' formation. Alternatively, in the so-called gas-collapse model, giant planets form from the gas's direct collapse in the proto-planetary disk due to gravitational instabilities. Scenarios of these kinds lead to different distributions of planet masses and distances from the host stars. Thus, making a census of planets around several types of stars and measuring their masses and orbital parameters are essential steps to validate planet formation models.

Unfortunately, there is no single method to find planets that allow probing the complete range of relevant masses and distances. For example, radial velocity shifts and transits are most efficient at finding massive planets within ~1 AU from their host stars. Thus, these methods do not allow to explore regions of the planetary systems that are particularly interesting for testing the scenarios of planet formation, like the so-called *habitable zone* (where liquid water can exist on the planet surface) or even more the region beyond the *snow line* (e.g. Lissauer 1987). This line marks the boundary distance from a star beyond which temperature drops below the point where water turns into ice. As an order of magnitude, this distance corresponds to ~1 AU.

As seen in the previous sections, microlensing is sensitive to planets near the host star's Einstein radius, $\sim r_E = D_L \theta_E$. For a solar mass star at $D_L \sim 4$ kpc and a source at $D_S \sim 8$ kpc, the Einstein ring radius is ~4 AU. This distance is similar to the semi-major axis of Jupiter's orbit (5.2 AU). Microlensing is thus complementary to the other methods to find planets. It allows us to explore a region in the planet mass vs. semi-major axis plane, which is difficult to investigate otherwise (Gaudi 2012). All the current microlensing events, which include planetary signatures, have been found at distances of several kilo-parsecs from the Sun (Tsapras 2018). Most of

---

[3]https://exoplanetarchive.ipac.caltech.edu.

the planets detected with other methods are instead confined within $\lesssim 1$ kpc. Since the Milky Way has a metallicity gradient (Daflon & Cunha 2004), microlensing detections are crucial to investigate planet formation in regions that are chemically different from the solar neighborhood.

The first detection of planetary microlensing is dated back to 2004, when the OGLE and MOA collaborations observed the event dubbed OGLE-2003-BLG-235/MOA-2003-BLG-53 toward the bulge of the Milky Way (Bond et al. 2004). The light-curve is a single-lens profile with a short (~ 7 days) deviation exhibiting the U-profile typical of a caustic crossing. The data are consistent with resonant caustic crossing event produced by a star-planet lens with $t_E = 61.5 \pm 1.8$ days, $q = 3.9^{+1.1}_{-0.7} \times 10^{-3}$, and $d = 1.120 \pm 0.007$. The host star's identification with HST allowed deriving other physical properties of this system (Bennett et al. 2006). The planet turned out to be a super Jupiter with mass $M_p = 2.6^{+0.8}_{-0.6} M_J$ orbiting around a M-dwarf star with mass $M = 0.63^{0.07}_{-0.09}$ $M_\odot$. The semi-major axis of the planet orbit is $a = 4.3^{+2.5}_{-0.8}$ AU.

The second planet discovered via microlensing (MOA-2007-BLG-400) is a super Jupiter planet orbiting around an M-dwarf star (Dong et al. 2009, Udalski et al. 2005). This discovery immediately raised the suspicion that such planetary systems are far from being rare. In fact, about $\sim 20\%$ of the microlensing detected planets are of this kind (Batista et al. 2011, Calchi et al. 2018, Han et al. 2013, Jung et al. 2015, Koshimoto et al. 2017; 2014, Poleski et al. 2014, Shin et al. 2016, Shvartzvald et al. 2014, Street et al. 2013, Udalski et al. 2005).

The third and the fourth planet detections with microlensing are mini-Neptune-like planets (Beaulieu et al. 2006, Gould et al. 2006). Given their low mass ($\sim$ few Earth masses), these planets are more difficult to detect compared to Jupiter-like planets. Thus, simple statistical arguments suggest that they are even more common than Jupiters. As new detections were made, it was possible to measure the mass ratio distribution of cold exoplanets (Sumi et al. 2010), which was found to scale as

$$\frac{dN}{d \log q} \propto q^{-0.7 \pm 0.2} \tag{4.114}$$

with a 95% confidence level upper limit of $n < -0.35$ (where $dN/d \log q \propto q^n$). This implies that Neptune-mass planets are at least three times more common than Jupiters at the 95% confidence level.

4.9   Python Applications

4.9.1   Standard Microlensing Light-Curve

In this example, we derive the standard light-curve in a microlensing event involving a point lens and a point source. We assume that the lens position is fixed. The source moves with a constant transverse velocity v relative to the lens (or relative proper motion $\mu_{rel} = \frac{v}{D_L}$), as shown in Fig. 4.2.

We define two python classes, one for the source and one for the lens. The first is very simple:

```python
from astropy import constants as const
from astropy import units as u
import numpy as np

class point_source(object):

    def __init__(self,flux=1.0,ds=10.0,vel=200.):
        """
        Initialize a point source.
        Parameters:
        - flux: baseline flux
        - ds: source distance
        - ds: source relative velocity
        """
        self.ds=ds
        self.flux=flux
        self.vel=vel
```

We create a `point_source` by specifying its baseline flux, distance, and relative velocity. Note that we imported some useful packages from `astropy`. We will use some constants defined in the module `constants` and make unit conversions using the module `units`.

The second class, `point_lens`, contains several methods implementing formulas discussed in Sects. 4.1 and 4.2. We initialize an instance of the class as follows:

```python
class point_lens(object):

    """
    Initialize a point lens.
    Parameters:
    - ps : point source
    - mass : lens mass
    - dl : lens distance
    - t0 : time of minimal distance from the source
           (magnification peak)
    - y0 : impact parameter
    """
    def __init__(self,ps,mass=1.0,dl=5.0,ds=8.0,t0=0.0,y0=0.1):
        self.M=mass
        self.dl=dl
        self.ps=ps
        self.y0=y0
        self.t0=t0
        self.tE=self.EinsteinCrossTime()
```

The first argument is the instance of the `point_source` class. The other parameters are the lens mass and distance, the time of the magnification peak, and the impact parameter. The initialization function contains a call to a function to calculate the Einstein crossing time, given in Eq. 4.29. This function calls another function to calculate the Einstein radius using Eq. 4.7:

```
# a function returning the Einstein radius
def EinsteinRadius(self):
    mass=self.M*const.M_sun
    G=const.G
    c=const.c
    # conversion factor: radian to arcsec
    aconv=np.rad2deg(1.0)*3600.0*u.arcsecond
    return((np.sqrt(4.0*(G*mass/c/c).to('kpc')*(self.ps.ds-self.dl)
                    /self.dl/self.ps.ds/u.kpc))*aconv)

# a function returning the Einstein radius crossing time
def EinsteinCrossTime(self):
    theta_e=self.EinsteinRadius()
    return(((theta_e.to('radian').value*self.dl*u.kpc).to('km')
            /self.ps.vel/u.km*u.s).to('day'))
```

We define a function to compute the source position relative to the lens at the time t. Function returns the two components of the vector $\vec{y}(t)$:

```
# a function returning the coordinates of the unlensed source
# at time t
def y(self,t):
    y1=(t-self.t0)/self.tE.value
    y2=np.ones(len(t))*self.y0
    return(y1,y2)
```

Finally, the function `mut` calculates the magnification as a function of time, using Eq. 4.33:

```
def mut(self,t):
    y0,y1=self.y(t)
    y=np.sqrt(y0**2+y1**2)
    return (self.ps.flux*(y**2+2)/y/np.sqrt(y**2+4))
```

We assume that the source is at a distance of $D_S = 8$ kpc and that its relative velocity is $v = 200$ km/s:

```
ps = point_source(flux=1.,ds=8.0,vel=200.)
```

We will display the light-curves for a variety of impact parameters y_0. The choice of t_0 is not important, because we will display the light-curves as a function of $(t - t_0)/t_E$.

```
# initialize the impact parameters
y0=np.linspace(1.0,0.1,10)
# passage at the minimum distance from the lens
t0=365 # days
```

We assume that the lens mass is $M = 0.3\ M_\odot$ and that the lens distance is $D_L = 4$ kpc. For each value of the impact parameter, we can now calculate the magnification as a function of time as follows:

```
# loop over the impact parameters and calculate
# the light-curves:

for i in range(y0.size):
    pl = point_lens(ps,mass=0.3,dl=4.0,t0=t0,y0=y0[i])
    t=pl.t0+np.linspace(-2,2,200)*pl.tE.value
    mut=pl.mut(t)
```

We show the resulting light-curves in Fig. 4.3.

4.9.2   Fitting the Standard Light-Curve

The microlensing light-curve is defined by the parameters t_0, y_0, and t_E. This last quantity depends also on M, v_{rel}, D_L, and D_S. Besides, the light-curve normalization depends on the baseline flux f. How precisely can we measure all these parameters?

Here, we set up the following experiment:

- we simulate the observation of a microlensing event and generate synthetic data, including measurement errors;
- we use the package lmfit to fit the data;
- we perform a Bayesian analysis using the package emcee to estimate the posterior probability distributions of the parameters.

We assume that the lens has a mass of $M = 0.3M_\odot$ and is at a distance of $D_L = 4$ kpc. We further assume that the source at a distance of $D_S = 8$ kpc. The source baseline flux is $f = 10$ (the units are arbitrary) and the impact parameter is $y_0 = 0.3$. The relative velocity of the source is $v = 210$ km/s. We assume that we could monitor the source star for a long period (2 years) and collect data regularly. This assumption is unrealistic, but we want to test an ideal situation. Also, we assume that the errors on the photometric measurement are at the level of 5%. The peak magnification occurs at $t_0 = 365$ days.

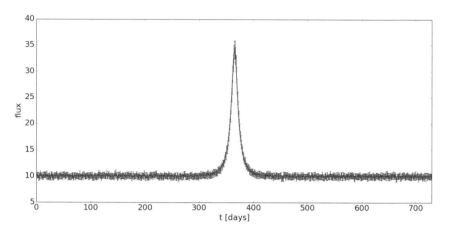

Fig. 4.23 Simulated light-curve of a microlensing event (blue points with error bars. The best fit to the data is shown in red

The code used to generate the synthetic data is here below. We use the classes `point_source` and `point_lens` from the previous example.

```
# input parameters for the source and the lens
# create a point source
ps = point_source(flux=10.,ds=8.0,vel=210.)
# create a point lens
pl = point_lens(ps,mass=0.3,dl=4.0,t0=365,y0=0.3)
# create a mock light-curve
t=np.linspace(0,730,730)
mut=pl.mut(t)+(np.random.randn(len(t))*0.05)

# we assign to the data some errors, which we assume to be a
# constant fraction of the measurement
emut=mut*0.05
```

The light-curve is displayed in Fig. 4.23 (blue points with error bars).

As said, to fit the data, we use the python package `lmfit`. This package allows building complex fitting models for non-linear least-squares problems. The implementation shown here was obtained by closely following the examples in the package documentation, which can be found at this link: http://cars9.uchicago.edu/software/python/lmfit_MinimizerResult/intro.html

We begin by setting up some initial guesses for the model parameters, storing them in a `lmfit.Parameter` object, including also some plausible ranges where the parameters can vary:

```
import lmfit

# initial guesses for the parameters:
# t0, M_lens, DL, DS, vel, y0, flux0
p = lmfit.Parameters()
```

```
p.add_many(('t0', 400.,True,0,720), ('M_lens', 1.0, True, 0.001, 100.0),
           ('DL', 5., True, 0.1, 10.), ('DS', 8., False, 5., 15.),
           ('vel',250,True,50.,300.), ('y0',0.8, True, 0.01,1.0),
           ('flux0',12, True, 8,12.0))
# For each parameter, we specify a initial value, a flag, two other values
# defining the search range. If the flag is True the parameter is free
# to vary, otherwise it is fixed to the initial value.
```

Then, we define a *cost* function to compare the data and the model:

```
def cost_function(p,t,mut,emut):
    ps = point_source(flux=p['flux0'],ds=p['DS'],vel=p['vel'])
    pl = point_lens(ps,mass=p['M_lens'],dl=p['DL'],
                    t0=p['t0'],y0=p['y0'])
    res=(pl.mut(t)-mut)/emut
    return (res)
```

This function returns the residuals between the model and the data.

The next step is to minimize the cost function (i.e. the residuals) to fit the data. Several algorithms are available in `lmfit`. Here, we perform the minimization using the Nelder-Mead optimization method:

```
mi = lmfit.minimize(cost_function, p, method='Nelder',
                    args=(t,mut,emut))
```

The red line in Fig. 4.23 shows the best-fit light-curve. Its parameters are:

```
[[Variables]]
    t0:       364.999736  (init = 400)
    M_lens:   0.48380390  (init = 1)
    DL:       4.85039494  (init = 5)
    DS:       8 (fixed)
    vel:      260.475326  (init = 250)
    y0:       0.30003253  (init = 0.8)
    flux0:    9.99984736  (init = 12)
```

The values of some of the input parameters are recovered almost perfectly: t_0, y_0, and f. Not surprisingly, the remaining parameters differ significantly from the truth. They are highly degenerate parameters, because they all concur to determine the value of t_E. Only the Einstein crossing time determines the shape of the light-curve.

We perform a Bayesian sampling of the posterior probability distribution of the parameters using the ensemble sampler for Markov chain Monte Carlo (MCMC) implemented in the Python package `emcee` (Foreman-Mackey et al. 2013). The log-posterior probability of the model parameters, p, given the data d, is

$$\ln P(p|d) \propto \ln P(d|p) + \ln P(p) , \tag{4.115}$$

where $\ln P(d|p)$ is the log-likelihood of the data given the model parameters and $\ln P(p)$ is the log-prior. We assume a uniform prior, meaning that $\ln P$ is zero if all the parameters are inside the bounds, and $-\infty$ if any of the parameters is outside its limits.

The log-likelihood function is

$$\ln P(d|p) = -\frac{1}{2} \sum_n \left[\frac{(model_n - data_n)^2}{s_n^2} + \ln 2\pi s_n^2 \right] , \qquad (4.116)$$

where s_n is the data uncertainty.

We calculate it as follows:

```
# log-likelihood function
def lnprob(p,t,mut,emut):
    from numpy import inf
    resid = cost_function(p,t,mut,emut)
    s = emut
    resid *= resid/s/s
    resid += np.log(2 * np.pi * s**2)
    lnp=-0.5 * np.sum(resid)
    return lnp
```

The next step is to call the function `minimize` using the method `emcee`. We use 100 walkers to explore the parameter space. We run a few *burn-in* steps in each MCMC chain to let the walkers get settled into the maximum of the density. Then, we do a production run of 2000 steps.

```
res = lmfit.minimize(lnprob, method='emcee',
                     nan_policy='omit',
                     nwalkers=100, burn=500, steps=40000,
                     params=mi.params,
                     progress=True,args=(t,mut,emut))
```

Finally, we visualize the posterior probability distributions using the `corner` package (Foreman-Mackey 2016). The function `corner` uses the samples stored in `res.flatchain` to draw two-dimensional histograms showing the probability density projected on planes defined by couples of parameters. These 2D-histograms are very useful to highlight the existing correlations between v_{rel}, D_L, and M. Besides, `corner` also plots the marginalized the one-dimensional density distributions for each parameter.

```
import corner
figure=corner.corner(res.flatchain, labels=[r"$t_0$",
                                            r"Mass",
                                            r"$D_L$",
                                            r"$v_{rel}$",
                                            r"$y_0$", r"f"],
                     truths=list(res.params.valuesdict().values()),
                     quantiles=[0.16,0.84],
                     show_titles=True,
                     title_kwargs={"fontsize": 16})
```

```
for ax in figure.get_axes():
    ax.tick_params(axis='both', labelsize=14)
    ax.tick_params(axis='both', labelsize=14)
    ax.xaxis.label.set_size(16)
    ax.yaxis.label.set_size(16)
```

We show the resulting plot in Fig. 4.24. The values reported in the `Minimizer Result` are the medians and the widths (estimated as half the difference between the 15.8 and 84.2 percentiles) of the marginalized probability distributions of each parameter:

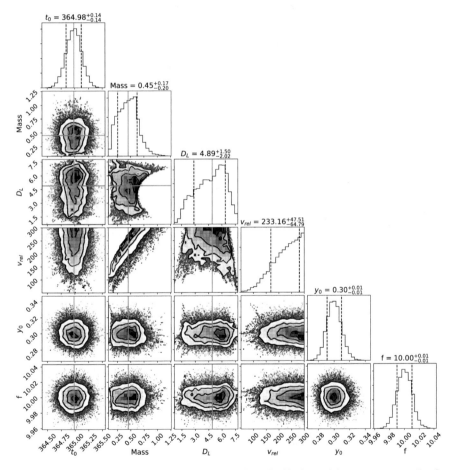

Fig. 4.24 Corner plot showing the posterior probability distributions of the parameters used to fit the light-curve in Fig. 4.23. We show the projections of the probability density on planes defined by each couple of parameters, as well as one-dimensional marginalized distributions. The blue and the dashed vertical lines in one-dimensional histograms indicate the medians and the 16th and 84th percentiles of the distributions

```
print("median of posterior probability distribution")
print('------------------------------------------------')
lmfit.report_fit(res.params)

median of the posterior probability distribution
----------------------------------------
[[Variables]]
    t0:      364.964216 +/- 0.14632837 (0.04%) (init = 364.9708)
    M_lens:  0.50785479 +/- 0.23382954 (46.04%) (init = 0.6573513)
    DL:      5.13183894 +/- 1.96273800 (38.25%) (init = 4.989481)
    DS:      8 (fixed)
    vel:     239.965356 +/- 53.2604941 (22.20%) (init = 299.7212)
    y0:      0.29986042 +/- 0.00610870 (2.04%) (init = 0.2990382)
    flux0:   9.98225154 +/- 0.01997525 (0.20%) (init = 9.981135)
[[Correlations]] (unreported correlations are < 0.100)
    C(M_lens, vel) =  0.638
    C(M_lens, DL)  =  0.392
```

Figure 4.25 shows the results of the fit to the measured light-curve with a model which depends only on `tE`, `t0`, and `y0`.

4.9.3 Distribution of Microlensing Event Timescale

In this example, we derive the distribution of microlensing event timescales expected from a given distribution of lenses. The procedure consists of drawing samples from Probability Distribution Functions (PDFs). For arbitrary distributions, we can do this using the *inverse transform sampling* method, which involves the following steps:

1. we compute the cumulative distribution function of the PDF we want to sample;
2. we invert the cumulative function;
3. we generate random numbers uniformly distributed between 0 and 1;
4. we read-off the corresponding values returned by the inverted cumulative function.

For example, for an exponential PDF,

$$p(x) = \lambda \exp(-\lambda x) , \qquad (4.117)$$

the cumulative distribution function is

$$P(x) = \int_0^x p(x)dx = 1 - \exp(-\lambda x) . \qquad (4.118)$$

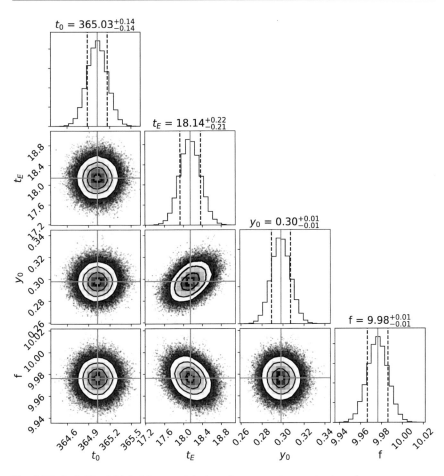

Fig. 4.25 As in Fig. 4.24, but for a model whose free parameters are t_E, t_0, and y_0

This can be readily inverted to find:

$$x = -\ln(1 - P)/\lambda . \qquad (4.119)$$

To obtain 10000 samples of this distribution (assuming a *rate* parameter $\lambda = 0.5$), we can do as follows:

```
# inverse cumulative function
def exp_icdf(p,lambd=1):
    return -np.log(1-p)/lambd


samples=10000
lambd=0.5
# generate uniformly distributed random numbers
p=np.random.random(samples)
```

```
# random samples:
x=exp_icdf(p,lambd=lambd)
```

The package numpy.random provides several functions to sample common PDFs, including the exponential PDF:

```
# draw samples from  an exponential PDF with rate parameter lambd:
x_exp=np.random.exponential(size=samples,scale=1.0/lambd)
# draw samples from  a Gaussian PDF with mean mu and
# standard deviation sigma:
mu=1.0
sigma=0.2
x_norm=np.random.normal(loc=mu,scale=sigma,size=samples)
# draw samples from a uniform distribution
a=0
b=2
x_unif=np.random.uniform(low=a,high=b,size=samples)
```

The purpose of this example is to show how some assumptions on the population of microlenses determine the expected distribution of observed microlensing event timescales.

Our assumptions are:

- the distribution of lens masses is a power-law or an exponential function;
- the distribution of source distances is a Gaussian with mean $D_S = 8$ kpc and standard deviation $\sigma = 0.3$ kpc;
- the distribution of lens distances is uniform between $D_L = 0$ and $min(D_S)$;
- the distribution of the lens-source relative velocities is a Gaussian with mean $v_{rel} = 220$ km/s and standard deviation $\sigma = 10$ km/s.

We begin by drawing the lens masses from a power-law distribution. The power function of numpy.random draws samples in the range [0, 1]. Thus, we re-scale the sampled values to cover the desired range of masses (e.g. [0, 10] $M_\odot$):

```
mmax=10.  # maximum mass
a = 0.5 # shape
# use the 'power' PDF in numpy.random
mass_pow = np.random.power(a, samples)*mmax
```

We draw a second sample of masses from an exponential distribution:

```
a = 0.5
# use the 'exponential' PDF in numpy.random
mass_exp = np.random.exponential(size=samples,scale=1.0/lambd)
mass_exp=mass_exp/mass_exp.max()*mmax
```

Then, we generate the source and lens distances:

```
# generate D_S
mu, sigma = 8.0, 0.3
ds=np.random.normal(loc=mu,scale=sigma,size=samples)
# generate D_L
dmin, dmax = 0.0, np.min(x_norm)
dl = np.random.uniform(low=dmin,high=dmax,size=samples)
```

Finally, we generate the velocities:

```
# generate vel
mu, sigma = 220, 10
vel=np.random.normal(loc=mu,scale=sigma,size=samples)
```

We show the distributions of the lens parameters, M, D_L, D_S, and v_{rel}, in Fig. 4.26. Assuming that these are not correlated, we use them to obtain the expected distribution of event timescales given the assumed model of lenses and sources:

```
ps=[point_source(ds=ds[i],vel=vel[i])
        for i in range(len(mass_pow))]
te_pow=[point_lens(ps=ps[i],mass=mass_pow[i],dl=dl[i]).tE.value
        for i in range(len(mass_pow))]
te_exp=[point_lens(ps=ps[i],mass=mass_exp[i],dl=dl[i]).tE.value
        for i in range(len(mass_exp))]
```

In Fig. 4.27, we can see that we obtain significantly different Einstein crossing time distributions, depending on the mass function used. This example is far from describing a realistic situation, but it serves to illustrate how information about the lens population is encoded in the statistics of the microlensing events (see, e.g. Gaudi 2012, Gould 1994, Sumi et al. 2011).

4.9.4 Astrometric Microlensing Effect

This example shows how the source apparent position changes during a microlensing event, the so-called astrometric microlensing. For this purpose, we add some additional functions to the point_lens class seen previously.

First, we define a new function to compute the source position at the time t. The function returns the two components of the vector $\vec{y}(t)$:

```
    # a function returning the coordinates of the unlensed source
    # at time t
    def y(self,t):
        y1=(t-self.t0)/self.tE.value
        y2=np.ones(len(t))*self.y0
        return(y1,y2)
```

In addition, we define two function returning the coordinates of the outer (xp) and of the inner (xm) image of the source. These functions implement Eqs. 4.42.

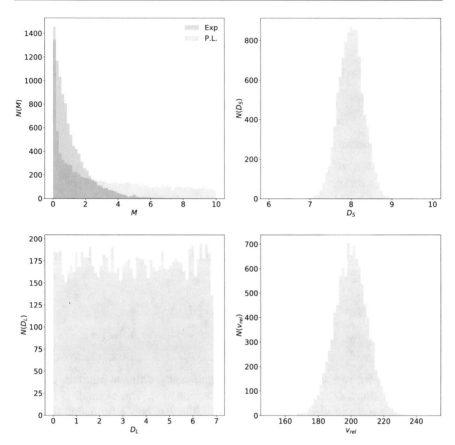

Fig. 4.26 From the upper left to the bottom right panels, distributions of the lens masses, of the source and of the lens distances, and of the relative velocities between the lenses and the sources. The distributions of the lens masses are modeled with a power-law (blue) and with an exponential (green) function

```
# a function returning the coordinates of the x_+ image at time t
def xp(self,t):
    y1, y2  = self.y(t)
    Q = np.sqrt(y1**2 + y2**2 +4)/(np.sqrt(y1**2 + y2**2))
    xp1= 0.5 *(1 + Q)* y1
    xp2= 0.5 *(1 + Q)* y2
    return(xp1, xp2)

# a function returning the coordinates of the x_- image at time t
def xm(self,t):
    y1, y2  = self.y(t)
    Q = np.sqrt(y1**2 + y2**2 +4)/(np.sqrt(y1**2 + y2**2))
    xm1= 0.5 *(1 - Q)* y1
```

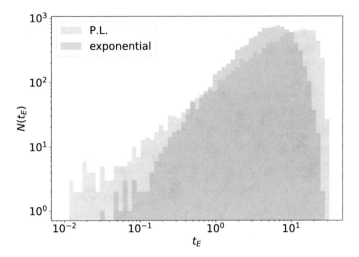

Fig. 4.27 Expected distribution of microlensing event timescales from the model outlined in Sect. 4.9.3. Results are shown for two mass distributions of lenses (power-law and exponential)

```
xm2= 0.5 *(1 - Q)* y2
return(xm1, xm2)
```

The following step is to compute the magnifications of the two images, using Eq. 4.18:

```
# the magnification of the x_+ image
def mup(self,t):
    y1, y2  = self.y(t)
    yy=np.sqrt(y1**2+y2**2)
    mup=0.5*(1+(yy**2+2)/yy/np.sqrt(yy**2+4))
    return (mup)

# the magnification of the x_- image
def mum(self,t):
    y1, y2  = self.y(t)
    yy=np.sqrt(y1**2+y2**2)
    mum=0.5*(1-(yy**2+2)/yy/np.sqrt(yy**2+4))
    return (mum)
```

Finally, the image centroid is computed using Eq. 4.44:

```
# a function returning the coordinate of the light centroid
def xc(self,t):
    xp=self.xp(t)
    xm=self.xm(t)
    xc=(xp*np.abs(self.mup(t))+xm*np.abs(self.mum(t)))/
    (np.abs(self.mup(t))+np.abs(self.mum(t)))
    return (xc)
```

Now, we can add some extra feature. For example, to draw Fig. 4.9, the source is assumed to have an extended size. The images of an extended source are also extended. To compute their shapes, we add another couple of functions to the class above:

```python
def xp_ext_source(self,t,r):
    phi=np.linspace(0.0,2*np.pi,360)
    dy1=r*np.cos(phi)
    dy2=r*np.sin(phi)
    y1,y2=self.y(t)
    yy1=y1+dy1
    yy2=y2+dy2
    Q=np.sqrt(yy1**2+yy2**2+4.0)/np.sqrt(yy1**2+yy2**2)
    xp1=0.5*(1+Q)*yy1
    xp2=0.5*(1+Q)*yy2
    return(xp1,xp2)

def xm_ext_source(self,t,r):
    phi=np.linspace(0.0,2*np.pi,360)
    dy1=r*np.cos(phi)
    dy2=r*np.sin(phi)
    y1,y2=self.y(t)
    yy1=y1+dy1
    yy2=y2+dy2
    Q=np.sqrt(yy1**2+yy2**2+4.0)/np.sqrt(yy1**2+yy2**2)
    xm1=0.5*(1-Q)*yy1
    xm2=0.5*(1-Q)*yy2
    return(xm1,xm2)
```

Let us assume that the source is a circle with radius r at position $\vec{y}$. Then, the points on its perimeter are at positions

$$\vec{y} + d\vec{y}_i,$$

where

$$d\vec{y}_i = r(\cos\phi, \sin\phi)$$

with $\phi \in [0, 2\pi]$. Each of these point sources produces two images $\vec{x}_{i,+}$ and $\vec{x}_{i,-}$, which can be found using the functions defined earlier. By doing this for all points on the source perimeter, we obtain the images' lensed contours. For example, to produce Fig. 4.9, we do as follows:

```python
# define a source and a lens
ps = point_source()
pl = point_lens(ps=ps, mass=1.0, dl=5.0, y0 = 0.2)

# define an array of times
t = np.linspace(-300,300,2000)
```

```
# compute the source, the image, and the centroid
# positions as function of time
y1, y2 = pl.y(t)
xp1, xp2= pl.xp(t)
xm1,xm2=pl.xm(t)
xc1,xc2=pl.xc(t)

# plot results
fig,ax=plt.subplots(1,1,figsize=(8,8))
ax.plot(y1,y2,'--',label='source traj.')
ax.plot(xp1,xp2,'--',label='image $x_+$')
ax.plot(xm1,xm2,'--',label='image $x_-$')
ax.plot(xc1,xc2,label='image centroid')

# define a coarser array of times
t_sparse=np.linspace(-60,60,19)
from matplotlib.pyplot import cm
color=iter(cm.rainbow(np.linspace(0,1,t_sparse.size)))
# at each time, compute the images of an extended
# circular source of radius r=0.05
for tt in t_sparse:
    c=next(color)
    xp1_e,xp2_e=pl.xp_ext_source(np.array([tt]),0.05)
    ax.plot(xp1_e,xp2_e,color=c,lw=2)
    xm1_e,xm2_e=pl.xm_ext_source(np.array([tt]),0.05)
    ax.plot(xm1_e,xm2_e,color=c,lw=2)

ax.set_xlim([-2,2])
ax.set_ylim([-1.8,2.2])
ax.plot([0.0],[0.0],'*',markersize=20,color='red')
circle=plt.Circle((0,0),1,color='black',fill=False)
ax.add_artist(circle)
ax.legend(fontsize=14)
ax.xaxis.set_tick_params(labelsize=20)
ax.yaxis.set_tick_params(labelsize=20)
ax.set_xlabel(r'$x_{||}$',fontsize=20)
ax.set_ylabel(r'$x_\perp$',fontsize=20)
```

Now, imagine to monitor a source over time. The source position is fixed until the microlensing event occurs. During the event the source moves, because the images light centroid is shifted. The amplitude of the shift is

$$\delta \vec{x}_c = \vec{x}_c - \vec{y} \ .$$

We wish to determine the trajectory of the source apparent position. First, we add this function to the class:

```
def deltaxc(self,t):
    y1,y2=self.y(t)
```

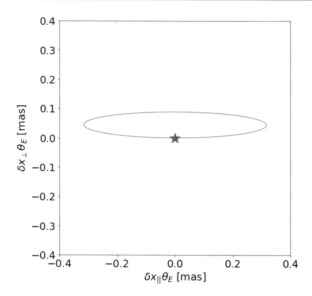

Fig. 4.28 Trajectory of a source centroid of light during a microlensing event produced by a star of mass $M = 1M_\odot$ at distance $D_{\rm L} = 5$ kpc, lensing a source at $D_{\rm S} = 10$ kpc. The impact parameter is $y_0 = 0.2$

```
yy=(y1**2+y2**2)
return(y1/(yy+2),y2/(yy+2))
```

Then, we use it to compute the coordinates of the centroid shift:

```
t=np.linspace(-5000,5000,5000)
```

```
dxc1,dxc2=pl.deltaxc(t)
fig,ax=plt.subplots(1,1,figsize=(8,8))
ax.plot(dxc1*pl.EinsteinRadius()*1000,dxc2*pl.EinsteinRadius()*1000)
ax.set_xlim([-0.4,0.4])
ax.set_ylim([-0.4,0.4])

ax.set_xlabel(r'$\delta x_{||}\theta_E$ [mas]',fontsize=20)
ax.set_ylabel(r'$\delta x_\perp \theta_E$ [mas]',fontsize=20)
ax.plot([0.0],[0.0],'*',markersize=20,color='red')
ax.xaxis.set_tick_params(labelsize=20)
ax.yaxis.set_tick_params(labelsize=20)
```

The result of this procedure is shown in Fig. 4.28. The red star indicates the unlensed position of the source. During the microlensing event, the light centroid follows an elliptical trajectory (blue solid line), as discussed in Sect. 4.4.

4.9.5   Critical Lines and Caustics of a Binary Lens

In the following examples we will consider binary lenses composed of two point masses M_1 and M_2. The two masses are related by $q = M_1/M_2$. They are placed at a distance d from each other (where d is in units of the lens Einstein radius). We choose the real axis to pass through the two point masses and assume that $z_2 = -z_1$. A sketch of such system is shown in Fig. 4.29.

The source is supposed to move behind the lens with a relative velocity v_{rel} along a linear trajectory forming an angle θ_s with the real axis. We measure the impact parameter y_0 relative to $z = 0$. The source reaches the minimal distance from $z = 0$ at $t = t_0$.

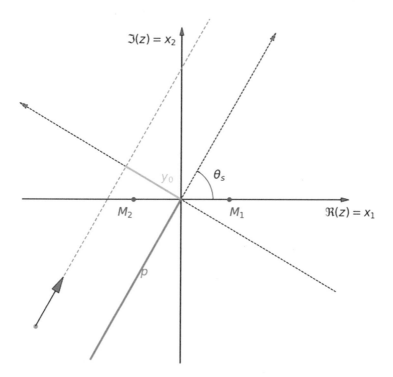

Fig. 4.29 Sketch of a binary lens system. The two lenses, dubbed M_1 and M_2, are indicated by red points. We choose the reference frame such that the origin is at the mid-point between the two masses. The real axis (or the axis x_1 on the lens plane) passes through the two lenses. The trajectory of the source is given by the blue dashed line. Its inclination relative to the real axis corresponds to the angle θ_s. We show the source position at time t with the orange point, and the blue arrow gives the velocity (relative to the lens). The impact parameter y_0 is measured with respect to the origin ($z = (x_1 + ix_2) = 0$). The source is assumed to pass at a distance y_0 from the origin at time t_0. In the frame rotated by the angle θ_s (dashed axes), the position of the source at time t is $z'_s = p + iy_0$, where $p = (t - t_0)/t_E$

The real and the imaginary parts of z_s are given by

$$\Re(z_s) = \cos(\theta_s)p - \sin(\theta_s)y_0 \tag{4.120}$$

and

$$\Im(z_s) = \sin(\theta_s)p + \cos(\theta_s)y_0, \tag{4.121}$$

where $p = (t - t_0)/t_E$. The effective Einstein radius, θ_E, and Einstein crossing time, t_E, are those of a point-mass lens with mass $M_{tot} = M_1 + M_2$.

As usual, D_L and D are the angular diameter distances to the lens and the source.

The critical lines and caustics of the binary lens are determined by solving numerically Eq. 4.93 for any $\phi \in [0, 2\pi)$. The equation can be turned into a fourth order complex polynomial, of which we shall find the roots. The polynomial is

$$p_4(z) = z^4 - z^2(2z_1^{*2} + e^{i\phi}) - zz_1^*2(m_1 - m_2)e^{i\phi} + z_1^{*2}(z_1^{*2} - e^{i\phi}) = 0. \tag{4.122}$$

For each ϕ there are up to 4 roots (critical points). By using the lens equation (Eq. 4.94), these can be mapped on the source plane to derive the caustics:

$$z_{cau} = z_{crit} - \frac{m_1}{z_{crit}^* - z_1^*} - \frac{m_2}{z_{crit}^* - z_2^*}. \tag{4.123}$$

The roots of the polynomial

$$p(x) = x^n + a_{n-1}x^{n-1} + \cdots + a_1x + a_0 \tag{4.124}$$

can be found by building its companion matrix

$$C = \begin{bmatrix} 0 & 0 & \cdots & 0 & -a_0 \\ 1 & 0 & \cdots & 0 & -a_1 \\ 0 & 1 & \cdots & 0 & -a_2 \\ \vdots & \vdots & \ddots & \vdots & \vdots \\ 0 & 0 & \cdots & 1 & -a_{n-1} \end{bmatrix} \tag{4.125}$$

and the characteristic polynomial

$$\det(xI - C) = p(x). \tag{4.126}$$

Thus, the eigenvalues of C are the roots of $p(x)$. Therefore, to find the critical points of the binary lens, we may proceed as follows:

- we build of the companion matrix of $p_4(z)$;
- we diagonalize the matrix and find the eigenvalues.

This procedure is implemented in the `numpy.roots` method, which will be used below.

We begin by importing numpy:

```
import numpy as np
```

We build a class for binary lenses. The class uses some functions from the `point_lens` class discussed in the previous sections.

We create an instance of the `binary_lens` class by setting the mass of the first lens, the mass ratio q and the separation between the lenses, d, in units of the equivalent Einstein radius. The other input parameters (D_L, D_S, t_0, y_0, θ_s, and v_{rel}) do not affect the results of this example.

```
class binary_lens(object):
    """
    Initialize a binary lens
    """
    def  __init__(self,dl=5.0,ds=8,m1=1.0,q=1.0,d=2.0,t0=0.0,y0=0.1,
                    theta=np.pi/4,vrel=200.0):
        # position of the first lens
        self.z1=complex(d/2.0,0.0)
        self.q=q
        self.dl=dl
        self.ds=ds
        # mass of the second lens
        m2=m1/q
        self.mtot=m1+m2
        # store masses in units of the total mass
        self.m1=m1/self.mtot
        self.m2=m2/self.mtot

        """
        we build a point_lens  to compute the Einstein radius
        and the Einstein crossing time. This requires defining a
        point source too.
        """
        ps = point_source(ds=ds,vel=vrel)
        pl = point_lens(ps=ps, mass=m1+m2, dl=dl)
        self.thetaE=pl.EinsteinRadius()
        self.tE=pl.EinsteinCrossTime()

        self.t0=t0
        self.y0=y0
        self.theta=theta
```

Then we add a function to find the lens critical lines and caustics using the method outlined above.

```
    def CritCau(self,ncpt=10000):
        # set the phase vector
        phi_=np.linspace(0,2.*np.pi,ncpt)
```

```
x=[]
y=[]
xs=[]
ys=[]

# we need to find the roots of our fourth order polynomial
# for each value of phi
for i in range(phi_.size):
    phi=phi_[i]
    # the coefficients of the complex polynomial
    coeff = [1.0,0.0,-2*np.conj(self.z1)**2-np.exp(1j*phi),
            -np.conj(self.z1)*2*(self.m1-self.m2)*np.exp(1j*phi),
            np.conj(self.z1)**2*(np.conj(self.z1)**2-np.exp(1j*phi))]
    # use the numpy function roots to find the roots of the
    # polynomial
    z=np.roots(coeff) # these are the critical points!

    # use the lens equation (complex form) to map the critical
    # points on the source plane
    zs=z-self.m1/(np.conj(z)-np.conj(self.z1))\
        -self.m2/((np.conj(z)-np.conj(-self.z1))) # these are the
    # caustics!

    # append critical and caustic points
    x.append(z.real)
    y.append(z.imag)
    xs.append(zs.real)
    ys.append(zs.imag)

return(np.array(x),np.array(y),np.array(xs),np.array(ys))
```

The function returns four numpy arrays, containing the coordinates of the lens
critical lines and caustics. The size of these arrays is defined by the input parameter
ncpt. By default, we sample the critical lines and the caustics by generating 10,000
values of the phase ϕ.

4.9.6   Solving the Lens Equation of the Binary Lens

To find the positions of the source images, we must solve Eq. 4.95. We can
use the same method adopted to find the critical points. The coefficients of the
complex polynomial are given in Eq. 4.96. We add the following function to the
binary_lens class:

```
def Images(self,ys1,ys2):
    # complex coordinate of the source
    zs=ys1+1j*ys2

    # coefficients of polynomial eq.
    m=0.5*(self.m1+self.m2)
```

```
Dm=(self.m2-self.m1)/2.0

c5=self.z1**2-np.conj(zs)**2
c4=-2*m*np.conj(zs)+zs*np.conj(zs)**2\
    -2*Dm*self.z1-zs*self.z1**2
c3=4.0*m*zs*np.conj(zs)+4.0*Dm*np.conj(zs)*self.z1\
    +2.0*np.conj(zs)**2*self.z1**2-2.0*self.z1**4
c2=4.0*m**2*zs+4.0*m*Dm*self.z1\
    -4.0*Dm*zs*np.conj(zs)*self.z1\
    -2.0*zs*np.conj(zs)**2\
    *self.z1**2+4.0*Dm*self.z1**3\
    +2.0*zs*self.z1**4
c1=-8.0*m*Dm*zs*self.z1\
    -4.0*Dm**2*self.z1**2\
    -4.0*m**2*self.z1**2\
    -4.0*m*zs*np.conj(zs)*self.z1**2\
    -4.0*Dm*np.conj(zs)*self.z1**3\
    -np.conj(zs)**2*self.z1**4\
    +self.z1**6
c0=self.z1**2*(4.0*Dm**2*zs\
                +4.0*m*Dm*self.z1\
                +4.0*Dm*zs*np.conj(zs)*self.z1+\
                2.0*m*np.conj(zs)*self.z1**2\
                +zs*np.conj(zs)**2*self.z1**2\
                -2*Dm*self.z1**3-zs*self.z1**4)

coefficients=[c5,c4,c3,c2,c1,c0]
# solutions of the lens equation
images=np.roots(coefficients)
        # drop spurious solutions
z2=-self.z1
deltazs=zs-(images-self.m1/(np.conj(images)-np.conj(self.z1))
            -self.m2/(np.conj(images)-np.conj(z2)))
return (np.array([images.real[np.abs(deltazs)<1e-3]]),
        np.array([images.imag[np.abs(deltazs)<1e-3]]))
```

We specify the source position in terms of the two coordinates ys1 and ys2, which are used to compute the complex coordinate zs. The polynomial $p_5(z)$ has five roots. However, the source can have 3 or 5 images, depending on whether the source is inside or outside the caustic. Therefore, sometimes two roots have to be discarded because they are spurious solutions of the lens equation. For doing this, we check which solutions satisfy the lens equation at the end of the function above. Once determined the polynomial roots, we insert them in the lens equation to verify if we obtain the input source position. We compute the quantity deltazs, i.e. the difference between the input source position and the result of the lens equation, and we discard those solutions for which deltazs is above some tolerance (10^{-3} in this example).

4.9.7   Light-Curve in a Binary Microlensing Event

In this example, we consider a source moving on a rectilinear trajectory with respect to the binary lens and derive its light-curve. The parameters defining the source trajectory (y0, t0, theta, vrel are used to initialize the binary_lens object, as shown in the previous examples.

We compute the source position as a function of time as follows:

```
def SourcePos(self,t):
    p=(t-self.t0)/self.tE.value
    zreal=np.cos(self.theta)*p-np.sin(self.theta)*self.y0
    zimag=np.sin(self.theta)*p+np.cos(self.theta)*self.y0
    return(zreal,zimag)
```

Then, the source position can be fed into the Images function to compute the corresponding images. For example, to produce Fig. 4.14, we used the following code:

```
times=np.linspace(-90,90,730)
bl=binary_lens(m1=1.0,q=1.0,d=1.,t0=0.0,y0=0.25,
               theta=np.pi/4.*3.0)
x1,x2,xs1,xs2=bl.CritCau()

color=iter(cm.rainbow(np.linspace(0,1,times.size)))

fig,ax=plt.subplots(1,2,figsize=(18,8))


for t in times:
    c=next(color)
    ys1,ys2=bl.SourcePos(t)
    xi1,xi2=bl.Images(ys1,ys2)
    ax[1].plot(ys1,ys2,'*',markersize=10,color=c)
    ax[0].plot(xi1,xi2,'o',markersize=10,color=c)

ax[0].plot(x1,x2,',',color='blue')
ax[1].plot(xs1,xs2,',',color='blue')
```

As in the case of microlensing by single lenses, multiple images are generally unresolved and the binary microlensing can be revealed only by means of how the magnification varies as a function of time. The magnification of each image can be computed using Eq. 4.102. The following function returns the Jacobian determinant, i.e. the inverse magnification at arbitrary positions z:

```
def detA(self,z):
    z2=-self.z1
    deta=1-np.abs(self.m1/(np.conj(z)-
            np.conj(self.z1))**2+
        self.m2/(np.conj(z)-np.conj(z2))**2)
    return(deta)
```

We apply it to the positions of each image at time t and sum the inverse of the resulting values to obtain the total magnification:

```
def Magnification(self,t):
    ys1,ys2=self.SourcePos(t)
    xi1,xi2=self.Images(ys1,ys2)
    images=xi1+1j*xi2
    mu=1.0/self.detA(images)
    return(np.abs(mu).sum())
```

By calling this function for different values of t, we obtain the light-curve:

```
def lightcurve(self,times):
    p=(times-self.t0)/self.tE.value
    mu=[]
    for t in times:
        mu.append(self.Magnification(t))
    return(p,mu)
```

The function `lightcurve(t)` returns the scaled time $p = (t - t_0)/t_E$ and the corresponding magnification values.

The code below can be used to produce Fig. 4.15.

```
p,mu=bl.LightCurve(times)

fig,ax=plt.subplots(1,2,figsize=(18,8))

color=iter(cm.rainbow(np.linspace(0,1,times.size)))
ys1,ys2=bl.SourcePos(times)
for i in range(times.size):
    c=next(color)
    ax[0].plot(ys1[i],ys2[i],'*',markersize=10,c=c)
    ax[1].plot([p[i]],[mu[i]],'o',markersize=10,c=c)

ax[0].plot(ys1,ys2,'--',color='blue')
ax[1].plot(p,mu,'-')
ax[1].set_ylim([0.0,np.max(mu)*1.1])
ax[0].set_xlim([xmin_,xmax_])
ax[0].set_ylim([ymin_,ymax_])

ax[0].xaxis.set_tick_params(labelsize=20)
ax[0].yaxis.set_tick_params(labelsize=20)

ax[0].set_xlabel('$y_1$',fontsize=20)
ax[0].set_ylabel('$y_2$',fontsize=20)

ax[1].xaxis.set_tick_params(labelsize=20)
ax[1].yaxis.set_tick_params(labelsize=20)
ax[1].set_xlabel('$(t-t_0)/t_E$',fontsize=20)
ax[1].set_ylabel('$\mu(t)$',fontsize=20)
ax[0].plot(xs1,xs2,',',color='blue')
```

References

Alard, C., Guibert, J., Bienayme, O., Valls-Gabaud, D., Robin, A. C., Terzan, A., & Bertin, E. (1995). The DUO programme: first results of a microlensing investigation of the Galactic disk and bulge conducted with the ESO Schmidt telescope. *The Messenger, 80*, 31–34.

Alcock, C., Allsman, R. A., Alves, D., Axelrod, T. S., Bennett, D. P., Cook, K. H., Sutherland, W. (1995). First observation of parallax in a gravitational microlensing event. *ApJL, 454*, L125. https://doi.org/10.1086/309783. eprint: astroph/9506114

Armitage, P. J. (2010). *Astrophysics of planet formation.*

Batista, V., Beaulieu, J.-P., Bennett, D. P., Gould, A., Marquette J.-B., Fukui, A., & Bhattacharya, A. (2015). Confirmation of the OGLE-2005BLG-169 planet signature and its characteristics with lens-source proper motion detection. *ApJ, 808*, 170. https://doi.org/10.1088/0004637X/808/2/170. arXiv: 1507.08914 [astro-ph.EP]

Batista, V., Gould, A., Dieters, S., Dong, S., Bond, I., Beaulieu, J. P., ... Street, R. A. (2011). MOA-2009-BLG-387Lb: A massive planet orbiting an M dwarf. *A & A, 529*, A102. https://doi.org/10.1051/00046361/201016111 arXiv: 1102.0558 [astro-ph.EP]

Beaulieu, J.-P., Bennett, D. P., Fouqué, P., Williams, A., Dominik, M., Jørgensen, U. G., ... Yoshioka, T. (2006). Discovery of a cool planet of 5.5 Earth masses through gravitational microlensing. *Nature, 439*, 437–440. https://doi.org/10.1038/nature04441 eprint: astroph/0601563

Bennett, D. P., Anderson, J., Bond, I. A., Udalski, A., & Gould, A. (2006). Identification of the OGLE-2003-BLG-235/MOA-2003-BLG-53 planetary host star. *ApJL, 647*, L171–L174. https://doi.org/10.1086/507585. eprint: astroph/0606038

Bennett, D. P., Bhattacharya, A., Anderson, J., Bond, I. A., Anderson, N., Barry R., Udalski, A. (2015). Confirmation of the planetary microlensing signal and star and planet mass determinations for event OGLE-2005-BLG-169. *ApJ, 808*, 169. https://doi.org/10.1088/0004637X/808/2/169 arXiv: 1507.08661 [astro-ph.EP]

Bond, I. A., Udalski, A., Jaroszyński, M., Rattenbury N. J., Paczyński, B., Soszyński, I., OGLE Collaboration. (2004). OGLE 2003-BLG-235/MOA 2003-BLG-53: A planetary microlensing event. *ApJL, 606*, L155–L158. https://doi.org/10.1086/420928 eprint: astroph/0404309

Bozza, V. (2010). Microlensing with an advanced contour integration algorithm: Green's theorem to third order error control, optimal sampling and limb darkening. *MNRAS, 408*(4), 2188–2200. https://doi.org/10.1111/j.13652966.2010.17265.x. arXiv: 1004.2796 [astro-ph.EP]

Bozza, V., Bachelet, E., Bartolić F., Heintz, T. M., Hoag, A. R., & Hundertmark, M. (2018). VBBINARYLENSING: A public package for microlensing light-curve computation. *MNRAS, 479*(4), 5157–5167. https://doi.org/10.1093/mnras/sty1791 arXiv: 1805.05653 [astro-ph.IM]

Calchi Novati, S., Suzuki, D., Udalski, A., Gould, A., Shvartzvald, Y., Bozza, V., Pogge, R. W. (2018). Spitzer microlensing parallax for OGLE-2016-BLG-1067: A sub-Jupiter orbiting an M-dwarf in the disk. ArXiv e-prints arXiv: 1801.05806 [astro-ph.EP]

Daflon, S., & Cunha, K. (2004). Galactic metallicity gradients derived from a sample of OB stars. *ApJ, 617*(2), 1115–1126. https://doi.org/10.1086/425607. arXiv: astroph/0409084 [astro-ph]

Dominik, M., & Sahu, K. C. (1998). Astrometric microlensing of stars. *ArXiv Astrophysics e-prints.* eprint: astroph/9805360

Dong, S., Bond, I. A., Gould, A., Kozłowski, S., Miyake, N., Gaudi, B. S., ... OGLE Collaboration. (2009). Microlensing event MOA-2007-BLG-40: Exhuming the buried signature of a Cool, Jovian-mass planet. *ApJ, 698*, 1826–1837. https://doi.org/10.1088/004637X/698/2/1826. arXiv: 0809.2997

Dong, S., Mérand, A., Delplancke-Ströbele, F., Gould, A., Chen, P., Post, R., ... Thompson, T. A. (2019). First resolution of microlensed images. *ApJ, 871*(1), 70. https://doi.org/10.3847/1538-4357/aaeffb. arXiv: 1809.08243 [astro-ph.SR]

Erdl, H., & Schneider, P. (1993). Classification of the multiple deflection two point-mass gravitational lens models and application of catastrophe theory in lensing. *A & A, 268*, 453–471.

Foreman-Mackey D. (2016). Corner.py: Scatterplot matrices in python. *The Journal of Open Source Software, 1*(2), 24. https://doi.org/10.21105/joss.00024

Foreman-Mackey D., Hogg, D. W., Lang, D., & Goodman, J. (2013). emcee: The MCMC hammer. *Proc.ASP, 125*(925), 306. https://doi.org/10.1086/670067. arXiv: 1202.3665 [astro-ph.IM]

Gaudi, B. S. (2012). Microlensing surveys for exoplanets. *Annual Review of Astronomy and Astrophysics, 50*, 411–453. https://doi.org/10.1146/annurevastro081811125518

Gaudi, B. S., Bennett, D. P., Udalski, A., Gould, A., Christie, G. W., Maoz, D., Macintosh, B. (2008). Discovery of a Jupiter/Saturn analog with gravitational microlensing. *Science, 319*, 927. https://doi.org/10.1126/science.1151947. arXiv: 0802.1920

Gaudi, B. S., & Gould, A. (1997). Planet parameters in microlensing events. *ApJ, 486*(1), 85–99. https://doi.org/10.1086/304491. arXiv: astroph/9610123 [astro-ph]

Gould, A. (1994). Proper motions of MACHOs. *ApJL 421* L71–L74. doi:10.1086/187190

Gould, A., & Loeb, A. (1992). Discovering planetary systems through gravitational microlenses. *ApJ, 396* 104–114. https://doi.org/10.1086/171700

Gould, A., Udalski, A., An, D., Bennett, D. P., Zhou, A.-Y., Dong, S., Swaving, S. C. (2006). Microlens OGLE-2005-BLG-169 implies that cool neptune-like planets are common. *ApJL, 644* L37–L40. https://doi.org/10.1086/505421 eprint: astroph/0603276

Gould, A., Udalski, A., Monard, B., Horne, K., Dong, S., Miyake, N., PLANET Collaboration. (2009). The extreme microlensing event OGLE-2007-BLG-224: Terrestrial parallax observation of a thick-disk brown dwarf. *ApJL, 698*, L147–L151. https://doi.org/10.1088/004-637X/698/2/L147. arXiv: 0904.0249 [astro-ph.GA]

Han, C. (2006). Properties of planetary caustics in gravitational microlensing. *ApJ, 638* 1080–1085. https://doi.org/10.1086/498937 eprint: astroph/0510206

Han, C., & Gaudi, B. S. (2008). A characteristic planetary feature in double-peaked, high-magnification microlensing events. *ApJ, 689*, 53–58. https://doi.org/10.1086/592723. arXiv: 0805.1103

Han, C., Jung, Y. K., Udalski, A., Sumi, T., Gaudi, B. S., Gould, A., RoboNet Collaboration. (2013). Microlensing discovery of a tight, low-mass-ratio planetary-mass object around an old field brown dwarf. *ApJ, 778*, 38. https://doi.org/10.1088/0004637X/778/1/38. arXiv: 1307.6335 [astro-ph.EP]

Hog, E., Novikov I. D., & Polnarev A. G. (1995). MACHO photometry and astrometry. *A & A, 294*, 287–294.

Jung, Y. K., Udalski, A., Bond, I. A., Yee, J. C., Gould, A., Han, C., MOA Collaboration. (2017a). OGLE-2016-BLG-1003: First resolved caustic-crossing binary-source event discovered by second-generation microlensing surveys. *ApJ, 841*, 75. https://doi.org/10.3847/15384357/aa7057. arXiv: 1705.01531 [astro-ph.SR]

Jung, Y. K., Udalski, A., Sumi, T., Han, C., Gould, A., Skowron, J., muFUN Collaboration. (2015). OGLE-2013-BLG-0102LA,B: Microlensing binary with components at star/brown dwarf and brown dwarf/planet boundaries. *ApJ, 798*, 123. https://doi.org/10.1088/0004637X/798/2/123. arXiv: 1407.7926 [astro-ph.SR]

Jung, Y. K., Udalski, A., Yee, J. C., Sumi, T., Gould, A., Han, C., MOA Collaboration. (2017b). Binary source microlensing event OGLE-2016-BLG-0733: Interpretation of a long-term asymmetric perturbation. *AJ, 153*, 129. https://doi.org/10.3847/15383881/aa5d07. arXiv: 1611.00775 [astro-ph.SR]

Kervella, P., Bersier D., Mourard, D., Nardetto, N., Fouqué, P., & Coudé du Foresto, V. (2004). Cepheid distances from infrared long-baseline interferometry III. Calibration of the surface brightness-color relations. *A & A, 428*, 587–593. https://doi.org/10.1051/00046361:20041416

Koshimoto, N., Shvartzvald, Y., Bennett, D. P., Penny M. T., Hundertmark, M., Bond, I. A., VST-K2C9 Team. (2017). MOA-2016-BLG-227Lb: A massive planet characterized by combining light-curve analysis and Keck AO imaging. *AJ, 154*, 3. https://doi.org/10.3847/15383881/aa72e0. arXiv: 1704.01724 [astro-ph.EP]

Koshimoto, N., Udalski, A., Sumi, T., Bennett, D. P., Bond, I. A., Rattenbury N., OGLE Collaboration. (2014). OGLE-2008-BLG-355Lb: A massive planet around a late-type star. *ApJ, 788*, 128. https://doi.org/10.1088/0004637X/788/2/128. arXiv: 1403.7005 [astro-ph.EP]

Lee, C.-H., Riffeser A., Seitz, S., & Bender R. (2009). Finite source effects in microlensing: A precise, easy to implement, fast and numerical stable formalism. *The Astrophysical Journal, 695*, 200–207. https://doi.org/10.1088/0004637X/695/1/200. arXiv: 0901.1316 [astro-ph.GA]

Lissauer J. J. (1987). Timescales for planetary accretion and the structure of the protoplanetary disk. *Icarus, 69*, 249–265. https://doi.org/10.1016/00191035(87)901047

Mao, S. (2008). Introduction to gravitational microlensing. ArXiv e-prints arXiv: 0811.0441

Mao, S., & Paczynski, B. (1991). Gravitational microlensing by double stars and planetary systems. *ApJL, 374*, L37. https://doi.org/10.1086/186066

Miyamoto, M., & Yoshii, Y. (1995). Astrometry for determining the MACHO mass and trajectory. *AJ, 110*, 1427. https://doi.org/10.1086/117616

Mollerach, S., & Roulet, E. (2002). *Gravitational lensing and microlensing* https://doi.org/10.1142/4890

Paczynski, B. (1986). Gravitational microlensing by the galactic halo. *ApJ, 304*, 1–5. https://doi.org/10.1086/164140

Paulin-Henriksson, S., Baillon, P., Bouquet, A., Carr B. J., Crézé, M., Evans, N. W., & POINT AGAPE Collaboration. (2003). The POINTAGAPE survey: 4 high signal-to-noise microlensing candidates detected towards M 31. *A & A, 405*, 15–21. https://doi.org/10.1051/00046361:20030519. eprint: astroph/0207025

Poleski, R., Udalski, A., Dong, S., Szymański, M. K., Soszyński, I., Kubiak, M., & Gould, A. (2014). Super-massive planets around late-type stars: The case of OGLE-2012-BLG-0406Lb. *ApJ, 782*, 47. https://doi.org/10.1088/0004637X/782/1/47. arXiv: 1307.4084 [astro-ph.EP]

Proft, S., Demleitner M., & Wambsganss, J. (2011). Prediction of astrometric microlensing events during the Gaia mission. *A & A, 536*, A50. https://doi.org/10.1051/0046361/201117663. arXiv: 1201.4000 [astro-ph.GA]

Rhie, S. H. (2001). Can a gravitational quadruple lens produce 17 images? ArXiv Astrophysics e-prints. eprint: astroph/0103463

Rhie, S. H. (2003). n-point gravitational lenses with 5(n-1) images. ArXiv Astrophysics e-prints. eprint: astroph/0305166

Sahu, K. C., Anderson, J., Casertano, S., Bond, H. E., Bergeron, P., Nelan, E. P., & Livio, M. (2017). Relativistic deflection of background starlight measures the mass of a nearby white dwarf star. *Science, 356*(6342), 1046–1050. https://doi.org/10.1126/science.aal2879. arXiv: 1706.02037 [astro-ph.SR]

Shin, I.-G., Ryu, Y-H., Udalski, A., Albrow M., Cha, S.-M., Choi, J.-Y., & Gould, A. (2016). A super-Jupiter microlens planet characterized by high-cadence KMTNeT microlensing survey observations of OGLE-2015-BLG-0954. *Journal of Korean Astronomical Society, 49*, 73–81. https://doi.org/10.5303/JKAS.2016.49.3.073. arXiv: 1603.00020 [astro-ph.EP]

Shvartzvald, Y., Maoz, D., Kaspi, S., Sumi, T., Udalski, A., Gould, A., & Pietrukowicz, P. (2014). MOA-2011-BLG-322Lb: A 'second generation survey' microlensing planet. *MNRAS, 439*, 604–610. https://doi.org/10.1093/mnras/stt2477. arXiv: 1310.0008 [astro-ph.EP]

Street, R. A., Choi, J.-Y., Tsapras, Y., Han, C., Furusawa, K., Hundertmark, & M., Surdej, J. (2013). MOA-2010-BLG-073L: An M-dwarf with a substellar companion at the planet/Brown Dwarf boundary. *ApJ, 763*, 67. https://doi.org/doi:10.1088/0004-637X/763/1/67. arXiv: 1211 3782 [astro-ph.EP]

Sumi, T., Bennett, D. P., Bond, I. A., Udalski, A., Batista, V., Dominik, M., & muFUN Collaboration. (2010). A cold neptune-mass planet OGLE-2007-BLG-368Lb: Cold Neptunes are common. *ApJ, 710*, 1641–1653. https://doi.org/10.1088/004637X/71/2/1641. arXiv: 0912.1171 [astro-ph.EP]

Sumi, T., Kamiya, K., Bennett, D. P., Bond, I. A., Abe, F., Botzler C. S., & Microlensing Observations in Astrophysics (MOA) Collaboration. (2011). Unbound or distant planetary mass population detected by gravitational microlensing. *Nature, 473*, 349–352. https://doi.org/10.1038/nature10092. arXiv: 1105.3544 [astro-ph.EP]

Tsapras, Y. (2018). Microlensing searches for exoplanets. *Geosciences, 8*(10), 365. https://doi.org/10.3390/geosciences8100365. arXiv: 1810.02691 [astro-ph.EP]

Udalski, A., Jaroszyński, M., Paczyński, B., Kubiak, M., Szymański, M. K., & Soszyński, I. (2005). A Jovian-Mass planet in microlensing event OGLE-2005-BLG-071. *ApJL, 628*(2), L109–L112. https://doi.org/10.1086/432795. arXiv: astroph/0505451 [astro-ph]

Witt, H. J. (1990). Investigation of high amplification events in light curves of gravitationally lensed quasars. *A & A, 236*, 311–322.

Witt, H. J., & Mao, S. (1994). Can lensed stars be regarded as pointlike for microlensing by MACHOs? *ApJ, 430*, 505–510. https://doi.org/10.1086/174426

Witt, H. J., & Mao, S. (1995). On the minimum magnification between caustic crossings for microlensing by binary and multiple stars. *ApJL, 447*, L105. https://doi.org/10.1086/309566

Yee, J. C., Udalski, A., Calchi Novati, S., Gould, A., Carey S., Poleski, R., & Wyrzykowski, Ł. (2015). First space-based microlens parallax measurement of an isolated star: Spitzer observations of OGLE-2014-BLG-0939. *ApJ, 802*, 76. https://doi.org/10.1088/0004637X/802/2/76. arXiv: 1410.5429 [astro-ph.SR]

Zurlo, A., Gratton, R., Mesa, D., Desidera, S., Enia, A., Sahu, K., & Roux, A. (2018). The gravitational mass of Proxima Centauri measured with SPHERE from a microlensing event. *MNRAS, 480*(1), 236–244. https://doi.org/10.1093/mnras/sty1805. arXiv: 1807.01318 [astro-ph.SR]

Extended Lenses

5

In this chapter, we review some properties of extended lenses, i.e. gravitational lenses, which can be described by extended, gravitationally bound, mass distributions. Cosmic structures like galaxies and galaxy clusters belong to this class of gravitational lenses. They produce the most spectacular lensing features observable in the sky, such as multiple images and gravitational arcs.

This chapter aims to understand how these effects depend on the specific properties of the lenses. We start by discussing circular, axially symmetric models and the impact of different mass profiles on their lens properties. Then, we introduce deviations from circular symmetry in the form of ellipticity and substructures. Finally, we consider the effects of the environment within which the lenses may reside.

5.1   Circular, Axially Symmetric Lenses

We begin with the most straightforward description of an extended lens, i.e. an axially symmetric or circular lens. For such a lens, the lensing potential is constant on circles centered on the lens center. Given the lens's symmetry properties, it is convenient to choose the reference frame's origin at the lens's center. Most of the equations, therefore, reduce to a one-dimensional form.

Deflection Angle
Let the lensing potential be

$$\hat{\Psi}(\vec{\theta}) = \hat{\Psi}(\theta), \tag{5.1}$$

© Springer Nature Switzerland AG 2021
M. Meneghetti, *Introduction to Gravitational Lensing*, Lecture Notes
in Physics 956, https://doi.org/10.1007/978-3-030-73582-1_5

where $\vec{\theta}$ is the usual vector (in angular units) on the lens plane. To use physical units, we have to multiply $\vec{\theta}$ by the angular diameter distance of the lens plane, $\vec{\xi} = D_L \vec{\theta}$.

As seen, in Sect. 3.2, the (reduced) deflection angle, $\vec{\alpha}(\vec{\theta})$, is the gradient of the lensing potential (in angular units). It is convenient to use polar coordinates. Then, the gradient operator can be written as

$$\vec{\nabla}_{\theta} \equiv D_L \left(\frac{\partial}{\partial \xi} \vec{e}_{\xi} + \frac{1}{\xi} \frac{\partial}{\partial \phi} \vec{e}_{\phi} \right) = \left(\frac{\partial}{\partial \theta} \vec{e}_{\theta} + \frac{1}{\theta} \frac{\partial}{\partial \phi} \vec{e}_{\phi} \right) , \tag{5.2}$$

where ϕ is the polar angle, $\vec{e}_{\xi} = \vec{e}_{\theta}$ and $\vec{e}_{\phi}$ are unit vectors, the first pointing in the radial direction and the second perpendicular to it.

Since the lensing potential does not depend on ϕ, for axially symmetric lenses the gradient of $\hat{\Psi}(\theta)$ is

$$\vec{\nabla}_{\theta} \hat{\Psi}(\vec{\theta}) = \hat{\Psi}'(\theta) \vec{e}_{\theta} = \vec{\alpha}(\vec{\theta}) = \alpha(\theta) \vec{e}_{\theta} . \tag{5.3}$$

Thus, the deflection angle is *central*, i.e. parallel to $\vec{\theta}$. In the equation above, we used the Lagrange notation for the derivatives, i.e. $\hat{\Psi}'(\theta) = \partial \hat{\Psi}(\theta)/\partial \theta$.

In addition, the Laplacian of the lensing potential is twice the convergence, as shown in Eq. 3.24. Writing the Laplace operator in polar coordinates, we obtain

$$\frac{1}{\theta} \frac{\partial}{\partial \theta} \left(\theta \frac{\partial}{\partial \theta} \right) \hat{\Psi}(\theta) = 2\kappa(\theta) . \tag{5.4}$$

From this equation, we see that

$$\alpha(\theta) = \frac{2 \int_0^{\theta} \kappa(\theta') \theta' d\theta'}{\theta}$$

$$= \frac{2 \int_0^{\theta} \Sigma(\theta') \theta' d\theta'}{\theta \Sigma_{cr}} \tag{5.5}$$

$$= \frac{D_{LS}}{D_S} \frac{4GM(\theta)}{c^2 D_L \theta} . \tag{5.6}$$

This formula is identical to that of the reduced deflection angle of the point-mass lens (e.g. Eq. 4.5), with the only difference that the mass M is substituted by the mass in a circle of radius θ, $M(\theta)$. This shows that, because of its symmetry, the lens properties are uniquely determined by the mass profile, $M(\theta)$, or alternatively by the profile of the surface-mass density, $\Sigma(\theta)$.

As usual, we can switch to the dimensionless notation by choosing a convenient arbitrary linear scale, ξ_0, which corresponds to the angular scale $\theta_0 = \xi_0/D_L$. From Eqs.3.6 and 3.17, the reduced deflection angle in the dimensionless form is

$$
\begin{aligned}
\alpha(x) &= \frac{D_L D_{LS}}{\xi_0 D_S}\hat{\alpha}(\xi_0 x) \\
&= \frac{D_L D_{LS}}{\xi_0 D_S}\frac{4GM(\xi_0 x)}{c^2 \xi} \\
&= \frac{M(\xi_0 x)}{\pi \xi_0^2 \Sigma_{cr}}\frac{1}{x} = \frac{m(x)}{x} \quad ,
\end{aligned}
\tag{5.7}
$$

where we have introduced the *dimensionless mass*

$$
m(x) \equiv \frac{M(\xi_0 x)}{\pi \xi_0^2 \Sigma_{cr}} \quad .
\tag{5.8}
$$

Note that

$$
\alpha(x) = \frac{2}{x}\int_0^x x' \kappa(x')\mathrm{d}x'
\tag{5.9}
$$

and

$$
m(x) = 2\int_0^x x' \kappa(x')\mathrm{d}x' \quad .
\tag{5.10}
$$

Lens Equation
Since $\vec{\alpha}(\vec{x})$ is parallel to $\vec{x}$, Eq. 3.7 can be written omitting the vector notation.
   Thus, using Eq. 5.10, we obtain

$$
y = x - \frac{m(x)}{x} \quad .
\tag{5.11}
$$

Convergence and Shear
From Eq. 5.4 and using the dimensionless form of the lensing potential in Eq. 3.14, we can easily find that the convergence profile is

$$
\kappa(x) = \frac{1}{2}\left[\Psi''(x) + \frac{\Psi'(x)}{x}\right] \quad .
\tag{5.12}
$$

Since

$$\Psi'(x) = \alpha(x) , \qquad (5.13)$$

we obtain that

$$\kappa(x) = \frac{1}{2}\left[\alpha'(x) + \frac{\alpha(x)}{x}\right] . \qquad (5.14)$$

Using Eq. 5.7, we also find that

$$\alpha'(x) = \frac{m'(x)}{x} - \frac{m(x)}{x^2} . \qquad (5.15)$$

Thus,

$$\kappa(x) = \frac{1}{2}\frac{m'(x)}{x} . \qquad (5.16)$$

The components of the shear are also readily derived by writing the partial derivative operators in polar coordinates,

$$\frac{\partial}{\partial x_1} = \cos\phi\,\frac{\partial}{\partial x} - \frac{\sin\phi}{x}\,\frac{\partial}{\partial \phi} ,$$
$$\frac{\partial}{\partial x_2} = \sin\phi\,\frac{\partial}{\partial x} + \frac{\cos\phi}{x}\,\frac{\partial}{\partial \phi} , \qquad (5.17)$$

where x_1 and x_2 are the dimensionless counterparts of the Cartesian coordinates ξ_1 and ξ_2 on the lens plane.

Since,

$$\alpha_1(x) = \alpha(x)\cos\phi ,$$
$$\alpha_2(x) = \alpha(x)\sin\phi , \qquad (5.18)$$

from Eqs. 3.35 and 3.36, we obtain that

$$\begin{aligned}
\gamma_1(x) &= \frac{1}{2}\left[\frac{\partial}{\partial x_1}\alpha_1(x) - \frac{\partial}{\partial x_2}\alpha_2(x)\right] \\
&= \frac{1}{2}\left[(\cos^2\phi - \sin^2\phi)\alpha'(x) - (\cos^2\phi - \sin^2\phi)\frac{\alpha(x)}{x}\right] \\
&= \frac{\cos 2\phi}{2}\left[\alpha'(x) - \frac{\alpha(x)}{x}\right] ,
\end{aligned} \qquad (5.19)$$

and

$$\gamma_2(x) = \frac{\partial}{\partial x_2}\alpha_1(x)$$

$$= \left[\sin\phi\cos\phi\alpha'(x) - \sin\phi\cos\phi\frac{\alpha(x)}{x}\right]$$

$$= \frac{\sin 2\phi}{2}\left[\alpha'(x) - \frac{\alpha(x)}{x}\right]. \tag{5.20}$$

The modulus of the shear is

$$\gamma(x) = \frac{1}{2}\left|\alpha'(x) - \frac{\alpha(x)}{x}\right|$$

$$= \frac{1}{2}\left|\frac{m'(x)}{x} - \frac{2m(x)}{x^2}\right|$$

$$= |\kappa(x) - \bar{\kappa}(x)|, \tag{5.21}$$

where $\bar{\kappa}(x)$ is the mean convergence within a circle of radius x:

$$\bar{\kappa}(x) = \frac{m(x)}{x^2} = 2\frac{\int_0^x x'\kappa(x')\mathrm{d}x'}{x^2}. \tag{5.22}$$

Lensing Jacobian

Using the results above and Eq. 3.46, the Jacobian matrix can be written as

$$A(x,\phi) = \left[1 - \frac{m'(x)}{2x}\right]I - \frac{1}{2}\left[\frac{m'(x)}{x} - \frac{2m(x)}{x^2}\right]\begin{pmatrix}\cos 2\phi & \sin 2\phi \\ \sin 2\phi & -\cos 2\phi\end{pmatrix}. \tag{5.23}$$

Its determinant is

$$\det A(x) = \frac{y}{x}\frac{\mathrm{d}y}{\mathrm{d}x} = \left[1 - \frac{\alpha(x)}{x}\right]\left[1 - \alpha'(x)\right]$$

$$= \left[1 - \frac{m(x)}{x^2}\right]\left[1 + \frac{m(x)}{x^2} - \frac{m'(x)}{x}\right]$$

$$= [1 - \bar{\kappa}(x)][1 + \bar{\kappa}(x) - 2\kappa(x)]$$

$$= [1 - \kappa(x) - \gamma(x)][1 - \kappa(x) + \gamma(x)]. \tag{5.24}$$

Finally, the magnification profile is $\mu(x) = \det A(x)^{-1}$.

Critical Lines and Caustics

Since the critical lines satisfy the condition $\det A(\vec{x}) = 0$, Eq. 5.24 implies that axially symmetric lenses with monotonically increasing $m(x)$ have at most two critical lines. These are circles, whose radii can be found by solving the equations

$$\frac{\alpha(x)}{x} = \frac{m(x)}{x^2} = \overline{\kappa}(x) = \kappa(x) + \gamma(x) = 1 \tag{5.25}$$

and

$$\alpha'(x) = \frac{m'(x)}{x} - \frac{m(x)}{x^2} = 2\kappa(x) - \overline{\kappa}(x) = \kappa(x) - \gamma(x) = 1 . \tag{5.26}$$

We can prove that vectors tangential to the first critical line, or normal to the second, are eigenvectors of the Jacobian matrix with zero eigenvalues. Thus, Eq. 5.25 defines the so-called *tangential critical line*. On the contrary, Eq. 5.26 defines the *radial critical line*. This can be seen as follows (Schneider et al. 1992). Consider the point with Cartesian coordinates $(x, 0)$ on the first critical line. This point is distant x from the lens center and has a phase $\phi = 0$. The Jacobian matrix at $(x, 0)$ is readily derived from Eq. 5.23:

$$A(x, 0) = I + \frac{m(x)}{x^2} \begin{pmatrix} 1 & 0 \\ 0 & -1 \end{pmatrix} - \frac{m'(x)}{x} \begin{pmatrix} 1 & 0 \\ 0 & 0 \end{pmatrix} . \tag{5.27}$$

Let consider a vector whose components are $(0, a)$ at $(x, 0)$. This vector is clearly tangential to the critical line at $(x, 0)$. Through the lens mapping, it is mapped onto

$$\begin{pmatrix} y_1 \\ y_2 \end{pmatrix} = A(x, 0) \begin{pmatrix} 0 \\ a \end{pmatrix}. \tag{5.28}$$

If $(x, 0)$ lays on the tangential critical line, then $[1 - m(x)/x^2] = 0$ and

$$\begin{pmatrix} y_1 \\ y_2 \end{pmatrix} = \left[1 - \frac{m(x)}{x^2} \right] \begin{pmatrix} 0 \\ a \end{pmatrix} = \begin{pmatrix} 0 \\ 0 \end{pmatrix} . \tag{5.29}$$

Thus any vector tangent to the critical line is an eigenvector of A with 0 eigenvalue.

Consider now a vector $(b, 0)$, normal to the critical line at $(x, 0)$. Mapping it to the source plane we obtain

$$\begin{pmatrix} y_1 \\ y_2 \end{pmatrix} = A(x, 0) \begin{pmatrix} b \\ 0 \end{pmatrix} = \left[1 + \frac{m(x)}{x^2} - \frac{m'(x)}{x} \right] \begin{pmatrix} b \\ 0 \end{pmatrix} . \tag{5.30}$$

If $(x, 0)$ lays on the radial critical line, then $[1 + m(x)/x^2 - m'(x)/x] = 0$, thus $(b, 0)$ is an eigenvector of A with 0 eigenvalue.

From the lens equation, it can be easily seen that all the points along the tangential critical line are mapped on the point $y = 0$ on the source plane. Indeed,

$$y = x \left[1 - \frac{m(x)}{x^2} \right] = 0 , \tag{5.31}$$

if x is the radius of the tangential critical line. Therefore, axially symmetric models have point-like tangential caustics. Instead, the radial critical points are mapped onto a circular caustic on the source plane.

Einstein Radius

The tangential critical line is the lens Einstein ring. Its angular size can be computed by solving the lens equation (in angular units)

$$\beta = \theta - \frac{D_{LS}}{D_S D_L} \frac{4GM(\theta)}{c^2\theta} , \tag{5.32}$$

for $\beta = 0$.

This leads to a formula for the Einstein radius which is identical to that in Eq. 4.7:

$$\theta_E = \sqrt{\frac{4GM(\theta_E)}{c^2} \frac{D_{LS}}{D_L D_S}} , \tag{5.33}$$

the only difference being that the total mass of the lens is here substituted by the mass enclosed by the Einstein ring. With little effort, we can see that

$$M(\theta_E) = \pi D_L^2 \theta_E^2 \Sigma_{cr} . \tag{5.34}$$

This is a quite important result, as it tells us that the average surface density within the Einstein ring is the critical surface density (see Eq. 5.25).

When using the dimensionless notation, a natural choice for ξ_0 is $\xi_0 = D_L \theta_E$, as done for the point mass in Chap. 4. In this case, the dimensionless mass $m(x)$ is the mass $M(\theta)$ in units of the mass $M(\theta_E)$,

$$m(x) = \frac{M(\theta)}{M(\theta_E)} . \tag{5.35}$$

Fig. 5.1 Illustration of the tangential and radial magnification of an infinitesimal source

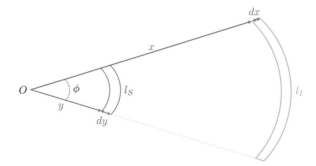

Tangential and Radial Magnification of the Images

The Jacobian matrix's eigenvalues are the inverse magnifications of the image in the tangential and radial directions (as seen above, the Jacobian matrix's eigenvectors are tangential and perpendicular to the tangential and the radial critical lines, respectively). This is illustrated in Fig. 5.1. An infinitesimal arc-like source at position y (shown in red) subtends an angle ϕ with respect to the lens center at O. We assume that its radial size is dy. Because the deflection angle is central for an axially symmetric lens, the source is mapped onto an arc-like image at position x (shown in green), whose radial size is dx and which subtends the same angle ϕ with respect to the center of the lens.

The radial magnification is obviously

$$\mu_r = \frac{dx}{dy} \; , \tag{5.36}$$

i.e. the inverse of the radial eigenvalue of the Jacobian matrix. On the contrary, the tangential magnification can be derived by comparing the tangential extension l_I and l_S of image and source. Since both subtend the same angle ϕ, we have that

$$\mu_t = \frac{l_I}{l_S} = \frac{x}{y} \; , \tag{5.37}$$

which is the inverse of the tangential eigenvalue of the Jacobian matrix.

5.2   Power-Law Lens

The formulas derived in the previous section are valid for any axially symmetric lens. Now, we consider a particular class of lenses, whose mass profile has a power-law form of the kind

$$m(x) = x^{3-n} \; . \tag{5.38}$$

Note that, for $x = 1$, the dimensionless mass is $m(1) = 1$. Thus, the radius x is in units of the lens Einstein radius, θ_E.

The convergence profile is derived from Eq. 5.16 and is given by

$$\kappa(x) = \frac{m'(x)}{2x} = \frac{3-n}{2} x^{1-n} . \tag{5.39}$$

Depending on n being larger or smaller than one, $\kappa(x)$ is a decreasing or an increasing function of x. Values of $n < 1$ are not interesting in the context of this Chapter, as we are focusing on gravitational lenses that have bound mass distributions. Therefore, they will be not considered further.

5.2.1 Lenses with $1 < n < 2$

We start from lenses with $1 < n < 2$. Their deflection angle is

$$\alpha(x) = \frac{m(x)}{x} = x^{2-n} . \tag{5.40}$$

Thus, this class of lenses have deflection angle profiles which monotonically increase with x and that are zero at the origin, $\alpha(0) = 0$.

The case $n = 1$ corresponds to a lens with constant convergence, $\kappa = 1$. For such a lens, $\alpha(x) = x$, implying $y(x) = 0$ for any x. Thus, this lens is perfectly convergent.

Critical Lines and Caustics
Because of the mass profile's chosen normalization, the power-law lens's tangential critical line is a circle with radius $x_t = 1$. As pointed out earlier, this is the Einstein ring, whose size in angular units was given in Eq. 5.33. The tangential caustic is a point at $y = 0$.

Instead, the size of the radial critical line in units of the Einstein radius depends on the power-law index n. Since the radial critical line forms where the condition $\alpha'(x) = 1$ is met, the critical radius can be found by solving the equation

$$(2-n)x_r^{1-n} = 1 , \tag{5.41}$$

from which we obtain

$$x_r = (2-n)^{1/(n-1)} . \tag{5.42}$$

In Fig. 5.2, we show how the size of the radial critical line (in units of the Einstein radius) varies as a function of n. By making the substitution $n' = 1/(n-1)$, we find that

$$x_r = \left(1 - \frac{1}{n'}\right)^{n'} . \tag{5.43}$$

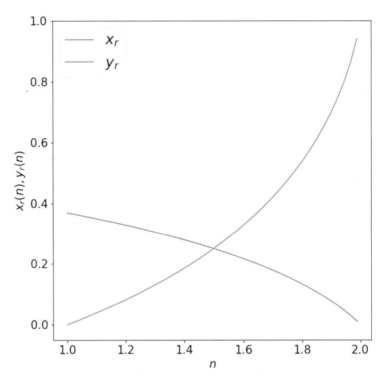

Fig. 5.2 Size of the radial critical line and caustic (in units of the Einstein radius) as a function of the power-law index n

Thus, for $n \to 1$, or $n' \to \infty$,

$$x_r = \lim_{n' \to \infty} \left(1 - \frac{1}{n'}\right)^{n'} = \frac{1}{e} \,. \tag{5.44}$$

The figure shows that the size of the radial critical line becomes smaller as n increases. Thus, lenses with steep density profiles have tiny radial critical lines.

In the same figure, we also show the size of the radial caustic. Contrary to the radial critical line's size, the radial caustic size grows as a function of n. In particular, for $n \to 2$, $y_r \to 1$.

Multiple Images
The number of multiple images that the power-law lens can produce can be determined by inspecting the so-called *image diagram*, shown in Fig. 5.3. The solid lines in the three panels show the curves $\alpha(x)$ corresponding to three values of the power index n, namely $n = 1.1$, 1.5, and 1.9.

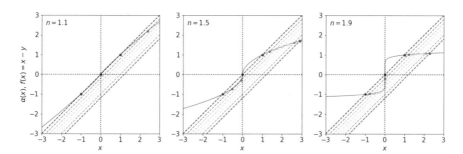

Fig. 5.3 Image diagram for power-law lenses with $n = 1.1$ (left panel), $n = 1.5$ (central panel), and $n = 1.9$ (right panel). The solid curves show the function $\alpha(x)$. The colored dashed lines show the function $f(x) = x - y$ for a range of values of $y \in [0, 1.2]$

Remark 5.1. Note that we can compute the deflection angle $\alpha(x)$ for negative values of x. Since the vector $\vec{\alpha}$ is pointing away from the lens, the deflection angle sign is positive on the positive x semi-axis and negative otherwise.

The lens equation states that the images of a source at position y form at the intersections of $\alpha(x)$ and $f(x) = x - y$. The latter is a line with a unit slope and intercept $-y$. Some examples, corresponding to values of y increasing from 0 to 1.2, are given by the colored dashed lines. We mark the intersections between $f(x)$ and $\alpha(x)$ with colored dots, thus identifying the locations of the source's multiple images when projected on the x axis.

As we can see, the power-law lens with $1 < n < 2$ can produce either three or one image of the background source, depending on whether y is smaller or larger than a particular value y_r. In fact, there exists a value of $y = y_r$ such that the line $f(x) = x - y_r$ is tangential to $\alpha(x)$. At the tangent point, two images of the source merge, and for $y > y_r$, they no longer exist. Obviously, y_r is the radius of a caustic, and the solution of the equation

$$\alpha(x_r) = x_r - y_r \tag{5.45}$$

gives the radius x_r of the corresponding critical line. Since $\alpha(x)$ is tangential to $f(x)$ at x_r, where $\alpha'(x_r) = 1$, we find that x_r is the radius of the radial critical line.

Thus, multiple images exist only if the source is inside the radial caustic, $0 < y \leq y_r$. One image forms on the positive x semi-axis, with $x > y$. Such an image is located outside the Einstein ring, as its distance from the lens center is > 1. Two additional images form inside the Einstein ring, on the negative x semi-axis. Of these two images, the inner one is inside the radial critical line, $|x| < x_r$. The other is between the radial and the tangential critical line.

Remark 5.2. For $y = 0$, the innermost image forms at $x = 0$. A source right behind a power-law lens with $1 < n < 2$ has a central image (in addition to the Einstein ring).

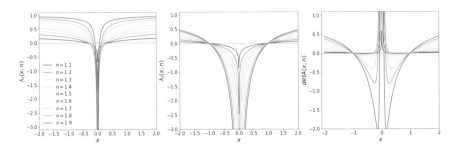

Fig. 5.4 Left and central panels: Radial and tangential eigenvalues of the Jacobian matrix as a function of x for different values of n. Right panel: determinant of the lensing Jacobian, resulting from the product of the curves in the first two panels

Image Magnification

The eigenvalues of the Jacobian matrix are

$$\lambda_t(x) = 1 - x^{1-n} \tag{5.46}$$

$$\lambda_r(x) = 1 - (2-n)x^{1-n} . \tag{5.47}$$

In the left and central panels of Fig. 5.4, their values are shown as a function of x. The sign of each eigenvalue changes from outside to inside the critical lines. Thus, the parity of the images changes accordingly. In particular, the outermost image always has positive parity. Both eigenvalues are positive at this image's position, meaning that it corresponds to a minimum of the time delay surface. The image forming between the radial and the tangential critical lines has negative parity ($\mu < 0$) because the eigenvalues have opposite signs. Thus, the image is located at a saddle point of the time delay surface. The innermost image has again positive parity, being the two eigenvalues both negative. This image is located at a local maximum of the time delay surface.

The right panel of Fig. 5.4 shows how the determinant of the lensing Jacobian matrix A varies as a function of x. For $|\det A| < 1$, the total magnification $\mu > 1$. Thus, the images outside the tangential critical line are magnified. Instead, those close to the lens center can be strongly de-magnified unless near the radial critical line.

In Fig. 5.5, we show the ratio of the tangential-to-radial magnifications for the three images of a source at $y = 0.05$. The outermost image is characterized by a tangential-to-radial magnification, which always exceeds one. Thus the overall distortion of this image is always tangential. The innermost image is predominantly radially distorted. In the case shown in the figure, the image is on the radial critical line for $n \sim 1.15$. For n smaller than this value, the source only has one image. Increasing n, the innermost image moves near the center (see Fig. 5.3), but so does the radial critical line. Consequently, the tangential-to-radial magnification ratio decreases also for large n. The intermediate image shows a transition from being

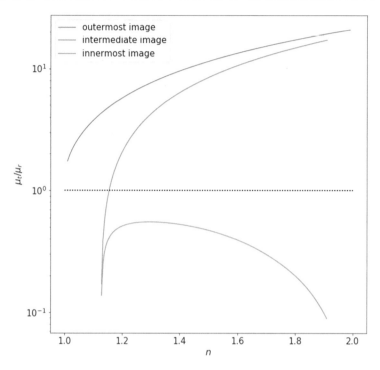

Fig. 5.5 Magnification ratio at the position of the multiple images of a source at $y = 0.2$

predominantly radially to tangentially distorted as n increases. The reason is again that the radial critical line shrinks as n increases.

In Fig. 5.6, the results discussed above are visualized with four examples, corresponding to lenses with $n = 1.05, 1.2, 1.4$, and 1.9 (from the upper left to the bottom right panels). We consider a circular source in each panel with a radius of $r = 0.02\theta_E$ (navy circle). The source is very close to the center of the lens projected onto the source plane ($y = 0.05$). We show the outermost, intermediate, and innermost source images in orange, green, and dark-red. The red and blue solid circles show the radial and the tangential critical lines of the lens. The dashed circle is the radial caustic. Note that the size of the caustic grows as the value of n increases. Therefore, the position of the source relative to the caustic changes as well. The source is outside the caustic in the upper left panel, meaning that it has only one image. Not only is the tangential magnification of this image large, but so is also the radial magnification, due to the small value of $n = 1.05$. For $n = 1.2$ (upper right panel) the source extends across the radial caustic. The innermost and the intermediate images merge across the corresponding critical line. For larger n (bottom panels), the outermost and the intermediate images' tangential magnification largely exceeds the radial magnification. Thus, they appear as long arcs with a length-to-width ratio significantly larger than one. In particular, as $n \to 2$, the radial magnification $\mu_r = \lambda_r^{-1} \to 1$, meaning that the radial sizes

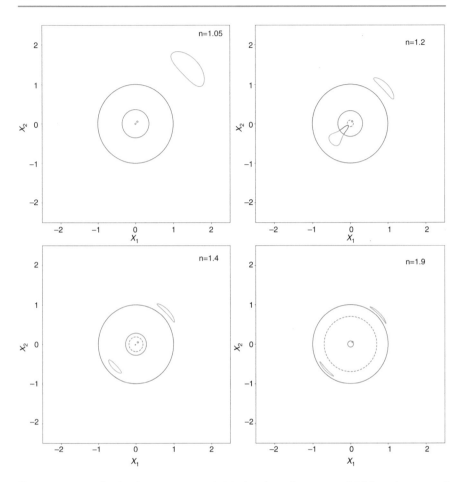

Fig. 5.6 Images of a circular source (navy circle) placed at a distance $y = 0.05$ from the center of the power-law lenses with index $n = 1.05$, 1.2, 1.4, and 1.9. In each panel, we show the tangential and the radial critical lines (blue and red solid lines) and the radial caustic (dashed red line). The outermost, intermediate, and innermost images are shown in orange, green, and dark-red. The source has a radius $r = 0.02\theta_E$

of the arcs approach the diameter of the source. The innermost image is severely de-magnified and barely visible only in the bottom left panel.

5.2.2 Lenses with $n > 2$

The power-law lenses with $n \geq 2$ have the peculiarity that the deflection angle profile $\alpha(x)$ is either flat ($n = 2$) or singular $n > 2$. We discuss the case $n = 2$ in the next section. Here, we briefly consider the case $n > 2$. In particular, we focus on image multiplicity. As shown in Fig. 5.7, such lenses always produce two

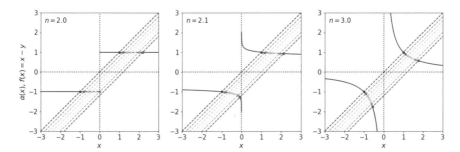

Fig. 5.7 Image diagram for power-law lenses with $n = 2$ (left panel), $n = 2.1$ (central panel), and $n = 3$ (right panel). The solid curves show the function $\alpha(x)$. The colored dashed lines show the function $f(x) = x - y$ for a range of values of $y \in [0, 1.2]$

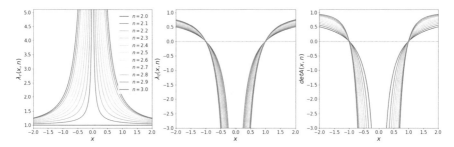

Fig. 5.8 As in Fig. 5.4, but for $n \geq 2$

images, one outside and one inside the Einstein radius. The image inside the Einstein radius approaches the lens center as y increases. For $y = 0$, the lens produces an Einstein ring. It can be easily seen that both images are radially de-magnified, being $\lambda_r(x) > 1$ for any x, as shown in the left panel of Fig. 5.8. The absolute tangential magnification, $|\mu_t|(x) = |\lambda_t^{-1}|(x)$, is larger than unity except near the lens center (central and right panels of Fig. 5.8). Thus, the observation of radially de-magnified tangential arcs may suggest that the lens has a steep surface density profile.

Remark 5.3. Note that, for power-law lenses with $n > 2$ do not have radial critical lines. The case $n = 3$ corresponds to the point-mass lens. Indeed, $m(x) = m =$ const. and $\alpha(x) = m/x$ for such lens.

5.2.3   Singular Isothermal Sphere

Due to its simplicity, one of the most widely used models for axially symmetric lenses is the Singular Isothermal Sphere (SIS hereafter). Its density profile is derived assuming that the matter content of the lens behaves as an ideal gas confined by a spherically symmetric gravitational potential. This gas is assumed to be in thermal and hydrostatic equilibria. One (three-dimensional) density profile satisfying these conditions is given by

$$\rho(r) = \frac{\sigma_v^2}{2\pi G r^2} , \tag{5.48}$$

where σ_v is the velocity dispersion of the "gas" particles and $r = \sqrt{\xi^2 + z^2}$ is the distance from the sphere center. As usual, ξ is the distance from the lens center on the lens plane and z is the coordinate along the line-of-sight. By projecting the three-dimensional density onto the lens plane, we obtain the surface density

$$\begin{aligned}
\Sigma(\xi) &= 2 \int_0^\infty \rho(\xi, z) dz \\
&= \frac{\sigma_v^2}{\pi G} \int_0^\infty \frac{dz}{\xi^2 + z^2} \\
&= \frac{\sigma_v^2}{\pi G} \frac{1}{\xi} \left[\arctan \frac{z}{\xi} \right]_0^\infty \\
&= \frac{\sigma_v^2}{2 G \xi} .
\end{aligned} \tag{5.49}$$

This density profile has a singularity at $\xi = 0$, where the density is infinite. By choosing

$$\theta_0 = 4\pi \left(\frac{\sigma_v}{c} \right)^2 \frac{D_{LS}}{D_S} ; \ \xi_0 = D_L \theta_0 \tag{5.50}$$

as the angular and length scales on the lens plane, respectively, we obtain

$$\Sigma(x) = \frac{1}{2x} \frac{c^2}{4\pi G} \frac{D_S}{D_L D_{LS}} = \frac{1}{2x} \Sigma_{cr} . \tag{5.51}$$

Thus, the convergence for the singular isothermal profile is

$$\kappa(x) = \frac{1}{2x} , \tag{5.52}$$

which shows that the SIS profile corresponds to the power-law lens with $n = 2$. Thus, the mass profile is

$$m(x) = x , \tag{5.53}$$

and the deflection angle is

$$\alpha(x) = \frac{x}{|x|} . \tag{5.54}$$

Since $|x| = 1$ for $y = 0$, θ_0 defined in Eq. 5.50 is the Einstein radius of the SIS, $\theta_E \equiv \theta_0$.

The lens equation reads

$$y = x - \frac{x}{|x|} . \tag{5.55}$$

As it can also be seen in the left panel of Fig. 5.7, if $y < 1$, two solutions of the lens equation exist. Their positions are $x_- = y - 1$ and $x_+ = y + 1$, on opposite sides with respect to the lens center. The corresponding angular positions of the images are

$$\theta_\pm = \beta \pm \theta_E . \tag{5.56}$$

The angular separation between the two images is always $\Delta(\theta) = 2\theta_E$.

On the other hand, if $y > 1$, Eq. 5.55 has a unique solution, $x_+ = y + 1$. Thus, the circle of radius $y = 1$ plays the same role of the radial caustic in the case of power-law lenses with $1 < n < 2$, separating the regions on the source plane corresponding to different image multiplicities. However, this circle is not a caustic, since $\alpha'(x) = 0$ for any x, implying that $\lambda_r = 1$. The circle of radius $y_{cut} = 1$ is a *pseudo-caustic* called the *cut*. We can see from the lens equation that

$$y_{cut} = \lim_{x \to 0} y(x) . \tag{5.57}$$

The shear follows from Eqs. 5.19 and 5.20. The modulus of γ is

$$\gamma(x) = \frac{1}{2x} , \tag{5.58}$$

i.e. the shear and the convergence profiles coincide. The shear components are

$$\gamma_1 = \frac{1}{2} \frac{\cos 2\phi}{x} , \tag{5.59}$$

$$\gamma_2 = \frac{1}{2} \frac{\sin 2\phi}{x} . \tag{5.60}$$

From Eq. 5.55, and given that $dy/dx = 1$, the magnification as a function of the image position is given by

$$\mu(x) = \frac{|x|}{|x| - 1} . \tag{5.61}$$

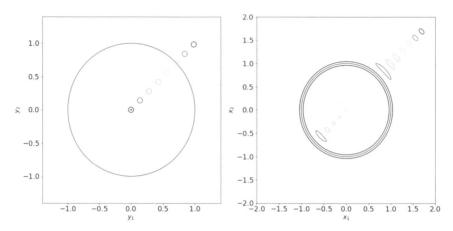

Fig. 5.9 Imaging of extended sources by a SIS lens

Images are only magnified in the tangential direction, since the radial eigenvalue of the Jacobian matrix is one everywhere.

If $y < 1$, the magnifications of the two images are

$$\mu_+(y) = \frac{y+1}{y} = 1 + \frac{1}{y} \; ; \; \mu_-(y) = \frac{|y-1|}{|y-1|-1} = \frac{-y+1}{-y} = 1 - \frac{1}{y}, \quad (5.62)$$

from which we see that for $y \to 1$, the second image becomes weaker and weaker until it disappears at $y = 1$. On the other hand, for $y \to \infty$, the source magnification obviously tends to unity: sources which are at large distance from the lens can only be weakly magnified by gravitational lensing.

Note that $\vec{x}_+$ is a minimum of the time delay surface, being both eigenvalues of the Jacobian matrix positive at this location (positive parity). Instead, $\vec{x}_-$, is a saddle point (negative parity), being the tangential eigenvalue negative.

In the left panel of Fig. 5.9, we show the cut (red circle) and the tangential caustic (blue point) of a SIS lens. In the same panel, we display several extended circular sources in different colors. We show the same sources' images in the right panel instead, illustrating the results anticipated above. The blue circle is the tangential critical line, a.k.a. the Einstein ring. Note that a source placed on the caustic is imaged onto an extended Einstein ring.

5.3   Softened (Non-singular) Isothermal Lenses

The lenses studied in the previous section have a central singularity in their surface density (or convergence). Now, we discuss the properties of cored lenses. More precisely, we introduce a core in the SIS, obtaining a Non-singular Isothermal Sphere (NIS) lens model.

The core is introduced in the surface density profile of the SIS as follows (see e.g. Kormann et al. 1994):

$$\Sigma(\xi) = \frac{\sigma_v^2}{2G}\frac{1}{\sqrt{\xi^2 + \xi_c^2}} = \frac{\Sigma_0}{\sqrt{1 + \xi^2/\xi_c^2}}\ . \tag{5.63}$$

With this modification, the profile reaches a constant density

$$\Sigma_0 = \frac{\sigma_v^2}{2G\xi_c} \tag{5.64}$$

for $\xi \ll \xi_c$.

As usual, we switch to dimensionless notation by choosing a length scale on the lens plane. We adopt the same scale ξ_0 given in Eq. 5.50, i.e. the Einstein radius of the SIS lens. Then, the convergence profile is

$$\kappa(x) = \frac{1}{2\sqrt{x^2 + x_c^2}}\ . \tag{5.65}$$

It follows that the mass profile is

$$m(x) = 2\int_0^x \kappa(x')x'dx' = \sqrt{x^2 + x_c^2} - x_c \tag{5.66}$$

and that the deflection angle profile is

$$\alpha(x) = \frac{m(x)}{x} = \sqrt{1 + \frac{x_c^2}{x^2}} - \frac{x_c}{x} \tag{5.67}$$

The convergence and the deflection angle profiles of the NIS lens model for different choices of x_c are shown in Fig. 5.10.

The shear can be derived from Eq. 5.21:

$$\gamma(x) = \frac{\sqrt{x^2 + x_c^2} - x_c}{x^2} - \frac{1}{2\sqrt{x^2 + x_c^2}}. \tag{5.68}$$

The radius of the tangential critical line can be calculated by solving the equation $1 - m(x)/x^2 = 0$, which gives

$$\sqrt{x^2 + x_c^2} - x_c = x^2\ , \tag{5.69}$$

Getting rid of the square root, the equation can be written as

$$x^2(x^2 + 2x_c - 1) = 0\ . \tag{5.70}$$

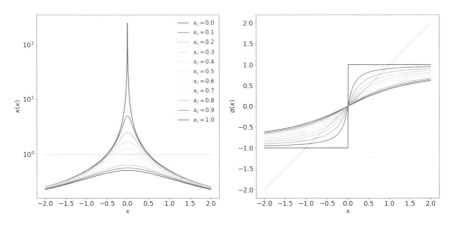

Fig. 5.10 Convergence and deflection angle profiles of the NIS lens model. Different colors correspond to different values of the core radius x_c

Discarding the solution $x = 0$, we find that

$$x_t = \sqrt{1 - 2x_c} \ . \tag{5.71}$$

Thus, the tangential critical line exists only for $x_c < 1/2$.

In angular units, the Einstein radius of the NIS is

$$\theta_E = \sqrt{\theta_0^2 - 2\theta_c\theta_0} \ , \tag{5.72}$$

where $\theta_0 = \xi_0/D_L$ and $\theta_c = \xi_c/D_L$.

The radius of the radial critical line is found by solving the equation $1 - \kappa(x) + \gamma(x) = 0$:

$$1 + \frac{\sqrt{x^2 + x_c^2} - x_c}{x^2} - \frac{1}{\sqrt{x^2 + x_c^2}} = 0 \tag{5.73}$$

which leads to

$$x_r^2 = \frac{1}{2}\left(2x_c - x_c^2 - x_c\sqrt{x_c^2 + 4x_c}\right) \ . \tag{5.74}$$

Note that $x_r^2 \geq 0$ for $x_c \leq 1/2$. Thus, the existence condition for the radial critical line is the same as for the tangential critical line.

While the tangential caustic is a point at $y_t = 0$, the radius of the radial caustic, y_r, can be obtained by inserting Eq. 5.74 into the lens equation.

The lens equation is

$$y = x - \frac{m(x)}{x} = x - \sqrt{1 + \frac{x_c^2}{x^2}} - \frac{x_c}{x} , \qquad (5.75)$$

which can be reduced to the third-order polynomial equation,

$$x^3 - 2yx^2 + (y^2 + 2x_c - 1)x - 2yx_c = 0 . \qquad (5.76)$$

Thus, the NIS can produce up to three images of a source at distance y from the lens. Whether the images are three or less depends on y and x_c. The right panel of Fig. 5.10 shows that, if $x_c > 1/2$, which implies $\alpha' < 1$ for any x, there is no line $f(x) = x - y$ which can intercept the function $\alpha(x)$ more than once. Thus, the lens cannot produce multiple images independently on the position of the source with respect to the lens. In addition, the left panel of Fig. 5.10 shows that for $x_c > 1/2$ the convergence never exceeds unity. This implies that the surface density $\Sigma(x)$ is always sub-critical, $\Sigma(x) < \Sigma_{cr}$.

On the contrary, for $x_c < 1/2$, multiple images exist if $y < y_r$. In the left panel of Fig. 5.11, some extended circular sources are placed at different distances y from the center of a NIS lens with $x_c = 0.1$. The blue point and the red circle show the tangential and the radial caustics. The right panel shows the corresponding critical lines (blue and red circles, respectively). We also show the images of the sources displayed in the left panel. One image is located on the same side of the source. This image, which corresponds to the minimum of the time delay surface, is external to the Einstein ring. When the source is inside the radial caustic, two additional images form, one inside and one outside the radial critical line. They are located on the opposite side of the lens. As discussed earlier, the innermost image

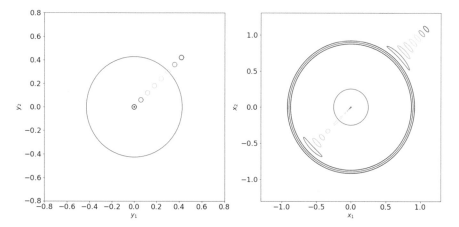

Fig. 5.11 Imaging of extended sources by a NIS lens

corresponds to the maximum of the time delay surface, while the other image to the saddle point. Thus the parity of this image is negative. If the source overlaps with the radial caustic, these two images merge at the radial critical line, forming a radial arc. Bringing the source closer to the tangential caustic causes the tangential distortion of the two outermost images to increase. When the source is exactly on the tangential caustic, the two images merge to form the Einstein ring. The central image instead is increasingly de-magnified and moves toward the center of the lens.

Remark 5.4. We have shown that, if $x_c > 1/2$, the lens

- is sub-critical, $\kappa(x) < 1$ for any x;
- does not develop critical lines, i.e. is not capable of producing large distortions;
- does not produce multiple images.

Under these circumstances, the lens is considered *weak*. On the contrary, a *strong* lens can produce large distortions and multiple images. As seen earlier, the number of these images depends on the relative positions of source and caustics. Sources inside the radial caustic produce three images. Sources outside the radial caustic have only one image. This is shown in Fig. 5.12. Since the tangential critical curve

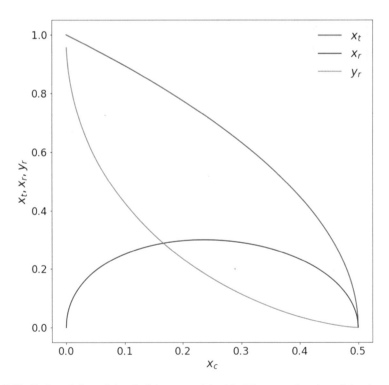

Fig. 5.12 Radius of the radial and of the tangential critical lines as a function of the size of the core radius x_c. The green line shows the size of the radial caustic

does not lead to a caustic curve, but the corresponding caustic degenerates to a single point $\vec{y} = 0$, the tangential critical curves do not influence the image multiplicity. Thus, pairs of images can only be created or destroyed if the radial critical curve exists (Fig. 5.11).

5.4   Elliptical Lenses

Having explored the effects of changing the density profile slope and including a central core, we consider how ellipticity affects the lens properties. Adding ellipticity removes the axial symmetry of the lens. Consequently, the tangential caustic is no longer a point at the center of the lens. Instead, it becomes an astroid-like kind of caustic, which can have two or four cusps.

5.4.1   Singular Isothermal Ellipsoid

As we have seen in the previous section, the singular isothermal profile is particularly tractable to derive its lensing properties. Similarly, it is quite straightforward to generalize to the elliptical case.

Convergence
As shown by Kormann et al. (1994), the Singular Isothermal Ellipsoid model (SIE) can be derived from the SIS by making the substitution

$$\xi \Rightarrow \sqrt{\xi_1^2 + f^2 \xi_2^2} \ . \tag{5.77}$$

Then, $\Sigma(\xi)$ becomes

$$\Sigma(\vec{\xi}) = \frac{\sigma_v^2}{2G} \frac{\sqrt{f}}{\sqrt{\xi_1^2 + f^2 \xi_2^2}} \ , \tag{5.78}$$

which is constant on ellipses with minor axis ξ and major axis ξ/f, oriented such that the major axis is along the ξ_2 axis. In the above formulas, f is the axis ratio of the ellipses, $0 < f \leq 1$. The normalization of the profile ensures that the mass within an elliptical iso-density contour for fixed Σ is independent on f.

By choosing $\xi_0 = \xi_{0,SIS}$ as reference scale and by repeating the same procedure outlined in Eq. 5.51, we obtain

$$\kappa(\vec{x}) = \frac{\sqrt{f}}{2\sqrt{x_1^2 + f^2 x_2^2}} \ . \tag{5.79}$$

Using polar coordinates, we obtain

$$\kappa(x, \varphi) = \frac{\sqrt{f}}{2x\,\Delta(\varphi)} \,, \tag{5.80}$$

where

$$\Delta(\varphi) = \sqrt{\cos\varphi^2 + f^2 \sin\varphi^2} \,. \tag{5.81}$$

This shows that the convergence is the product of the convergence profile of the SIS lens and of the function $\sqrt{f}/\Delta(\varphi)$.

Lensing Potential

The lensing potential can be found by solving the Poisson equation:

$$\frac{\partial^2 \Psi}{\partial x^2} + \frac{1}{x}\frac{\partial \Psi}{\partial x} + \frac{1}{x^2}\frac{\partial^2 \Psi}{\partial \varphi^2} = 2\kappa = \frac{\sqrt{f}}{x\,\Delta(\varphi)} \,. \tag{5.82}$$

Making the ansatz $\Psi(x, \varphi) := x\tilde{\Psi}(\varphi)$ we find

$$\tilde{\Psi}(\varphi) + \frac{d^2}{d\varphi^2}\tilde{\Psi}(\varphi) = \frac{\sqrt{f}}{\Delta(\varphi)} \,. \tag{5.83}$$

This differential equation can be solved using Green's method (Kormann et al. 1994). The resulting lensing potential is

$$\Psi(x, \varphi) = x\frac{\sqrt{f}}{f'}\left[\sin\varphi \arcsin\left(f'\sin\varphi\right) + \cos\varphi\, \text{arcsinh}(f'/f\cos\varphi)\right] \,. \tag{5.84}$$

In the formula above we introduced $f' = \sqrt{1 - f^2}$.

In Fig. 5.13, we compare the map of the convergence and of the potential for a SIE lens. The contours of constant convergence are indeed ellipses with axis ratio $f = 0.6$. The iso-potential contours are much rounder: they have axis ratios $f_\Psi \sim 0.84 = 1.4f$.

Deflection Angle

The deflection angle can be derived as usual by taking the gradient of the lensing potential. It is convenient to operate in polar coordinates, so that

$$\frac{\partial}{\partial x_1} = \cos\varphi\frac{\partial}{\partial x} - \frac{\sin\varphi}{x}\frac{\partial}{\partial \varphi} \tag{5.85}$$

and

$$\frac{\partial}{\partial x_2} = \sin\varphi\frac{\partial}{\partial x} + \frac{\cos\varphi}{x}\frac{\partial}{\partial \varphi} \,. \tag{5.86}$$

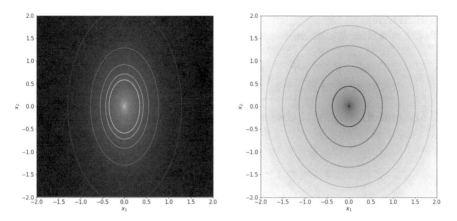

Fig. 5.13 Maps of the convergence and of the lensing potential for a SIE lens with $f = 0.6$

Thus, the components of the deflection angle are

$$\alpha_1(\vec{x}) = \frac{\sqrt{f}}{f'} \operatorname{arcsinh}\left(\frac{f'}{f}\cos\varphi\right)$$

$$\alpha_2(\vec{x}) = \frac{\sqrt{f}}{f'} \arcsin(f'\sin\varphi). \tag{5.87}$$

As found for the SIS, the deflection angle of the SIE does not depend on x.

Shear

The shear components can be obtained from the derivatives of the deflection angle:

$$\gamma_1(\vec{x}) = \frac{1}{2}\left(\frac{\partial\alpha_1}{\partial x_1} - \frac{\partial\alpha_2}{\partial x_2}\right),$$

$$\gamma_2(\vec{x}) = \frac{\partial\alpha_1}{\partial x_2}. \tag{5.88}$$

Using the differential operators in polar coordinates and the results above, we find that the shear components are

$$\gamma_1(\vec{x}) = -\frac{\sqrt{f}}{2x\,\Delta(\varphi)}\cos 2\varphi = -\kappa(\vec{x})\cos 2\varphi$$

$$\gamma_2(\vec{x}) = -\frac{\sqrt{f}}{2x\,\Delta(\varphi)}\sin 2\varphi = -\kappa(\vec{x})\sin 2\varphi\,, \tag{5.89}$$

and that $\gamma = \sqrt{\gamma_1^2 + \gamma_2^2} = \kappa$, as for the SIS.

Critical Lines

We can now compute the lensing Jacobian, which is

$$A = \begin{bmatrix} 1 - \kappa - \gamma_1 & -\gamma_2 \\ \gamma_2 & 1 - \kappa + \gamma_1 \end{bmatrix} = \begin{bmatrix} 1 - 2\kappa \sin^2 \varphi & \kappa \sin 2\varphi \\ \kappa \sin 2\varphi & 1 - 2\kappa \cos^2 \varphi \end{bmatrix} . \tag{5.90}$$

The tangential and radial eigenvalues of A are

$$\lambda_t(\vec{x}) = 1 - \kappa(\vec{x}) - \gamma(\vec{x}) = 1 - 2\kappa(\vec{x})$$
$$\lambda_r(\vec{x}) = 1 - \kappa(\vec{x}) + \gamma(\vec{x}) = 1 . \tag{5.91}$$

It turns out that, similarly to the SIS lens, the SIE does not have a radial critical line, being the radial magnification always unity. Instead, the tangential critical line is the ellipse defined by the condition

$$\kappa(\vec{x}) = \frac{1}{2} . \tag{5.92}$$

The coordinates of the points on the ellipse can be obtained by inverting Eq. 5.80:

$$\vec{x}_t(\varphi) = \frac{\sqrt{f}}{\Delta(\varphi)} [\cos \varphi, \sin \varphi] . \tag{5.93}$$

The critical lines of two SIE lenses with $f = 0.6$ and $f = 0.2$ are shown with dashed red lines in the left and in the right panels of Fig. 5.14.

Caustic and Cut

The points on the critical line can be mapped onto the source plane using the lens equation, to obtain the corresponding points on the tangential caustics:

$$y_{t,1}(\varphi) = \frac{\sqrt{f}}{\Delta(\varphi)} \cos \varphi - \frac{\sqrt{f}}{f'} \text{arcsinh} \left(\frac{f'}{f} \cos \varphi \right) ,$$
$$y_{t,2}(\varphi) = \frac{\sqrt{f}}{\Delta(\varphi)} \sin \varphi - \frac{\sqrt{f}}{f'} \arcsin(f' \sin \varphi) . \tag{5.94}$$

The resulting curves for the same lenses with $f = 0.6$ and $f = 0.2$ are given by the solid red lines in Fig. 5.14. Introducing the ellipticity breaks the axial symmetry of the SIS lens, transforming the point caustic into an extended curve with four cusps and four folds.

As for the SIS (see Eq. 5.57), we can determine the *cut*, whose points are defined as

$$\vec{y}_c(\varphi) = \lim_{x \to 0} \vec{y}(x, \varphi) = -\vec{\alpha}(\varphi) . \tag{5.95}$$

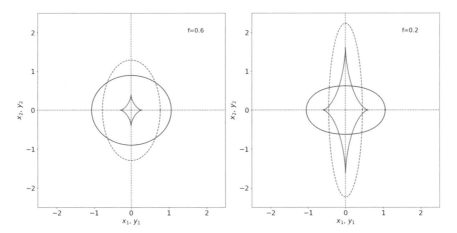

Fig. 5.14 Critical line (dashed red), caustic (solid red), cut (solid blue) for two SIE lenses with $f = 0.6$ (left panel) and $f = 0.2$ (right panel)

Thus, we obtain

$$y_{c,1}(\varphi) = -\frac{\sqrt{f}}{f'}\operatorname{arcsinh}\left(\frac{f'}{f}\cos\varphi\right)$$

$$y_{c,2}(\varphi) = -\frac{\sqrt{f}}{f'}\arcsin(f'\sin\varphi)\,. \tag{5.96}$$

The cuts of the two SIE lenses with $f = 0.6$ and $f = 0.2$ are given by the blue solid curves in each panel of Fig. 5.14.

The cut and the caustic intercept the y_1 and the y_2 axes at points that are symmetric with respect to the center of the lens. These points have coordinates

$$s_{1,\pm,c} = [y_{c,1}(\varphi = 0, \pi), 0]\,,$$

$$s_{2,\pm,c} = [0, y_{c,2}(\varphi = \pi/2, -\pi/2)] \tag{5.97}$$

for the cut, and

$$s_{1,\pm,t} = [y_{t,1}(\varphi = 0, \pi), 0]\,,$$

$$s_{1,\pm,t} = [0, y_{t,2}(\varphi = \pi/2, -\pi/2)] \tag{5.98}$$

for the caustic. The sign $\pm$ indicates whether the points are on the positive or the negative semi-axes.

Figure 5.15 shows how s_1 and s_2 (on the positive y_1 and y_2 axes) vary as a function of f for both the cut and the caustic. Clearly, $s_{1,c} > s_{1,t}$ for any f, meaning that the cut is always external to the tangential caustic along the y_1 axis, i.e. along

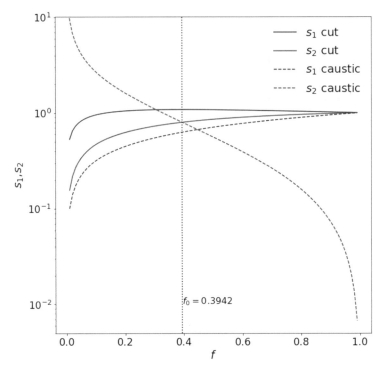

Fig. 5.15 Intercepts between the cut (solid lines) and the caustics (dashed lines) with the positive y_1 and y_2 semi-axes (blue and red lines, respectively) as a function of f

the minor axis of the lens. On the contrary, there exists a value, $f = f_0 = 0.3942$, such that

$$s_{2,c} \leq s_{2,t} \text{ for } f \leq f_0 \,,$$
$$s_{2,c} > s_{2,t} \text{ for } f > f_0 \,. \tag{5.99}$$

Therefore, for large ellipticities, corresponding to small values of f, the tangential caustic extends outside the cut along the y_2 axis. The cusps which are not contained within the cut are called *naked*.

Multiple Images
Finding the multiple images of a source lensed by the SIE model requires using numerical methods. We discuss the implementation of one of such algorithms in python in Sect. 5.11.1.

It is useful to see how the pieces of the tangential critical line are mapped onto the caustic. In the left panel of Fig. 5.16, we divide the critical line into four parts, corresponding to the quadrants in the lens plane. We use different colors to indicate the critical points with polar angles in the ranges $[0, \pi/2)$, $[\pi/2, \pi)$, $[\pi, 3/2\pi)$,

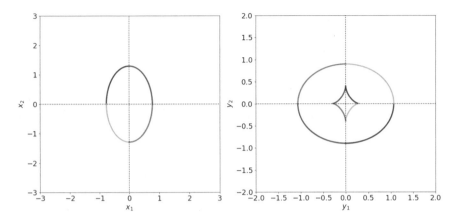

Fig. 5.16 Mapping of points on the critical line onto the caustic. Different colors correspond to different quadrants on the lens plane. The mapping is shown also for the cut, which would become the radial caustic in presence of a cores

and $[3/2\pi, 2\pi)$ (see Eq. 5.93). In the right panel, the corresponding points on the caustic, obtained using Eq. 5.94, are shown using the same colors. We use the same color-code to visualize the points on the cut, depending on the polar angle φ.

We can see that the mapping of the critical points follows a left–right rule: the first and the fourth quadrants in the lens plane ($\varphi \in [0, \pi/2)$ and $\varphi \in [3/2\pi, 2\pi)$) are mapped onto the second and the third quadrants of the source plane, and vice versa.

The points on the cut, instead, follow a diagonal rule. For example, the portion of the cut in the first quadrant in the source plane corresponds to $\varphi \in [\pi, 3/2\pi)$.

Remark 5.5. As seen earlier, in lenses with a radial critical line, the cut becomes the radial caustic. Thus the same diagonal rule applies for the mapping of the radial critical points onto the radial caustic.

With these results in mind, we can easily guess the position of a source whose images are near some portion of the critical line. For example, if we observe images near the red portion of the critical line in Fig. 5.16, then the corresponding source position is near the red portion of the caustic.

The multiplicity of the images produced by the SIE lens depends on the caustic and cut shapes and the source's position with respect to these two curves. The following rules apply:

- A source far away from the lens, i.e. outside of both the cut and the caustic, has only one image.
- The number of images increases by one, if the source is inside the cut.
- The number of images increases by two if the source is inside the caustic.

The last two rules do not exclude each other, i.e. if the source is inside the cut AND the caustic, the number of images increases by three. Thus, a SIE lens can produce up to four images of a single source. More precisely,

- If $f > f_0$, the lens can produce either 1, 2, or 4 images. It depends on whether the source is outside the cut, between the cut and the caustic, or inside the caustic.
- If $f \leq f_0$, the lens can produce 1, 2, 3, or 4 images. Indeed, for such lenses, a source can be located within the caustic but outside the cut. Thus, up to three images can form in this case.

Figure 5.17 shows the images of circular sources at different positions relative to the caustic and the cut. The SIE lens used in these examples has $f = 0.6$. The solid and the dashed lines in the left panels indicate the caustics and the cut, respectively. The colored circles mark the positions of several circular sources with radii $r = 0.05\xi_0$. We show the corresponding images in the right panels. The solid black lines indicate the tangential critical lines.

The upper panels show how the image geometry changes when the source is brought closer to the lens center by crossing the cut and the fold of the caustic. Sources outside the cut have only one image located in the same quadrant of the image plane onto which the source position is projected. When the source crosses the cut, one additional image appears near the center of the lens. This is not surprising, given that the cut is given by Eq. 5.95. As the source is brought closer to the caustic, the central image moves away from the lens center. It is located in the lens plane's quadrant opposite to that where the source is projected. In this case, the abovementioned diagonal rule applies. At the caustic crossing, two images appear on opposite sides of the critical line. Note that if the source is near the fold in the first quadrant of the source plane, these two images appear in the second quadrant of the lens plane, following the left–right mapping rule, which applies to the tangential critical points. As the source is brought even closer to the center of the lens, the images have a symmetric configuration called *Einstein cross*.

The middle and bottom panels refer to the same lens. We move the source from outside the cut toward the center of the lens, passing through the cusps of the tangential caustic. At the caustic crossing, three images merge at the critical line. Again, the left–right rule applies.

Figure 5.18 is analogous to Fig. 5.17, but refers to a SIE lens with $f = 0.2$. Note that the middle panels show the image configurations for a source moved across the naked cusp. When the source is on the cusp, we find again that three images merge. When the source is inside the caustic but still outside the cut, the image multiplicity is three, as anticipated earlier.

Distortion and Parity of the Images

We can now comment on the magnification and parity. From Eqs. 5.80 and 5.89, we obtain that

$$\mu = \frac{1}{1 - 2\kappa} \, , \tag{5.100}$$

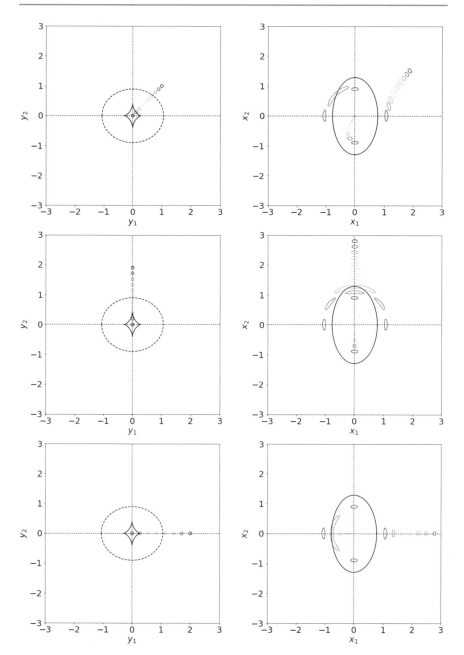

Fig. 5.17 Lensing of circular sources by a SIE lens with $f = 0.6$. In the left panels, we use different colors to indicate sources at several angular separations from the lens center. The caustic and the cut are given by the black solid and dashed lines. The images (colored contours) and the critical line (solid black line) are shown in the right panels

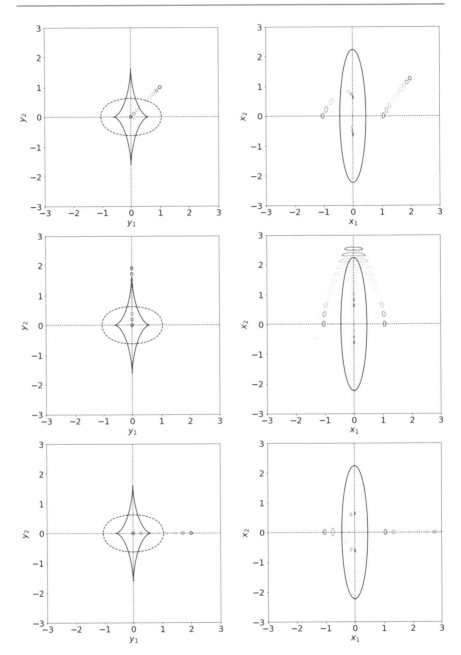

Fig. 5.18 As in Fig. 5.17, but for a SIE lens with $f = 0.2$

meaning that $\mu > 0$ for $\kappa < 0.5$. Therefore, the parity of the images is positive
outside the critical line, negative otherwise.

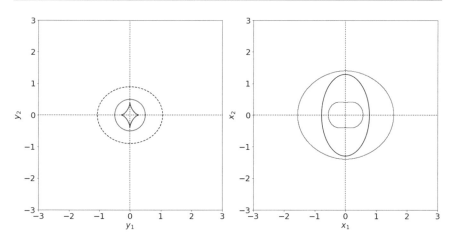

Fig. 5.19 Lensing of an extended source of radius $r = 0.5$ (shown in red) by a SIE lens with $f = 0.6$. The black solid and dashed lines in the left panel show the caustic and the cut of the lens. The black solid line in the right panel is the critical line

Given the singularity of the lensing potential near the center, the images are either minima or saddle points of the time delay surface. The images inside the critical line are saddle points. Those outside the critical line are minima.

The magnification of the images near the center of the lens is minimal ($|\mu| \sim 0$), being the lensing potential discontinuous at the lens's center. Since the radial magnification is always unity, the images near the lens center are tangentially demagnified. Of course, the magnification diverges near the critical line.

Figures 5.17 and 5.18 show that the images near the critical line are distorted tangentially leading to the formation of gravitational arcs. The individual images are elongated and merge at the critical line. The largest gravitational arcs result from the merger of three images of sources near the caustic cusps.

In the examples shown in Figs. 5.17 and 5.18, the source size is $r = 0.05$ in units of ξ_0. In Fig. 5.19, we show how a source of radius $r = 0.5$ at the caustic center is distorted to form a complete Einstein ring. Thus, the observed distortions depend on the size of the source relative to the caustic size. If the source size is large compared to the caustic size, the effect of ellipticity is merely detectable.

Figure 5.20 illustrates how the image parity changes moving from inside to outside the critical line. For each source in the left panel, we mark four characteristic points in their contours. We indicate them with circles, triangles, squares, and stars in counter clockwise order. We can see that the same symbol sequence is preserved in images outside the critical line in the right panel. In the images inside the critical line, circles and squares are interchanged, following the left–right rule introduced above.

Note that half of the orange source is within the caustic. Therefore, only part of this source has four images, while the remaining portion only has two images. For

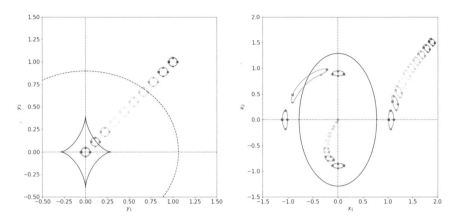

Fig. 5.20 Lensing of an extended source of radius $r = 0.05$ by a SIE lens with $f = 0.6$ moved across the cut and the fold of the caustic. Four points on the source contours are marked with symbols. The same symbols are mapped onto the image plane on the image contours

example, the points marked with a circle and triangle in this source do not appear in the fold arc in the second quadrant of the lens plane.

5.4.2   Softened (Non-singular) Elliptical Models

We can introduce a core also in the SIE lens model. In this case, however, the model becomes not easily tractable analytically. A discussion on the properties of the Non-singular Isothermal Ellipsoid (NIE) is given by Kormann et al. (1994). Similar complications arise from changing the slope of the density profile. A detailed description of the elliptical power-law lens model can be found in Tessore and Metcalf (2015).

Here, we only summarize some properties of the NIE model, whose surface density is written as

$$\Sigma(\vec{\xi}) = \frac{\sigma^2}{2G} \frac{\sqrt{f}}{\sqrt{\xi_1^2 + f^2\xi_2^2 + \xi_c^2}} \; , \tag{5.101}$$

where ξ_c is the core radius. With the usual choice of $\xi_0 = \xi_{0,SIS}$, the convergence profile becomes

$$\kappa(\vec{x}) = \frac{\sqrt{f}}{2\sqrt{x_1^2 + f^2x_2^2 + x_c^2}} \; . \tag{5.102}$$

We show some examples of critical lines and caustics in Fig. 5.21. Depending on the values of f and of x_c, the lens can have two separate critical lines and caustics

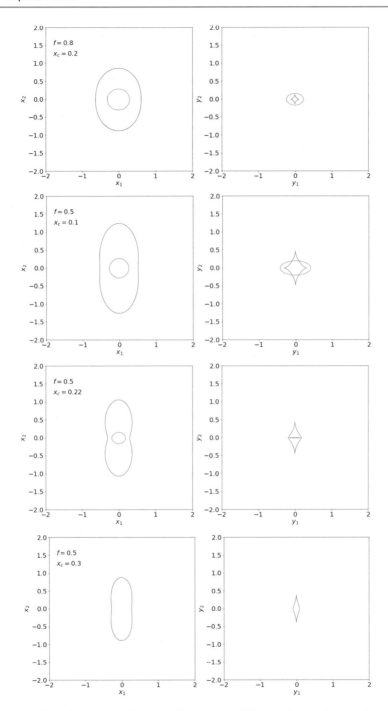

Fig. 5.21 Critical lines and caustics for NIE lenses with different values of f and x_c. Examples of each topology are shown

(one radial and one tangential), only one tangential critical line and caustic, or no critical lines and caustics at all. In particular,

- If $x_c < f^{3/2}/2$ there are two distinguished critical lines and caustics. One caustic is the tangential caustic and has four cusps. The other is the radial caustic and has no cusps. The tangential caustic is completely contained within the radial caustic if the ellipticity is small (f is large). Even in the case of mildly elliptical lenses, the radial caustic contains the tangential caustic if the core is small enough.
- If $f^{3/2}/2 < x_c < f^{3/2}/(1 + f)$, the radial caustic is contained within the tangential caustic. Besides, both the radial and the tangential caustics have only two cusps.
- If $f^{3/2}/(1 + f) < x_c < f^{1/2}/(1 + f)$, the lens has only a tangential critical line and caustic. The radial critical lines and caustics disappear for $x_c = f^{3/2}/(1+f)$.
- Even the tangential caustic disappears if $x_c = f^{1/2}/(1 + f)$. Thus, for $x_c > f^{1/2}/(1 + f)$ the lens does not have critical lines and caustics.

Depending on the caustics structure, the NIE can produce 1, 3, or 5 images of a source. These cases are illustrated in Fig. 5.21.

5.4.3   Pseudo-Elliptical Models

Pseudo-elliptical lenses are lenses whose iso-potential contours are ellipses with axis ratio f.

Let us consider a circular symmetric potential $\Psi(x)$, where x is, as usual, the distance from the center of the lens in dimensionless units. As previously done for the surface density, circular iso-potential contours can be transformed into ellipses by making the substitution

$$x \rightarrow x = \sqrt{x_1^2 + f^2 x_2^2} \, . \tag{5.103}$$

The ellipses have minor axis x and major axis x/f, and their major axes are aligned with the x_2 axis.

The great advantage of introducing the ellipticity in the potential rather in the projected density is that all lens properties can be obtained more easily by means of derivatives of circular symmetric quantities. For example, the components of the deflection angle are

$$\alpha_1(\vec{x}) = \frac{\partial \Psi(x)}{\partial x_1} = \Psi'(x) \frac{\partial x}{\partial x_1} = \tilde{\alpha}(x) \frac{\partial x}{\partial x_1} \, ,$$

$$\alpha_2(\vec{x}) = \frac{\partial \Psi(x)}{\partial x_2} = \Psi'(x) \frac{\partial x}{\partial x_2} = \tilde{\alpha}(x) \frac{\partial x}{\partial x_2} \, . \tag{5.104}$$

where $\tilde{\alpha}(x)$ is the deflection angle of the circular symmetric lens whose potential is $\Psi(x)$. Convergence and shear are linear combinations of the second derivatives of the lensing potential, or, equivalently, of the first derivatives of the components of the deflection angle (Sect. 3.3):

$$\Psi_{11}(\vec{x}) = \frac{\partial \alpha_1(\vec{x})}{\partial x_1} = \tilde{\alpha}'(x)\left(\frac{\partial x}{\partial x_1}\right)^2 + \tilde{\alpha}(x)\frac{\partial^2 x}{\partial x_1^2} ,$$

$$\Psi_{22}(\vec{x}) = \frac{\partial \alpha_2(\vec{x})}{\partial x_2} = \tilde{\alpha}'(x)\left(\frac{\partial x}{\partial x_2}\right)^2 + \tilde{\alpha}(x)\frac{\partial^2 x}{\partial x_2^2} ,$$

$$\Psi_{12}(\vec{x}) = \frac{\partial \alpha_1(\vec{x})}{\partial x_2} = \tilde{\alpha}'(x)\frac{\partial x}{\partial x_1}\frac{\partial x}{\partial x_2} + \tilde{\alpha}(x)\frac{\partial^2 x}{\partial x_1 x_2} . \tag{5.105}$$

Thus, the calculation of the lens properties is fast even when done with numerical methods.

The disadvantage of introducing ellipticity in the potential rather than in the surface density is that the resulting iso-surface density contours are not elliptical. They have a characteristic dumbbell shape, as shown in Fig. 5.22 (see e.g. Kassiola

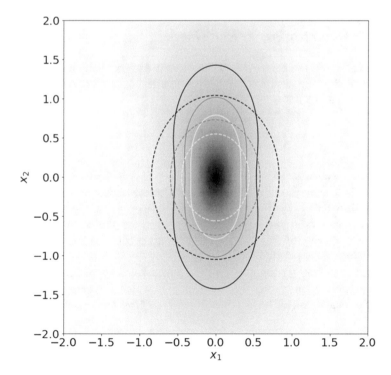

Fig. 5.22 Convergence map of a pNIE lens with $f = 0.7$. The solid lines show some contours of constant convergence. For comparison, the dashed lines correspond to NIE lens with the same value of f

& Kovner 1993). The figure displays the convergence map of a pseudo-NIE (pNIE) model with lensing potential

$$\Psi(\vec{x}) = \sqrt{x_1^2 + f^2 x_2^2 + x_c^2} \tag{5.106}$$

and axis ratio $f = 0.7$. The solid lines are contours of constant convergence. The dashed lines show equivalent contours for a NIE lens with the same axis ratio f. Note that the lens potential is rounder than the corresponding mass distribution.

In the case of large ellipticities, pseudo-elliptical potentials correspond to nonphysical mass distributions (e.g. with negative convergence) (Golse & Kneib 2002, Kassiola & Kovner 1993).

5.5 Other Profiles

A large variety of mass profiles has been used in the literature to model gravitational lenses. A relatively comprehensive catalog of such models is presented in Keeton (2001). Here, we briefly discuss two of them. The first is the Navarro–Frenk–White model, and the second is the dual Pseudo-Isothermal Elliptical model.

5.5.1 The Navarro–Frenk–White Model

Navarro et al. (1997) found that the density profile of dark matter halos numerically simulated in the framework of Cold-Dark-Matter cosmogony can be very well described by the radial function

$$\rho(r) = \frac{\rho_s}{(r/r_s)(1 + r/r_s)^2} , \tag{5.107}$$

within the wide mass range $3 \times 10^{11} \lesssim M_{vir}/(h^{-1} M_\odot) \lesssim 3 \times 10^{15}$. The logarithmic slope of this density profile changes from -1 at the center to -3 at large radii. Therefore, it is flatter than that of the SIS in the inner part of the halo and steeper in the outer part. The two parameters r_s and ρ_s are the scale radius and the characteristic density of the halo. The profile in Eq. 5.107 is known as Navarro–Frenk–White density profile (NFW hereafter).

Navarro et al. (1997) parameterized dark matter halos by their masses M_{200}, i.e. the masses enclosed in spheres with radius r_{200} in which the average density is 200 times the critical density. The relationship between M_{200} and r_{200} is given by

$$r_{200} = 1.63 \times 10^{-2} \left(\frac{M_{200}}{h^{-1} M_\odot} \right)^{1/3} \left[\frac{\Omega_0}{\Omega(z)} \right]^{-1/3} (1 + z)^{-1} h^{-1} \text{ kpc} . \tag{5.108}$$

This definition depends on the redshift z at which the halo is identified as well as on the background cosmological model. In particular, Ω_0 and $\Omega(z)$ are the density parameters of the universe at redshift 0 and z, respectively.

From the former definition of r_{200}, the *concentration*, $c \equiv r_{200}/r_s$, and the characteristic density are linked by the relation,

$$\rho_s = \frac{200}{3}\rho_{\mathrm{cr}}\frac{c^3}{[\ln(1+c) - c/(1+c)]} . \tag{5.109}$$

Numerical simulations show that the scale radii of dark matter halos at any redshift z systematically change with mass in such a way that concentration is a characteristic function of M_{200} (Bhattacharya et al. 2013, De Boni et al. 2013, Diemer & Kravtsov 2015, Dolag et al. 2004, Duffy et al. 2008, Dutton & Macciò 2014, Ludlow et al. 2014, Meneghetti et al. 2014).

The lensing properties of halos with NFW profiles are discussed in several papers (e.g. Bartelmann 1996, Meneghetti et al. 2003, Wright and Brainerd 2000). If we take $\xi_0 = r_s$, the density profile 5.107 implies the surface-mass density

$$\Sigma(x) = \frac{2\rho_s r_s}{x^2 - 1}f(x) , \tag{5.110}$$

with

$$f(x) = \begin{cases} 1 - \dfrac{2}{\sqrt{x^2-1}}\arctan\sqrt{\dfrac{x-1}{x+1}} & (x > 1) \\[2mm] 1 - \dfrac{2}{\sqrt{1-x^2}}\operatorname{arctanh}\sqrt{\dfrac{1-x}{1+x}} & (x < 1) \\[2mm] 0 & (x = 1) \end{cases} . \tag{5.111}$$

The lensing potential is given by

$$\Psi(x) = 4\kappa_s g(x) , \tag{5.112}$$

where

$$g(x) = \frac{1}{2}\ln^2\frac{x}{2} + \begin{cases} 2\arctan^2\sqrt{\dfrac{x-1}{x+1}} & (x > 1) \\[2mm] -2\operatorname{arctanh}^2\sqrt{\dfrac{1-x}{1+x}} & (x < 1) \\[2mm] 0 & (x = 1) \end{cases} , \tag{5.113}$$

and $\kappa_s \equiv \rho_s r_s \Sigma_{\mathrm{cr}}^{-1}$. This implies the deflection angle

$$\alpha(x) = \frac{4\kappa_s}{x}h(x) , \tag{5.114}$$

with

$$
h(x) = \ln\frac{x}{2} +
\begin{cases}
\frac{2}{\sqrt{x^2-1}}\arctan\sqrt{\frac{x-1}{x+1}} & (x > 1) \\
\frac{2}{\sqrt{1-x^2}}\operatorname{arctanh}\sqrt{\frac{1-x}{1+x}} & (x < 1) \\
1 & (x = 1)
\end{cases}
\tag{5.115}
$$

It is an important feature of the NFW lensing potential (Eq. 5.112) that its radial profile is considerably less curved near the center than the SIS profile (Eq. 5.84). Since the curvature of Ψ determines the local imaging properties, this immediately implies substantial changes to the lensing properties.

The convergence can be written as

$$
\kappa(x) = \frac{\Sigma(x)}{\Sigma_{\mathrm{cr}}} = 2\kappa_s \frac{f(x)}{x^2 - 1} \;,
\tag{5.116}
$$

from which we obtain the dimensionless mass,

$$
m(x) = 2\int_0^x \kappa(x')x'dx' = 4k_s h(x) \;.
\tag{5.117}
$$

Finally, the shear profile can be derived from Eq. 5.21.

We can solve the lens equation for this kind of lens model using numerical methods also in the case of circular symmetry (e.g. employing the image diagram). At fixed halo mass, the NFW lens's critical curves are closer to its center than for a SIS lens because of its flatter density profile. There, the potential is less curved. Thus the image magnification is larger and decreases more slowly away from the critical curves. Therefore NFW lenses are less efficient in image splitting than SIS lenses but comparably efficient in image magnification. An essential feature of the NFW model is that it can have a radial critical line, and it is thus capable of reproducing radial arcs observed in several galaxy clusters.

Due to the relative complexity of this lens model, ellipticity is usually introduced in the lensing potential (e.g. Golse & Kneib 2002, Meneghetti et al. 2003) as discussed in Sect. 5.4.3.

5.5.2 The Dual Pseudo-Isothermal Mass Distribution

The dual Pseudo-Isothermal model is a variation of the Isothermal model discussed earlier, including both a core and a cut radius in the density profile. Together with its simplicity, this feature makes it useful to describe mass distributions both on the scales of galaxies and galaxy clusters.

The profile of the three-dimensional density is given by

$$\rho(r) = \frac{\rho_0}{(1 + r^2/r_{core}^2)(1 + r^2/r_{cut}^2)} , \qquad (5.118)$$

where r_{core} and r_{cut} are the core and the cut radii, respectively, with $r_{cut} > r_{core}$. The profile is isothermal, $\rho \propto r^{-2}$, for $r_{core} < r < r_{cut}$. At small radii, $r < r_{core}$ it reaches a plateau with central density ρ_0. Finally, for $r \gg r_{cut}$, the density falls off very steeply as $\rho \propto r^{-4}$.

The central density ρ_0 is related to the 1D-central velocity dispersion, σ_0, as (Eliasdóttir et al. 2007, Limousin et al. 2005)

$$\rho_0 = \frac{\sigma_0^2}{2\pi G} \frac{r_{cut} + r_{core}}{r_{core}^2 r_{cut}} . \qquad (5.119)$$

In the limit, $r_{core} \to 0$ and $r_{cut} \to \infty$, Eq. 5.118 reduces to Eq. 5.48 with $\sigma_v = \sigma_0$.

By integrating the three-dimensional density along the line-of-sight, we obtain the surface density

$$\Sigma(\xi) = \frac{\sigma_0^2}{2G} \frac{r_{cut}}{r_{cut} - r_{core}} \left(\frac{1}{\sqrt{\xi^2 + r_{core}^2}} - \frac{1}{\sqrt{\xi^2 + r_{cut}^2}} \right) . \qquad (5.120)$$

This shows a nice property of the model, namely that the surface density profile is the difference between two NIS profiles (see Eq. 5.63). This feature makes possible to compute the properties of the lens model using the Equations discussed in Sect. 5.3.

It is important to note that the mass of a dual Pseudo-Isothermal lens is finite. Indeed, the projected mass profile is

$$M(\xi) = 2\pi \int_0^\xi \Sigma(\xi')\xi' d\xi'$$

$$= \frac{\pi \sigma_0^2}{G} \frac{r_{cut}}{r_{cut} - r_{core}} \left(\sqrt{r_{core}^2 + \xi^2} - r_{core} - \sqrt{r_{cut}^2 + \xi^2} + r_{cut} \right) , \qquad (5.121)$$

and, in the limit $\xi \to \infty$, we obtain

$$M_{tot} = \frac{\pi \sigma_0^2 r_{cut}}{G} . \qquad (5.122)$$

5.6   External Perturbations

It is often necessary to embed a lens into an external shear to account for matter in the lens surroundings. A useful approach is to model this shear by means of a potential Ψ_γ, which must satisfy the following conditions:

$$\gamma_1 = \frac{1}{2}(\Psi_{11} - \Psi_{22}) = \text{const.}$$

$$\gamma_2 = \Psi_{12} = \text{const.}$$

$$\kappa = \frac{1}{2}(\Psi_{11} + \Psi_{22}) = \text{const.} \tag{5.123}$$

If $\Psi_{11} \pm \Psi_{22}$ are required to be constant, Ψ_{11} and Ψ_{22} must separately be constants, thus

$$\Psi_\gamma(\vec{x}) = C x_1^2 + C' x_2^2 + D x_1 x_2 + E . \tag{5.124}$$

By differentiating, we obtain that

$$\frac{1}{2}(\Psi_{11} - \Psi_{22}) = C - C' = \gamma_1$$

$$\Psi_{12} = D = \gamma_2$$

$$\frac{1}{2}(\Psi_{11} + \Psi_{22}) = C + C' = \kappa. \tag{5.125}$$

Imposing $\kappa = 0$, we obtain

$$C = -C' \Rightarrow C = \frac{\gamma_1}{2} . \tag{5.126}$$

Therefore,

$$\Psi_\gamma(\vec{x}) = \frac{\gamma_1}{2}(x_1^2 - x_2^2) + \gamma_2 x_1 x_2 . \tag{5.127}$$

If ϕ_γ is the angle defining the direction of the external shear (or better the direction of the eigenvectors of the shear with eigenvalue γ, see Sect. 3.3), i.e.

$$\gamma_1 = \gamma \cos 2\phi_\gamma ,$$

$$\gamma_2 = \gamma \sin 2\phi_\gamma , \tag{5.128}$$

then, in polar coordinates,

$$\Psi_\gamma(x, \phi) = \frac{\gamma}{2} x^2 \cos 2(\phi - \phi_\gamma) . \tag{5.129}$$

Likewise, if the lens is embedded in a sheet of constant surface-mass density producing no shear, from Eq. 5.125 we find

$$\Psi_\kappa(\vec{x}) = \frac{\kappa}{2}(x_1^2 + x_2^2) = \frac{\kappa}{2}x^2 \ . \tag{5.130}$$

Irrelevant constants have been omitted in the equations above.

The deflection angle of a sheet of constant surface-mass density is

$$\vec{\alpha}(\vec{x}) = \vec{\nabla}\Psi_\kappa(\vec{x}) = \kappa\vec{x} \ . \tag{5.131}$$

Thus, the lens equation reads, in this case,

$$\vec{y} = \vec{x} - \vec{\alpha}(\vec{x}) = \vec{x}(1 - \kappa) \ . \tag{5.132}$$

If $\kappa = 1$, $y = 0$ for all images, i.e. this sheet focuses all light rays exactly on the origin. This gravitational lens thus has a well-defined focal point.

The potential Ψ_γ is a particular case of a more general class of perturbations described by potentials of the kind

$$\Psi_{ext}(x, \phi) = \frac{\epsilon}{m}x^n \cos m(\phi - \phi_\epsilon) \ . \tag{5.133}$$

For example, the external shear corresponds $m = 2$ and $n = 2$. Higher-order perturbations can be modeled with such potentials (see e.g. Meneghetti et al. 2017, Oguri et al. 2013).

5.7   Multiple Mass Components

We often cannot describe gravitational lenses with smooth mass distributions composed of a single mass clump. Indeed, many lenses show a hierarchy of mass components. The largest mass components, which we may refer to as *macro-lenses*, are responsible for the largest-scales lensing effects, often the easiest to identify and measure. On the contrary, the smallest mass clumps within the macro-lenses may be referred to as *substructures* and act as perturbers of the macro-lens.

The superposition principle allows to compute the total potential of the lens as the sum of the potentials of the individual mass components. Thus, the total potential of a lens composed of n_{smooth} smooth large-scale mass components and of n_{sub} small-scale mass clumps can be written as

$$\Psi(\vec{x}) = \sum_{i=1}^{n_{smooth}} \Psi_{smooth,i}(\vec{x} - \vec{x}_{smooth,i}) + \sum_{i=1}^{n_{sub}} \Psi_{sub,i}(\vec{x} - \vec{x}_{sub,i}) \ , \tag{5.134}$$

where $\Psi_{smooth,i}$ and $\Psi_{sub,i}$ are the lensing potentials of the i-th large- and small-scale mass clumps at positions $\vec{x}_{smooth,i}$ and $\vec{x}_{sub,i}$, respectively.

In case of lenses embedded in some external perturbation described by a potential Ψ_{ext}, we can further add it to the potential in Eq. 5.134:

$$\tilde{\Psi}(\vec{x}) = \Psi(\vec{x}) + \Psi_{ext}(\vec{x}) . \tag{5.135}$$

We can compute lens properties such as deflection angles, convergence, and shear as usual by taking the first and second derivatives of the lensing potential and by combining them.

Multi-modality in the mass distribution of gravitational lenses is very relevant, particularly in galaxy clusters and galaxy groups. In the hierarchical model of structure formation, galaxy clusters are the largest and the youngest structures in the universe. They form through the assembly of smaller-scale structures, and for this reason, they are often observed during mergers. Several studies have shown that multi-modal mass distributions are very efficient gravitational lenses. Indeed, the shear of each mass component causes the critical lines and the caustics of the lens to extend into regions of low convergence (see, e.g. Meneghetti et al. 2010, Torri et al. 2004).

Substructures act as lenses within larger lenses. They can significantly affect the appearance and the positions of images of sources lensed by the primary lens where they reside. The magnitude of these effects depends on the relative size of the substructure and source. If the latter is much more extended than the Einstein ring of the substructure, then the substructure manifests as a small-scale perturbation of the lensed images' surface brightness. Suppose instead that the source is sufficiently small. In that case, the substructure can induce significant magnification variations, often seen as "anomalies."

The so-called "flux-ratio anomalies"İ represent a long-standing problem in strong lensing (Kochanek 1991). In many multiply imaged quasars, the image positions can be well reproduced with lens models having smooth mass distributions, but relative flux-ratios of multiple images cannot. In particular, there are image configurations where simple mathematical relations between the image fluxes must be satisfied. For example, in the case of a source near the cusp of the tangential caustic of an elliptical lens, three images form near the tangential critical line, as shown in the middle and bottom panels of Figs. 5.17 and 5.18. Let us dub these images as 1, 2, and 3, respectively. In this situation, the image fluxes should satisfy the so-called cusp-relation:

$$R_{cusp} = \frac{\mu_1 + \mu_2 + \mu_3}{|\mu_1| + |\mu_2| + |\mu_3|} \rightarrow 0 , \text{ for } \mu_{tot} \rightarrow \infty , \tag{5.136}$$

i.e. the sum of the signed magnifications of the cusp images approaches zero as the source moves toward the cusp (Blandford & Narayan 1986, Schneider et al. 1992). Similarly, a source near the fold of the caustic has two images, one inside and one outside the critical line, that must satisfy the so-called fold-relation:

$$R_{fold} = \frac{\mu_1 + \mu_2}{|\mu_1| + |\mu_2|} \to 0 \,, \text{ for } \mu_{tot} \to \infty \,, \tag{5.137}$$

i.e. the sum of the signed magnifications of the fold images approaches zero as the source moves toward the fold. It has been suggested that substructures in galaxies could be among the possible causes why these relations are often not satisfied (Amara et al. 2006, Mao & Schneider 1998, Metcalf & Madau 2001, Metcalf & Zhao 2002, Xu et al. 2015).

Consider the mass distribution in Fig.5.23, obtained by combining several mass clumps modeled as NIE lenses. In addition to a large-scale mass component, the lens contains 100 randomly distributed substructures, all having the same mass. This toy lens model could be a representative of a massive galaxy cluster. In Fig. 5.24, we show some examples of strong lensing effects by substructures. Massive substructures such as cluster galaxies can split large gravitational arcs into smaller arclets (Desprez et al. 2018, Kassiola et al. 1992, Rivera-Thorsen et al.

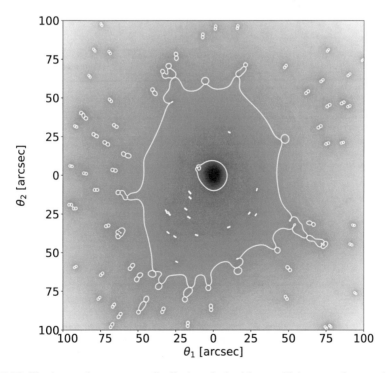

Fig. 5.23 The image shows a mass distribution obtained by combining several mass clumps modeled as NIE lenses. In addition to a large-scale mass component, the lens contains 100 randomly distributed substructures, all having the same mass. The yellow and white lines show the tangential and radial critical lines, respectively

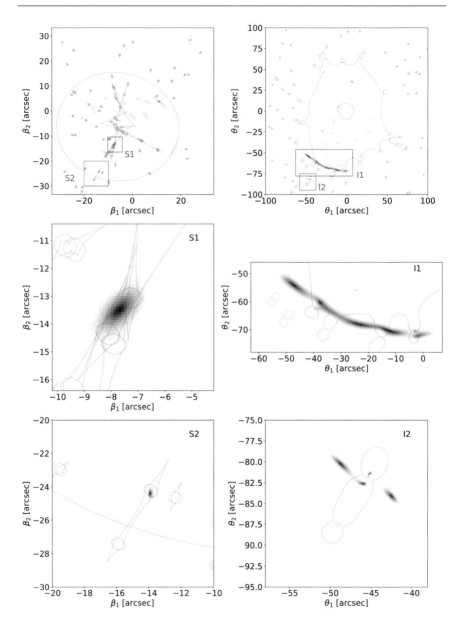

Fig. 5.24 Strong lensing effects by substructures. In the upper left and right panels, we show the caustics and the critical lines of the lens in Fig. 5.23. Two sources are located within the blue rectangles S1 and S2 in the upper left panel. We zoom over the same regions in the middle and bottom left panels, and we show the corresponding images I1 and I2 in the right panels

2019), as in the case of the image indicated as I1 in the upper right panel. The corresponding source is labeled S1 in the upper left panel. The primary lens and the substructures have their critical lines and caustics, shown in red in the right and left panels. The source S1 is within the cusp of the tangential caustic of the primary lens. For this reason, three of its images merge into the tangential arc in panel I1. Besides, the source in S1 is also located at the intersection of several small substructure caustics, as it can be better seen in the middle left panel. The source parts which overlap with these caustics are imaged several additional times around the substructure critical lines. Consequently, the arc appears broken in several small pieces and is locally curved around the substructures. We show a zoom-in of the arc in the middle right panel. If the source is sufficiently small to be fully contained into a secondary caustic, as the source S2, it produces sets of multiple distinct images around the corresponding secondary critical line. For example, we show the images I2 of S2 in the bottom right panel. It has been estimated that substructures account for $\sim 30\%$ of the efficiency of galaxy clusters to produce strong lensing effects (Meneghetti et al. 2007).

5.8   Time Delays

As seen in Sect. 3.6.1, the time delay of a source at $\vec{y}$ is

$$
\begin{aligned}
t(\vec{x}) &= \frac{(1+z_{\mathrm{L}})}{c} \frac{D_{\mathrm{L}} D_{\mathrm{S}}}{D_{\mathrm{LS}}} \frac{\xi_0^2}{D_{\mathrm{L}}^2} \left[\frac{1}{2}(\vec{x} - \vec{y})^2 - \Psi(\vec{x}) \right] \\
&= \frac{(1+z_{\mathrm{L}})}{c} \frac{D_{\mathrm{L}} D_{\mathrm{S}}}{D_{\mathrm{LS}}} \tau(\vec{x}) = (1+z_{\mathrm{L}}) \frac{D_{\Delta t}}{c} \tau(\vec{x}) .
\end{aligned}
\tag{5.138}
$$

We cannot measure the time delay of individual images in an absolute sense. However, we can measure the relative time delay between images if the source is intrinsically variable. In this case, the same variability-induced features should appear in the light-curves of its multiple images at different times, depending on the delay accumulated by light along each line-of-sight.

We consider an axially symmetric lens with a power-law mass profile given in Eq. 5.38. The profile of the deflection angle is then given in Eq. 5.40. Given that the deflection angle is the gradient of the lensing potential, it turns out that the latest is

$$
\Psi(x) = \frac{x^{3-n}}{3-n} .
\tag{5.139}
$$

If $n < 2$, the lens can produce up to three multiple images of a source, depending on whether y is smaller or larger than the radius of the radial critical line, y_r.

Let us assume that there are three multiple images, denoted as A, B, C, where A is the image located at the minimum of the time delay surface. We may use this image as the reference to measure the relative time delays of the other two images. Let us further assume that images B and C correspond to the saddle point and the maximum time delay surface.

Inserting the proper form of the lensing potential in the time delay function, we obtain that the Fermat potential of the i-th image is

$$\tau(x_i) = \frac{\xi_0^2}{D_{\rm L}^2} \left[\frac{1}{2} x_i^{2(2-n)} - \frac{1}{3-n} x_i^{3-n} \right] . \tag{5.140}$$

Thus, the time delay relative to image A is

$$\Delta t_{iA} \propto \Delta \tau_{iA} = \frac{\xi_0^2}{D_{\rm L}^2} \left[\frac{1}{2} \left(x_i^{2(2-n)} - x_A^{2(2-n)} \right) - \frac{1}{3-n} \left(x_i^{3-n} - x_A^{3-n} \right) \right] , \tag{5.141}$$

where $i \in [B, C]$.

Let us assume that the observed spatial offset between images i and A is $\Delta x_{iA} = x_A - x_i$. Then, the relative time delay can be written as a function of the image position A as

$$\Delta t_{iA} \propto \Delta \tau_{iA} = \frac{\xi_0^2}{D_{\rm L}^2} \left\{ \frac{1}{2} \left[(x_A - \Delta x_{iA})^{2(2-n)} - x_A^{2(2-n)} \right] \right.$$
$$\left. - \frac{1}{3-n} \left[(x_A - \Delta x_{iA})^{3-n} - x_A^{3-n} \right] \right\} . \tag{5.142}$$

We may consider the case of a source quite close to the lens tangential caustic. In this case, the images A and B will be located close to the Einstein ring, $\Delta x_{BA} \sim 2$, independently on the value of n. Instead, image C will be located near the center of the lens, so that $\Delta x_{CA} \sim 1$.

As the central image is usually undetected, we focus our attention on image B. In Fig. 5.25, we can see how that the larger is the logarithmic slope n, the larger is the relative time delay between this image and image A. This example shows that steeper and, generally speaking, more compact lenses produce longer time delays between multiple images of the same source.

Of course, the exact location of the images relative to the lens will impact the time delay's magnitude. In the case discussed above, if the source is too close to the tangential caustic, the images A and B will have short time delays independently on the slope of the mass profile, given the nearly perfect symmetry of the time delay surface (see Sect. 3.6.3). On the contrary, the larger is the offset between the lens and the source (provided that the source is still within the radial caustic or the cut, otherwise no multiple images would be produced), the longer is the time delay between the image pair.

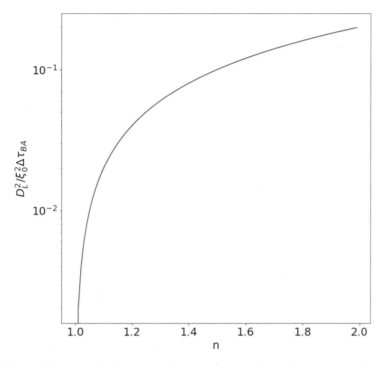

Fig. 5.25 Relative time delay between the images of a source lensed by a power-law lens. The results are shown as a function of the power-law index n for a source position slightly offset from the center of the lens, assuming that the images A and B, corresponding to the minimum and to the saddle point of the time delay surface, respectively, have a separation $\Delta x_{BA} \sim 1$

5.9 Mass-Sheet Degeneracy

We consider a lens with potential $\Psi(\vec{x})$ placed on a constant sheet of matter. From Eqs. 5.130 and 5.135 the total potential is

$$\tilde{\Psi}(\vec{x}) = \Psi(\vec{x}) + \frac{1}{2}\kappa_{ext}x^2 . \tag{5.143}$$

The lens equation can then be written as

$$\vec{y} = \vec{x} - \vec{\nabla}\tilde{\Psi}(\vec{x}) = (1 - \kappa_{ext})\vec{x} - \vec{\nabla}\Psi(\vec{x}) . \tag{5.144}$$

This shows that, if we re-scale the potential $\Psi(\vec{x})$ and the source position $\vec{y}$ by the factor $(1 - \kappa_{ext})$, we obtain a new lens equation which has the same solutions as the lens equation of the isolated lens (i.e. without the mass sheet):

$$\vec{y} = \vec{x} - \vec{\nabla}\Psi(\vec{x}) . \tag{5.145}$$

More generally, any transformation of the potential of the kind

$$\Psi(\vec{x}) \rightarrow \tilde{\Psi}(\vec{x}) = \frac{1}{2}(1 - \lambda)x^2 + \lambda\Psi(\vec{x}), \tag{5.146}$$

where λ plays the role of $(1 - \kappa_{ext})$ and leaves the solutions of the lens equation unaltered. This degeneracy is called mass-sheet degeneracy and the potential transformation in Eq. 5.146 is called *mass-sheet transformation* .

By taking the second derivatives of the potential in Eq. 5.146, we find that, under mass-sheet transformations, the convergence and the shear change as

$$\tilde{\kappa}(\vec{x}) = (1 - \lambda) + \lambda\kappa(x) , \tag{5.147}$$

$$\tilde{\gamma}(\vec{x}) = \lambda\gamma(x) . \tag{5.148}$$

Therefore, the eigenvalues of the Jacobian matrix transform as

$$\tilde{\lambda}_t(\vec{x}) = 1 - \tilde{\kappa}(\vec{x}) - \tilde{\gamma}(\vec{x}) = \lambda[1 - \kappa(\vec{x}) - \gamma(\vec{x})] , \tag{5.149}$$

$$\tilde{\lambda}_r(\vec{x}) = 1 - \tilde{\kappa}(\vec{x}) + \tilde{\gamma}(\vec{x}) = \lambda[1 - \kappa(\vec{x}) + \gamma(\vec{x})] . \tag{5.150}$$

Thus, mass-sheet transformations keep the critical lines of the lens unchanged.

Also the reduced shear is note altered by the mass-sheet transformation, because

$$\tilde{g}(\vec{x}) = \frac{\tilde{\gamma}(\vec{x})}{1 - \tilde{\kappa}(\vec{x})} = \frac{\lambda\gamma(\vec{x})}{\lambda(1 - \kappa(\vec{x}))} = g(\vec{x}) . \tag{5.151}$$

On the contrary, the determinant of the Jacobian transforms as

$$\tilde{\det} A(\vec{x}) = [1 - \tilde{\kappa}(\vec{x})]^2 - \tilde{\gamma}(\vec{x})^2 = \lambda^2\{[1 - \kappa(\vec{x})]^2 - \gamma(\vec{x})^2\} = \lambda^2 \det A(\vec{x}) . \tag{5.152}$$

Thus, the magnification changes as

$$\tilde{\mu}(\vec{x}) = \lambda^{-2}\mu(\vec{x}) . \tag{5.153}$$

Unless the source size is known in a statistical sense, the magnification is not directly observable. We can compare the magnifications of multiple images of the same source by means of the magnification ratios, which, however, are invariant under mass-sheet transformations.

The mass-sheet transformation of the time delay surface can be derived as follows:

$$\tilde{t}(\vec{x}) \propto \frac{1}{2}(\vec{x} - \tilde{\vec{y}})^2 - \tilde{\Psi}(\vec{x})$$

$$= \frac{(\vec{x} - \lambda\vec{y})^2}{2} - \lambda\Psi(\vec{x}) - \frac{1 - \lambda}{2}x^2$$

$$= \lambda \left[\frac{1}{2}(\vec{x} - \vec{y})^2 - \Psi(\vec{x}) \right] + \frac{1}{2}\lambda(1 + \lambda)y^2$$

$$= \lambda t(\vec{x}) + \frac{1}{2}\lambda(1 + \lambda)y^2 . \tag{5.154}$$

The absolute time delay of an image is not measurable, but we can measure the relative time delays between images, in case of intrinsically variable sources. Since the source position is the same, the additive term in the second member of the equation above cancels out. Thus,

$$\Delta \tilde{t}_{ij} = \lambda \Delta t_{ij} . \tag{5.155}$$

Thus, time delays seem to offer a possibility to break the mass-sheet degeneracy. However, this is true only if the time delay distance $D_{\Delta t}$ is known.

5.10 Multiple Lens Planes

While propagating from a distant source to an observer, the light might have encounters with multiple gravitational lenses at different distances. We may assume that each lens lies on a different lens plane. If the angular separation between the lenses on the sky is not too large so that the sky's curvature can be neglected, the lens planes can be assumed to be parallel. We can study how the light propagates through the sequence of lens planes using the thin-screen approximation, as discussed in Chap. 3, and work out a new multiplane lens equation.

To begin, we consider the case of two lens planes. The angular diameter distances to the two planes are D_i with $i \in [1, 2]$. The first plane is the closest to the observer. We assume that the source lays on a third plane at a distance D_S. We show a sketch of this system in Fig. 5.26. The light is emitted by a source at angular position $\vec{\beta}$ relative to an arbitrary line-of-sight chosen as reference. It intercepts the second lens plane at position $\vec{\theta}_2$, where it is deflected by the angle $\hat{\vec{\alpha}}_2(\vec{\theta}_2)$. Then it intercepts the first lens plane at position $\vec{\theta}_1$, where it is deflected by the angle $\hat{\vec{\alpha}}_1(\vec{\theta}_1)$. Finally, it reaches the observer, who will see the image subtending the angle $\vec{\theta}_1$ on the sky.

From the figure, we can see that

$$D_S\vec{\beta} = D_S\vec{\theta}_1 - D_{1S}\hat{\vec{\alpha}}_1(\vec{\theta}_1) - D_{2S}\hat{\vec{\alpha}}_2(\vec{\theta}_2) , \tag{5.156}$$

where D_{iS} is the angular diameter distance between the i-th lens plane and the source. Dividing by D_S and introducing the reduced deflection angles, $\vec{\alpha}_i = D_{iS}/D_S\hat{\vec{\alpha}}_i$, we obtain

$$\vec{\beta} = \vec{\theta}_1 - \sum_{i=1}^{2} \vec{\alpha}_i(\vec{\theta}_i) . \tag{5.157}$$

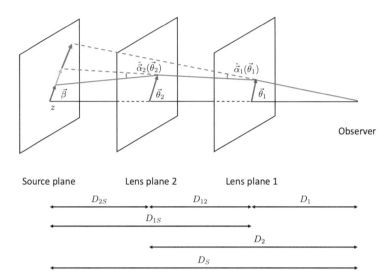

Fig. 5.26 Sketch of a sequence of multiple lens planes

The vectors $\vec{\theta}_1$ and $\vec{\theta}_2$ are related. Indeed, if the source was at position $\vec{\theta}_2$, the observer would see an image at $\vec{\theta}_1$ such that

$$\vec{\theta}_2 = \vec{\theta}_1 - \frac{D_{12}}{D_2}\hat{\vec{\alpha}}_1(\vec{\theta}_1)$$

$$= \vec{\theta}_1 - \frac{D_{12}}{D_2}\frac{D_S}{D_{1S}}\vec{\alpha}_1(\vec{\theta}_1), \tag{5.158}$$

where D_{12} is the angular diameter distance between the first and the second lens planes.

Equations 5.157 and 5.158 can be generalized to the case of N lens planes:

$$\vec{\beta} = \vec{\theta}_1 - \sum_{i=1}^{N}\vec{\alpha}_i(\vec{\theta}_i)\,, \tag{5.159}$$

$$\vec{\theta}_i = \vec{\theta}_1 - \sum_{j=1}^{i-1}\frac{D_{ji}}{D_i}\frac{D_S}{D_{jS}}\vec{\alpha}_j(\vec{\theta}_j)\,. \tag{5.160}$$

The effects of masses along the line-of-sight can be substantial and, in several applications, they must be taken into account (Dalal et al. 2005, Fassnacht et al. 2006, Rusu et al. 2017). Image positions, fluxes, and even time delays can be affected (Chirivì et al. 2018, D'Aloisio & Natarajan 2011, D'Aloisio et al. 2014, Inoue 2016, Xu et al. 2012). The strong lensing cross-sections can be boosted by the presence of structures along the line-of-sight. Thus, galaxies and galaxy clusters selected because of their strong lensing strength could be biased populations, i.e. preferentially aligned with other cosmic structures (see, e.g. Bayliss et al. 2014, Faure et al. 2009, Puchwein & Hilbert 2009).

5.11   Python Applications

5.11.1 Numerical Solution of the Lens Equation

In most cases, the lens models' complexity does not allow to solve the lens equation for a given source position analytically. Then, we have to find the images using numerical methods. Several algorithms can be used for this purpose. Here, we discuss two of them.

Multiple Images by a SIE Lens
Kormann et al. (1994) propose a method to solve the lens equation for SIE lenses (see Sect.5.4.1). We begin with the two coordinates of the source,

$$y_1 = x_1 - \alpha_1(\vec{x}) \tag{5.161}$$

$$y_2 = x_2 - \alpha_2(\vec{x}) . \tag{5.162}$$

Multiplying Eq. 5.161 by $\cos\varphi$ and Eq. 5.162 by $\sin\varphi$, we obtain

$$y_1 \cos\varphi = x_1 \cos\varphi - \alpha_1(\vec{x}) \cos\varphi = x \cos^2\varphi - \alpha(x, \varphi) \cos^2\varphi , \tag{5.163}$$

$$y_2 \sin\varphi = x_2 \sin\varphi - \alpha_2(\vec{x}) \sin\varphi = x \sin^2\varphi - \alpha(x, \varphi) \sin^2\varphi . \tag{5.164}$$

We remind that

$$\alpha(x, \varphi) = \tilde{\psi}(\varphi) \tag{5.165}$$

for the SIE lens.

Equations 5.163 and 5.164 can be combined to compute the distance from the lens center as a function of φ:

$$x(\varphi) = y_1 \cos\varphi + y_2 \sin\varphi + \tilde{\psi}(\varphi) . \tag{5.166}$$

By reinserting Eq. 5.166 into the lens equation, we finally obtain

$$F(\varphi) = \left[y_1 + \frac{\sqrt{f}}{f'} \operatorname{arcsinh}\left(\frac{f'}{f} \cos\varphi \right) \right] \sin\varphi$$

$$- \left[y_2 + \frac{\sqrt{f}}{f'} \arcsin(f' \sin\varphi) \right] \cos\varphi = 0. \tag{5.167}$$

Now the problem of finding the images of a source at position (y_1, y_2) reduces to the problem of finding the zeros of $F(\varphi)$. Once φ has been determined, it can be inserted in Eq. 5.166 above to obtain x.

The solutions cannot be found analytically: a root-finding algorithm such as Brent's method must be employed (Brent 1972).

The following code implements a class for the SIE lens. An instance of the class has five input parameters, namely the redshifts of the lens and the source zl and zs, the velocity dispersion sigmav, the axis ratio f, and the position angle pa. The latest is the angle between the major axis of the lens and the axis x_2. An astropy cosmology object, co, must also be provided. The class includes methods to compute several quantities as a function of the coordinates on the lens plane, like the deflection angle, the convergence, the shear, and the lensing potential. It also includes functions to compute the critical line, the cut, and the caustic.

```python
from astropy.constants import c, G
import numpy as np
from scipy.optimize import brentq
from astropy.cosmology import FlatLambdaCDM


class sie_lens(object):
    """
    class SIE
    """
    def __init__(self,co, zl=0.3,zs=2.0,sigmav=200,f=0.6,pa=45.0):
        """
        Initialize the SIE object.
        Dimensionless units are used. The scale angle is given by the
        SIS Einstein radius.
        """

        self.sigmav=sigmav # velocity dispersion
        self.co=co # cosmological model
        self.zl=zl # lens redshift
        self.zs=zs # source redshift
        self.f=f # axis ratio
        self.pa=pa*np.pi/180.0 # position angle
        # compute the angular diameter distances:
        self.dl=self.co.angular_diameter_distance(self.zl)
        self.ds=self.co.angular_diameter_distance(self.zs)
        self.dls=self.co.angular_diameter_distance_z1z2(self.zl,self.zs)
        # calculates the Einstein radius of the SIS lens
        # in arcsec
        self.theta0=np.rad2deg((4.0*np.pi*sigmav**2/(c.to("km/s"))**2*
                        self.dls/self.ds).value)*3600.0

    def delta(self,f,phi):
        return np.sqrt(np.cos(phi-self.pa)**2+
                        self.f**2*np.sin(phi-self.pa)**2)

    def kappa(self,x,phi):
```

```
        """
        Convergence for the SIE lens at position (x,phi) in polar
        coordinates.
        """
        return(np.sqrt(self.f)/2.0/x/self.delta(self.f,phi))

    def gamma(self,x,phi):
        """
        Shear for the SIE lens at position (x,phi) in polar coordinates.
        """
        return(-self.kappa(x,phi)*np.cos(2.0*phi-self.pa),
               -self.kappa(x,phi)*np.sin(2.0*phi-self.pa))

    def mu(self,x,phi):
        """
        Magnification for the SIE lens at position (x,phi) in polar
        coordinates.
        """
        ga1,ga2=self.gamma(x,phi)
        ga=np.sqrt(ga1*ga1+ga2*ga2)
        return 1.0/(1.0-self.kappa(x,phi)-ga)/(1.0-self.kappa(x,phi)+ga)

    def psi_tilde(self,phi):
        """
        angular part of the lensing potential at the polar angle phi
        """
        if (self.f < 1.0):
            fp=np.sqrt(1.0-self.f**2)
            return np.sqrt(self.f)/fp*\
                    (np.sin(phi-self.pa)*np.arcsin(fp*np.sin(phi-self.pa))+
                     np.cos(phi-self.pa)*np.arcsinh(fp/f
                     *np.cos(phi-self.pa)))
        else:
            return(1.0)

    def psi(self,x,phi):
        """
        Lensing potential at polar coordinates x,phi
        """
        psi=x*self.psi_tilde(phi)
        return psi

    def alpha(self,phi):
        """
        Deflection angle as a function of the polar angle phi
        """
        fp=np.sqrt(1.0-self.f**2)
        a1=np.sqrt(self.f)/fp*np.arcsinh(fp/self.f*np.cos(phi))
        a2=np.sqrt(self.f)/fp*np.arcsin(fp*np.sin(phi))
        return a1,a2
```

```
def cut(self,phi_min=0,phi_max=2.0*np.pi,nphi=1000):
    """
    Coordinates of the points on the cut.
    The arguments phi_min, phi_max, nphi define the range of
    polar angles used.
    """
    phi=np.linspace(phi_min,phi_max,nphi)
    y1_,y2_=self.alpha(phi)
    y1 = y1_ * np.cos(self.pa) - y2_ * np.sin(self.pa)
    y2 = y1_ * np.sin(self.pa) + y2_ * np.cos(self.pa)
    return -y1,-y2


def tan_caustic(self,phi_min=0,phi_max=2.0*np.pi,nphi=1000):
    """
    Coordinates of the points on the tangential caustic.
    The arguments phi_min, phi_max, nphi define the range of
    polar angles used.
    """
    phi=np.linspace(phi_min,phi_max,nphi)
    delta=np.sqrt(np.cos(phi)**2+self.f**2*np.sin(phi)**2)
    a1,a2=self.alpha(phi)
    y1_=np.sqrt(self.f)/delta*np.cos(phi)-a1
    y2_=np.sqrt(self.f)/delta*np.sin(phi)-a2
    y1 = y1_ * np.cos(self.pa) - y2_ * np.sin(self.pa)
    y2 = y1_ * np.sin(self.pa) + y2_ * np.cos(self.pa)
    return y1,y2

def tan_cc(self,phi_min=0,phi_max=2.0*np.pi,nphi=1000):
    """
    Coordinates of the points on the tangential critical line.
    The arguments phi_min, phi_max, nphi define the range of
    polar angles used.
    """
    phi=np.linspace(phi_min,phi_max,nphi)
    delta=np.sqrt(np.cos(phi)**2+self.f**2*np.sin(phi)**2)
    r=np.sqrt(self.f)/delta
    x1=r*np.cos(phi+self.pa)
    x2=r*np.sin(phi+self.pa)
    return(x1,x2)
```

The method to solve Eq. 5.167 is implemented in the function phi_ima below. First, we evaluate the function F on an array of polar angles φ. Then, we loop over the elements of the array, checking whether F changes sign between two consecutive polar angles. When this happens, a zero of the function F is bracketed by these polar angles, and it is determined by using Brent's method. This is implemented in the scipy.optimize module, for example.

```
def x_ima(self,y1,y2,phi):
    """
```

```
        Distance of the image from the lens center
        """
        x=y1*np.cos(phi)+y2*np.sin(phi)+(self.psi_tilde(phi+self.pa))
        return x

    def phi_ima(self,y1,y2,checkplot=True,eps=0.001,nphi=100):
        """
        Solve the lens Equation for a given source position (y1,y2)
        """
        # source position in the frame where the lens major axis is
        # along the £x_2£ axis.
        y1_ = y1 * np.cos(self.pa) + y2 * np.sin(self.pa)
        y2_ = - y1 * np.sin(self.pa) + y2 * np.cos(self.pa)

        # This is Eq.\,\ref{eq:ffunct}
        def phi_func(phi):
            a1,a2=self.alpha(phi)
            func=(y1_+a1)*np.sin(phi)-(y2_+a2)*np.cos(phi)
            return func

        # Evaluate phi_func and the sign of phi_func on an array of
        # polar angles
        U=np.linspace(0.,2.0*np.pi+eps,nphi)
        c = phi_func(U)
        s = np.sign(c)
        phi=[]
        xphi=[]
        # loop over polar angles
        for i in range(len(U)-1):
            # if two polar angles bracket a zero of phi_func,
            # use Brent's method to find exact solution
            if s[i] + s[i+1] == 0: # opposite signs
                u = brentq(phi_func, U[i], U[i+1])
                z = phi_func(u)
                if np.isnan(z) or abs(z) > 1e-3:
                    continue
                x=self.x_ima(y1_,y2_,u)
                # append solution to a list if it corresponds to radial
                # distances x>0; discard otherwise (spurious solutions)
                if (x>0):
                    phi.append(u)
                    xphi.append(x)

        # convert lists to numpy arrays
        xphi=np.array(xphi)
        phi=np.array(phi)

        # returns radii and polar angles of the images. Add position angle
        # to go back to the rotated frame of the lens.
        return xphi,phi+self.pa
```

The function `phi_ima` returns the coordinates of the multiple images in polar coordinates (x_i, φ_i).

In the following example, we use the functions above to find the images of a source at position ($y_1 = 0.2$, $y_2 = 0.2$):

```
# define a SIE lens with sigmav=200 km/s, f=0.3, pa=0.0
sigmav=200.0
f=0.3
pa=0.0
co = FlatLambdaCDM(Om0=0.3,H0=70.0)
sie=sie_lens(co,sigmav=sigmav,f=f,pa=pa)

# source coordinates
y1=0.2
y2=0.2

# set up a figure

fig,ax=plt.subplots(1,2,figsize=(18,8))

# polar coordinates of the images
x,phi=sie.phi_ima(y1,y2)

# convert to Cartesian coordinates
x1_ima=x*np.cos(phi)
x2_ima=x*np.sin(phi)

# compute cut, caustic, and critical line
y1_cut,y2_cut=sie.cut()
y1_cau,y2_cau=sie.tan_caustic()
x1_cc,x2_cc=sie.tan_cc()
```

The results are shown in Fig. 5.27. The functions $F(\varphi)$ and $x(\varphi)$ are given by the blue and the orange lines in the left panel. The zeros of $F(\varphi)$ found with the method outlined above are marked with green dots. The solid and the dashed lines in the right panel show the caustic and the cut of the lens, respectively. The solid blue line is the lens critical line. The source's position is indicated by the orange dot, while the images are shown in green.

5.11.2 Triangle Mapping

The triangle mapping method (see e.g. Bartelmann 2003, Schneider et al. 1992) is more generic and can be used with any lens model, including non-analytic mass distributions.

The method is grid-based. Consider a regular grid covering the lens plane, with cell corners at positions x_{ij}, with $i, j \in [0, N)$. We could find the images of a given source by checking if that source falls into one or more of the grid cells when these are mapped onto the source plane via ray-tracing (see, e.g., Sect. 3.7.1).

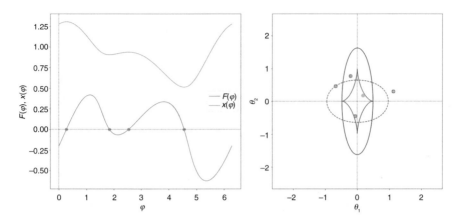

Fig. 5.27 Solution of the lens equation for a SIE lens. Left panel: the blue and the orange lines show the functions $F(\varphi)$ and $x(\varphi)$. The green dots indicate the zeros of $F(\varphi)$. The lens caustic and cut are shown by the solid and dashed lines in the right panel. The solid blue line gives the lens critical line. The positions of the source and its multiple images computed using the method in Sect. 5.11.1 are marked with orange and green points, respectively. The yellow crosses overlapped with the green dots show the image positions determined using the triangle mapping method discussed in Sect. 5.11.2

However, a complication arises from the fact that, while the grid cells on the lens plane are squares (or rectangles, if the grid is rectangular), their shapes can be highly distorted when mapped onto the source plane. For example, near the lens caustics, the opposite corners of the original grid cells can be interchanged, making it difficult to decide whether a given point in the source plane is inside or outside the mapped grid cell.

This problem can be solved by splitting each grid cell on the lens plane into two triangles. When mapped onto the source plane, these remain triangles. If one or more of these triangles contain the source, then the source can be associated with the image positions at the center of the corresponding triangles on the lens plane.

To check whether a source point is inside or outside a triangle, the following method can be employed. Let $\vec{d}_k$ be the vectors connecting the triangle's vertices to the source point, with $k \in [1..3]$. It can be shown that the source point falls within the triangle if the three vector products

$$\vec{d}_1 \times \vec{d}_2 \; ; \vec{d}_1 \times \vec{d}_3 \; ; \vec{d}_2 \times \vec{d}_3 \tag{5.168}$$

are all positive. The method is implemented in the Python function below.

```
def find_images(self,ys1_,ys2_,xmin=-2,xmax=2,npix=2048):
    ys1 = ys1_ * np.cos(self.pa) + ys2_ * np.sin(self.pa)
    ys2 = - ys1_ * np.sin(self.pa) + ys2_ * np.cos(self.pa)
    # create a mesh in the search region:
    x=np.linspace(xmin,xmax,npix)
    grid_pixel = x[1]-x[0]
```

```
x1,x2 = np.meshgrid(x,x)

# compute deflections at each grid point
phi=np.arctan2(x2,x1)
x=np.sqrt(x1*x1+x2*x2)
a1,a2=self.alpha(phi)

# ray-trace the deflector grid onto the source plane
y1 = x1 - a1
y2 = x2 - a2

# convert to pixel units
xray = y1 / grid_pixel + (len(x)-1) / 2.0
yray = y2 / grid_pixel + (len(x)-1) / 2.0

y1s = ys1 / grid_pixel + (len(x)-1) / 2.0
y2s = ys2 / grid_pixel + (len(x)-1) / 2.0

# shift the maps by one pixel
xray1 = np.roll(xray, 1, axis=1)
xray2 = np.roll(xray1, 1, axis=0)
xray3 = np.roll(xray2, -1, axis=1)
yray1 = np.roll(yray, 1, axis=1)
yray2 = np.roll(yray1, 1, axis=0)
yray3 = np.roll(yray2, -1, axis=1)


"""

for each pixel on the lens plane, build two triangles.
Using ray-tracing, these are mapped onto the source plane
into the other two triangles. Compute the distances of the
vertices of the triangles on the source plane from the source
and check using cross-products if the source is inside one of
the two triangles
"""


x1 = y1s - xray
y1 = y2s - yray

x2 = y1s - xray1
y2 = y2s - yray1

x3 = y1s - xray2
y3 = y2s - yray2

x4 = y1s - xray3
y4 = y2s - yray3

prod12 = x1 * y2 - x2 * y1
prod23 = x2 * y3 - x3 * y2
prod31 = x3 * y1 - x1 * y3
prod13 = -prod31
```

```python
prod34 = x3 * y4 - x4 * y3
prod41 = x4 * y1 - x1 * y4

image = np.zeros(xray.shape)
image[((np.sign(prod12) == np.sign(prod23)) &
       (np.sign(prod23) == np.sign(prod31)))] = 1
image[((np.sign(prod13) == np.sign(prod34)) &
       (np.sign(prod34) == np.sign(prod41)))] = 2

# first kind of images (first triangle)
images1 = np.argwhere(image == 1)
xi_images_ = images1[:, 1]
yi_images_ = images1[:, 0]
xi_images = xi_images_[(xi_images_ > 0) & (yi_images_ > 0)]
yi_images = yi_images_[(xi_images_ > 0) & (yi_images_ > 0)]

# compute the weights
w = np.array([1. / np.sqrt(x1[xi_images, yi_images] ** 2 +
                           y1[xi_images, yi_images] ** 2),
              1. / np.sqrt(x2[xi_images, yi_images] ** 2 +
                           y2[xi_images, yi_images] ** 2),
              1. / np.sqrt(x3[xi_images, yi_images] ** 2 +
                           y3[xi_images, yi_images] ** 2)])
xif1, yif1 = self.refineImagePositions(xi_images, yi_images, w, 1)

# second kind of images (second triangle)
images1 = np.argwhere(image == 2)
xi_images_ = images1[:, 1]
yi_images_ = images1[:, 0]
xi_images = xi_images_[(xi_images_ > 0) & (yi_images_ > 0)]
yi_images = yi_images_[(xi_images_ > 0) & (yi_images_ > 0)]

# compute the weights
w = np.array([1. / np.sqrt(x1[xi_images, yi_images] ** 2 +
                           y1[xi_images, yi_images] ** 2),
              1. / np.sqrt(x3[xi_images, yi_images] ** 2 +
                           y3[xi_images, yi_images] ** 2),
              1. / np.sqrt(x4[xi_images, yi_images] ** 2 +
                           y4[xi_images, yi_images] ** 2)])
xif2, yif2 = self.refineImagePositions(xi_images, yi_images, w, 2)

xi = np.concatenate([xif1, xif2])
yi = np.concatenate([yif1, yif2])

xi = (xi - 1 - (len(x)-1) / 2.0) * grid_pixel
yi = (yi - 1 - (len(x)-1) / 2.0) * grid_pixel

xi_ = xi * np.cos(self.pa) - yi * np.sin(self.pa)
yi_ = xi * np.sin(self.pa) + yi * np.cos(self.pa)
return (xi_, yi_)
```

```
def refineImagePositions(self, x, y, w, typ):
    """
    Image positions are calculated as the weighted mean of the
    positions of the triangle vertices. The weights are the distances
    between the vertices mapped onto the source plane and the source
    position.
    """
    if (typ == 2):
        xp = np.array([x, x + 1, x + 1])
        yp = np.array([y, y, y + 1])
    else:
        xp = np.array([x, x + 1, x])
        yp = np.array([y, y + 1, y + 1])
    xi = np.zeros(x.size)
    yi = np.zeros(y.size)
    for i in range(x.size):
        xi[i] = (xp[:, i] / w[:, i]).sum() / (1. / w[:, i]).sum()
        yi[i] = (yp[:, i] / w[:, i]).sum() / (1. / w[:, i]).sum()
    return (xi, yi)
```

This method works well if the image separation is larger than the size of grid cells on the image plane. Therefore, it might fail to find images of sources just inside the caustics. Indeed, such sources have pairs or triplets of close images forming on the opposite sides of the critical lines. To solve this problem, one can implement adaptive meshes to refine the number of grid cells near these regions.

The function find_images can be included in the class SIE and used to re-compute the multiple images of the source in the example discussed in Sect. 5.11.1. The resulting image positions are shown by the yellow crosses overlapped with the green dots in Fig. 5.27 and are identical to those computed with the previous method.

5.11.3 SIS Lens in an External Shear

How do the properties of a SIS lens change when we embed it in an external shear? In this example, we answer this question.

We begin by implementing a class for the external shear lens model. The class will contain methods to compute several quantities, namely the lensing potential, the deflection angle, and the shear. Eqs. 5.129 and 5.128 give the lensing potential and the components of a constant external shear. By construction, the convergence is zero. We obtain the deflection angle components by differentiating the lensing potential,

$$
\alpha_1(x, \phi) = \cos\phi \frac{\partial \Psi_\gamma(x, \phi)}{\partial x} - \frac{\sin\phi}{x} \frac{\partial \Psi_\gamma(x, \phi)}{\partial \phi} ,
$$

$$
\alpha_2(x, \phi) = \sin\phi \frac{\partial \Psi_\gamma(x, \phi)}{\partial x} + \frac{\cos\phi}{x} \frac{\partial \Psi_\gamma(x, \phi)}{\partial \phi} ,
$$

which gives

$$\alpha_1(x, \phi) = \gamma x \cos(2\phi_\gamma - \phi) , \qquad (5.169)$$

$$\alpha_2(x, \phi) = \gamma x \sin(2\phi_\gamma - \phi) . \qquad (5.170)$$

The class is implemented as follows:

```python
class ext_shear(object):
    def __init__(self,g,phi_g):
        """
        Initialize an external shear using
        the amplitude g and the angle phi_g (in degrees)
        """
        self.g = g
        self.phi_g = np.deg2rad(phi_g)

    def psi(self,x,phi):
        """
        Returns the lensing potential at polar
        coordinates x, phi
        """
        return 0.5*self.g*x**2*np.cos(2*(phi-self.phi_g))

    def alpha(self,x,phi):
        """
        Returns the components of the deflection
        angle at polar coordinates x, phi
        """
        a1=self.g*x*np.cos(2*self.phi_g-phi)
        a2=self.g*x*np.sin(2*self.phi_g-phi)
        return a1, a2

    def gamma(self):
        """
        Returns the components of the shear
        at polar coordinates x, phi
        """
        g1=self.g*np.cos(2*self.phi_g)
        g2=self.g*np.sin(2*self.phi_g)
        return g1, g2
```

We assume an external shear of amplitude $\gamma = 0.1$ and direction $\phi_\gamma = 45\,\text{deg}$ relative to the x_2 axis:

```python
eg=ext_shear(g=0.1,phi_g=45.0)
```

We construct an instance of a SIS lens by using the `sie_lens` class from the previous example, setting $f = 1$:

```
# define a SIS lens with sigmav=200 km/s (i.e. a SIE with f=1.0, pa=0.0)
sigmav=200.0
f=1.0
pa=0.0
# assume a flat LCDM cosmology with Om0=0.3
co = FlatLambdaCDM(Om0=0.3,H0=70.0)
sie=sie_lens(co,sigmav=sigmav,f=f,pa=pa)
```

To embed the SIS in the external shear, we use the superposition principle. The total potential is the sum of the potentials of the SIS and external shear. We define a regular grid on the lens plane and compute the polar coordinates:

```
# define a grid on the lens plane
fov=3.0
x_ = np.linspace(-fov/2.,fov/2.,1000)
x1,x2 = np.meshgrid(x_,x_)
# compute phases and distances from the center of the SIS,
# assumed to be at the center of the grid, in (0,0)
phi=np.arctan2(x2,x1)
x=np.sqrt(x1*x1+x2*x2)
```

The total convergence at any position on the lens plane is simply the convergence of the SIS:

```
# SIS kappa map:
kappa_SIS = sie.kappa(x,phi)
```

We show this map in panel A of Fig. 5.28. Instead, we obtain the total potential as follows:

```
# SIS potential map:
psi_SIS=sie.psi(x,phi)
# external shear potential map:
psi_eg=eg.psi(x,phi)
# total lensing potential
psi_tot=psi_SIS+psi_eg
```

In panel B of Fig. 5.28, we show the total lensing potential of the SIS embedded in the external shear. The solid contours show levels of constant potential. These contours are elongated in the direction of the external shear. Thus, the lensing potential is no longer axially symmetric. On the contrary, it is similar to that of an elliptical lens, although the contours are not elliptical and their elongation is not constant. For comparison, we also show the contour levels of the SIS potential without external shear using dotted lines.

To derive the critical lines, we compute the map of the tangential eigenvalue of the lensing Jacobian. For this purpose, we need to calculate the components of the

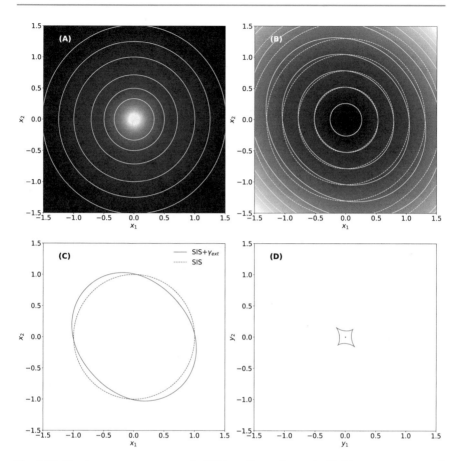

Fig. 5.28 Panel **a**: convergence map of a SIS lens. Even when embedded in a constant external shear, the convergence is axially symmetric, as shown by the yellow contours. Indeed, the external shear does not contribute to the convergence. Panel **b**: lensing potential map of a SIS lens embedded in an external shear with amplitude $\gamma_{ext} = 0.1$ and direction $\phi_\gamma = 45$ deg. The solid yellow lines show some level contours. For comparison, we show the same level contours for the SIS lens without external shear (dashed lines). Panel **c**: tangential critical lines of the SIS embedded in the external shear and for the SIS alone (red solid and blue dashed lines, respectively). Panel **d**: as in panel **c**, but for the lens caustics

shear first:

```
# SIS shear components:
g1_SIS,g2_SIS=sie.gamma(x,phi)
# external shear components:
g1_eg, g2_eg = eg.gamma()
# total shear components
g1_tot=g1_SIS+g1_eg
g2_tot=g2_SIS+g2_eg
```

We obtain the tangential eigenvalue map as follows:

```
# tangential eigenvalue:
lambdat_tot = 1.0 - kappa_SIS - np.sqrt(g1_tot**2+g2_tot**2)
```

The tangential critical line is the level-zero contour of this map. We show it in panel C of Fig. 5.28 (solid red line). Since the lensing potential is no longer axially symmetric, the lens critical line is not circular as for the SIS lens (blue dashed line). Similar to the potential contours, it is elongated in the direction of ϕ_γ. Mapping the critical points onto the source plane, we obtain the lens caustic, which is not a single point, as for the SIS. It has an astroid-like shape with cusps and folds, similar to elliptical lenses. Since the caustic is not point-like, the SIS lens embedded in the external shear produces four images if the source is within the caustic.

5.11.4   Multiple Lens Planes

In this example, we implement a ray-tracing algorithm to propagate light rays through multiple lens planes. One of the lens planes contains the projected mass distribution of a galaxy cluster. This massive object called *Ares* has a virial mass of $M_{vir} \sim 10^{15}$ $M_\odot$ and a redshift $z_2 = 0.5$. It consists of two large clumps of matter and many smaller-scale sub-halos. A detailed description of this lens can be found in Meneghetti et al. (2017). We show its mass distribution in panel B of Fig. 5.29.

The other two lens planes are at redshifts $z_1 = 0.25$ and $z_3 = 0.75$ and contain some smaller galaxy scale mass concentrations. They are shown in panels A and C of Fig. 5.29. All lens planes have sizes of 300×300 arcsec.

We work on a flat ΛCDM cosmological model with $\Omega_{m,0} = 0.3$:

```
from astropy.cosmology import FlatLambdaCDM
cosmo = FlatLambdaCDM(H0=70, Om0=0.3)
```

We compute the deflection angle maps of each lens plane as shown in Sect. 3.7.1. We do not repeat the calculations here. The deflection angle maps are saved in .fits

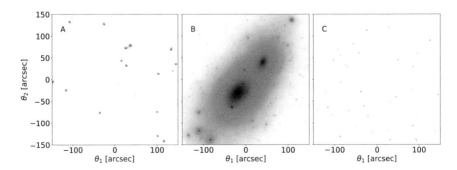

Fig. 5.29   Mass distributions on three lens planes at redshift 0.25, 0.5, and 0.75 (panels **a**, **b**, and **c**, respectively). The lens in panel **b** is a massive galaxy cluster (Meneghetti et al. 2017)

files, whose headers contain all the information needed for the calculations below. To handle these files, we create a class called `lensplane`:

```python
import astropy.io.fits as pyfits
import numpy as np
from scipy.ndimage import map_coordinates

class lensplane(object):
    def __init__(self,filename):
        # read the deflection angle maps from the fits file
        myhd=pyfits.open(filename)
        self.a1=myhd[0].data
        self.a2=myhd[1].data
        # read lens plane parameters from header of the fits file:
        self.z=myhd[0].header['ZLENS']
        self.xmin=myhd[0].header['XMIN']
        self.xmax=myhd[0].header['XMAX']
        self.ymin=myhd[0].header['YMIN']
        self.ymax=myhd[0].header['YMAX']
        self.omega=myhd[0].header['OMEGA']
        self.lambd=myhd[0].header['LAMBDA']
        self.hubble=myhd[0].header['H']
        # compute the pixel scale:
        self.px=(self.xmax-self.xmin)/(myhd[0].header['NAXIS1']-1)

    def angles(self,theta1,theta2):
        """
        This function returns the interpolated values of the deflection
        angle components at theta1, theta2
        """
        theta1pix=(theta1-self.xmin)/self.px
        theta2pix=(theta2-self.ymin)/self.px
        a1 = map_coordinates(self.a1,[theta2pix,theta1pix],order=2)
        a2 = map_coordinates(self.a2,[theta2pix,theta1pix],order=2)
        return(a1,a2)

# create a list of lens planes
listplane=['lensplane_z=0.25.fits','deflAnglesAres.fits',
                'lensplane_z=0.75.fits']
lp=[]
for lenspl in listplane:
    lp.append(lensplane(lenspl))
```

In addition to the initialization function, the class contains a method, called `angles` to interpolate the deflection angle maps at arbitrary positions.

We propagate a bundle of 60×60 light rays from the observer position through a regular grid covering the first lens plane:

```python
fov=300.0 # FOV in arcsec
theta=np.linspace(-fov/2.,fov/2..,60)
theta1,theta2=np.meshgrid(theta,theta)
```

For the purpose of this example, we will use only comoving positions and angular distances:

$$\xi_c = (1+z)\xi \, , \qquad\qquad\qquad (5.171)$$

$$T = (1+z)D \, . \qquad\qquad\qquad (5.172)$$

The comoving angular distances can be computed using the function `comoving_transverse_distance` of `astropy.cosmology`. This function returns the transverse comoving distance in Mpc at a given redshift corresponding to an angular separation of 1 radian. Thus, the comoving positions of the light rays on the first plane are

```
x1_1=np.deg2rad(theta1/3600.0)*cosmo.comoving_transverse_distance(lp[0].z)
x2_1=np.deg2rad(theta2/3600.0)*cosmo.comoving_transverse_distance(lp[0].z)
```

After being deflected on the first lens plane, the light rays propagate toward the second lens plane. We calculate their arrival positions as follows:

```
# difference between the transverse distances at the redshifts
# of the second and first planes
T_last = cosmo.comoving_transverse_distance(lp[1].z)
T_before = cosmo.comoving_transverse_distance(lp[0].z)
deltaT=T_last-T_before
# deflection angle components at the ray positions on the first plane:
alpha1,alpha2=lp[0].angles(theta1,theta2)
# angle between the deflected ray and the optical axis
Talpha1_1=theta1-alpha1
Talpha2_1=theta2-alpha2
# arrival positions on the second lens plane
x1_2=x1_1+np.deg2rad(Talpha1_1/3600.0)*deltaT
x2_2=x2_1+np.deg2rad(Talpha2_1/3600.0)*deltaT
```

In the last equations, we sum the comoving coordinates on the first lens plane (translated to the second lens plane) and the shift between the first and the second plane corresponding to the angle `(Talpha1_1,Talpha2_1)`. This is the angle between the deflected ray and the optical axis.

The angular positions of the light rays on the second lens plane (in arcsec) are given by

```
theta1_2=np.rad2deg((x1_2/T_last).value)*3600.0
theta2_2=np.rad2deg((x2_2/T_last).value)*3600.0
```

Now, we move to the third lens plane:

```
# difference between the transverse distances at the redshifts
# of the third and second planes
T_last = cosmo.comoving_transverse_distance(lp[2].z)
T_before = cosmo.comoving_transverse_distance(lp[1].z)
deltaT=T_last-T_before
```

```
# deflection angle components at the ray positions on the second plane:
alpha1,alpha2=lp[1].angles(theta1_2,theta2_2)
# angle between the deflected ray and the optical axis
Talpha1_2=Talpha1_1-alpha1
Talpha2_2=Talpha2_1-alpha2
# arrival positions on the third lens plane
x1_3=x1_2+np.deg2rad(Talpha1_2/3600.0)*deltaT
x2_3=x2_2+np.deg2rad(Talpha2_2/3600.0)*deltaT
# arrival positions in arcsec
theta1_3=np.rad2deg((x1_3/T_last).value)*3600.0
theta2_3=np.rad2deg((x2_3/T_last).value)*3600.0
```

Finally, we compute the light rays arrival positions on the source plane, which we assume to be at redshift $z_s = 3.0$:

```
# source redshift
zs=3.0
# difference between the transverse distances at the redshifts
# of the source and third planes
T_source = cosmo.comoving_transverse_distance(zs)
T_before = cosmo.comoving_transverse_distance(lp[2].z)
deltaT=T_source-T_before
# deflection angle components at the ray positions on the third plane:
alpha1,alpha2=lp[2].angles(theta1_3,theta2_3)
# angle between the deflected ray and the optical axis
Talpha1_3=Talpha1_2-alpha1
Talpha2_3=Talpha2_2-alpha2
# arrival positions on the source plane
x1_s=x1_3+np.deg2rad(Talpha1_3/3600.0)*deltaT
x2_s=x2_3+np.deg2rad(Talpha2_3/3600.0)*deltaT
# arrival positions in arcsec
beta1=np.rad2deg((x1_s/T_source).value)*3600.0
beta2=np.rad2deg((x2_s/T_source).value)*3600.0
```

We show in Fig.5.30 the arrival positions of the light rays on each lens plane and on the final source plane. As it is clear, the largest deflections occur on the second lens plane, which contains the cluster mass distribution.

As seen, the computations for each lens plane are recursive. Thus, the whole procedure can be written in a much more compact and efficient way. We do that in the class multiplane below:

```
class multiplane(object):

    def __init__(self,cosmo,lp,zs):
        """
        Initialize a multiplane instance using a list of lens plane and
        a source redshift
        """
        self.lp=lp
        self.cosmo = cosmo
        zl_=[]
```

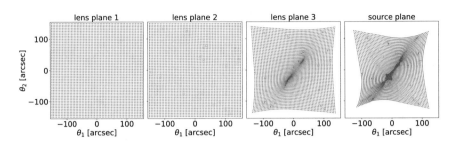

Fig. 5.30 Arrival positions of the light rays on each lens plane shown in Fig. 5.29 and on the source plane at redshift $z_S = 2$

```python
# create list of lens plane redshifts.
# Consider only the lens planes at
# redshift smaller than the source redshift
self.nlens=0
for i in range(len(lp)):
    if lp[i].z < zs:
        self.nlens+=1
        zl_.append(lp[i].z)
# append the source plane redshift
zl_.append(zs)

# loop over the lens planes to initialize the distances
# differences between transverse distances on planes i,j
self.deltaT_list=[]
# distances between the observer and lens planes
self.T_list=[]

# start from redshift z=0
z_before = 0.0
for idx in range(self.nlens):
    z_lens = zl_[idx]
    # calculate the differences between the transverse distances in
    # consecutive pairs of lens planes
    T_last=self.cosmo.comoving_transverse_distance(z_lens)
    T_before=self.cosmo.comoving_transverse_distance(z_before)
    delta_T = T_last-T_before
    self.deltaT_list.append(delta_T.value)
    # save transverse comoving distance for lens plane
    self.T_list.append(T_last.value)
    z_before = z_lens
# add deltaT between source plane and last lens plane
T_source = self.cosmo.comoving_transverse_distance(zs)
T_before = self.cosmo.comoving_transverse_distance(z_before)
delta_T = T_source-T_before
self.deltaT_list.append(delta_T.value)
# save the comoving transverse distance at the source redshift
self.T_source = T_source.value
```

```python
def com2rad_source(self, x_1, x_2):
    """
    compute angular positions of light rays on the source plane
    """
    T = self.T_source
    theta_1 = x_1 / T
    theta_2 = x_2 / T
    return theta_1, theta_2

def com2rad(self, x_1, x_2, idex):
    """
    compute angular positions of light rays on the lens plane
    """
    T = self.T_list[idex]
    theta_1 = x_1 / T
    theta_2 = x_2 / T
    return theta_1, theta_2

def raytrace(self, theta_1, theta_2):
    """
    propagate the light rays from the first lens plane toward
    the sources
    """
    x1 = np.zeros_like(theta_1)
    x2 = np.zeros_like(theta_2)
    alpha_1 = theta_1
    alpha_2 = theta_2
    i = 0
    for i in range(self.nlens):
        delta_T = self.deltaT_list[i]
        x1, x2 = self.next_step(x1, x2, alpha_1, alpha_2, delta_T)
        alpha_1, alpha_2 = self.Talpha(x1, x2, alpha_1, alpha_2, i)
    delta_T = self.deltaT_list[i+1]
    x1, x2 = self.next_step(x1, x2, alpha_1, alpha_2, delta_T)
    beta_1, beta_2 = self.com2rad_source(x1, x2)
    return beta_1, beta_2


def next_step(self, x1, x2, alpha_1, alpha_2, delta_T):
    """
    Computes arrival position of light ray on current plane
    """
    x1_ = x1 + alpha_1 * delta_T
    x2_ = x2 + alpha_2 * delta_T

    return x1_, x2_

def Talpha(self, x1, x2, alpha_1, alpha_2, idex):
    """
```

```
     Computes Talpha at step idex
     """
     theta_1, theta_2 = self.com2rad(x1, x2, idex)
     alpha_1_arcsec, alpha_2_arcsec = \
     self.lp[idex].angles(np.rad2deg(theta_1)*3600.0,
                          np.rad2deg(theta_2)*3600.0)
     Talpha_1 = alpha_1 - np.deg2rad(alpha_1_arcsec/3600.0)
     Talpha_2 = alpha_2 - np.deg2rad(alpha_2_arcsec/3600.0)

     return Talpha_1, Talpha_2

 def alpha(self,theta1,theta2,beta1,beta2):
     a1,a2=theta1-beta1,theta2-beta2
     return (a1,a2)
```

```
# initialize the distances between planes
theta=np.linspace(-150.,150.,1000)
theta1,theta2=np.meshgrid(theta,theta)

mp=multiplane(cosmo,lp,zs)
beta1_,beta2_=mp.raytrace(np.deg2rad(theta1/3600.0),
                          np.deg2rad(theta2/3600.0))
beta1_=np.rad2deg(beta1_)*3600.0
beta2_=np.rad2deg(beta2_)*3600.0
```

The light rays' effective deflection angles are the difference between their starting positions on the first lens plane and their arrival positions on the source plane:

```
a1=theta1-beta1_
a2=theta2-beta2_
```

By taking the first derivatives of a1 and a2, we compute the effective convergence as

```
def convergence(a1,a2,px):
    a12,a11=np.gradient(a1/px)
    a22,a21=np.gradient(a2/px)
    ka=0.5*(a11+a22)
    g1=0.5*(a11-a22)
    g2=a21
    return (ka,g1,g2)

px=theta[1]-theta[0]
ka,g1,g2=convergence(a1,a2,px)
```

We show the effective convergence maps for source redshifts $z_S = 0.4, 0.65$, and 3 in Fig. 5.31. We also show the lens critical lines in yellow. The lens planes at redshifts higher than the source redshift do not contribute to the effective deflection angles. Thus, the mass clumps on these lens planes are not visible in the effective

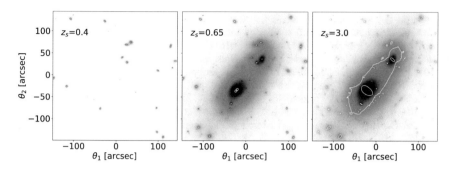

Fig. 5.31 Effective convergence for several source redshifts. Only the lens planes at redshifts lower than the source redshift deflect the light rays. Therefore, some massive structures become visible in the effective convergence maps only for large source redshifts. We show the lens critical lines in yellow

convergence maps until a sufficiently high source redshift is chosen. For example, the massive cluster does not appear in the convergence map for $z_S = 0.4$, but it is visible in the convergence maps for $z_S > 0.5$. Note that the masses on the highest redshift lens planes appear distorted by the lower redshift lens planes' lensing effects.

References

Amara, A., Metcalf, R. B., Cox, T. J., & Ostriker J. P. (2006). Simulations of strong gravitational lensing with substructure. *MNRAS, 367*(4), 1367–1378. https://doi.org/10.1111/j13652966. 200610053.x. arXiv: astroph/0411587 [astro-ph]

Bartelmann, M. (1996). Arcs from a universal dark-matter halo profile. *A & A 313*, 697–702. arXiv: astroph/9602053 [astro-ph]

Bartelmann, M. (2003). Numerical methods in gravitational lensing. arXiv e-prints, astro-ph/0304162. arXiv: astroph/0304162 [astro-ph]

Bayliss, M. B., Johnson, T., Gladders, M. D., Sharon, K., & Oguri, M. (2014). Line-of-sight structure toward strong lensing galaxy clusters. *ApJ, 783*(1), 41. https://doi.org/10.1088/0004-637X/783/1/41. arXiv: 1312.3637 [astro-ph.CO]

Bhattacharya, S., Habib, S., Heitmann, K., & Vikhlinin, A. (2013) Dark matter halo profiles of massive clusters: Theory versus observations. *ApJ, 766*(1), 32. https://doi.org/10.1088/0004-637X/766/1/32. arXiv: 1112.5479 [astro-ph.CO]

Blandford, R. D., & Narayan, R. (1986). Fermat's principle, caustics, and the classification of gravitational lens images. *ApJ, 310*, 568. https://doi.org/10.1086/164709

Brent, R. P. (1972). *Algorithms for minimization without derivatives (Prentice-Hall series in automatic computation)*. Prentice-Hall. Retrieved from https://www.xarg.org/ref/a/0130223352/

Chirivì, G., Suyu, S. H., Grillo, C., Halkola, A., Balestra, I., Caminha, G. B., & Rosati, P. (2018). MACS J0416.1-2403: Impact of line-of-sight structures on strong gravitational lensing modelling of galaxy clusters. *A & A, 614*, A8. https://doi.org/10.1051/00046361/201731433. arXiv: 1706.07815 [astro-ph.CO]

D'Aloisio, A., & Natarajan, P. (2011). Cosmography with cluster strong lenses: The influence of substructure and line-of-sight haloes. *MNRAS, 411*(3), 1628–1640. https://doi.org/10.1111/j13652966.2010.17795.x. arXiv: 1010.0004 [astro-ph.CO]

D'Aloisio, A., Natarajan, P., & Shapiro, P. R. (2014). The effect of large-scale structure on the magnification of high-redshift sources by cluster lenses. *MNRAS, 445*(4), 3581–3591. https://doi.org/10.1093/mnras/stu1931. arXiv: 1311.1614 [astro-ph.CO]

Dalal, N., Hennawi, J. F., & Bode, P. (2005). Noise in strong lensing cosmography. *ApJ, 622*(1), 99–105. https://doi.org/10.1086/427323. arXiv: astroph/0409028 [astro-ph]

De Boni, C., Ettori, S., Dolag, K., & Moscardini, L. (2013). Hydrodynamical simulations of galaxy clusters in dark energy cosmologies II. c-M relation. *MNRAS, 428*(4), 2921–2938. https://doi.org/10.1093/mnras/sts235. arXiv: 1205.3163 [astro-ph.CO]

Desprez, G., Richard, J., Jauzac, M., Martinez, J., Siana, B., & Clément, B. (2018). Galaxy-galaxy lensing in the outskirts of CLASH clusters: Constraints on local shear and testing mass-luminosity scaling relation. *MNRAS, 479*(2), 2630–2648. https://doi.org/10.1093/mnras/sty1666. arXiv: 1806.08120 [astro-ph.GA]

Diemer B., & Kravtsov A. V. (2015). A universal model for Halo concentrations. *ApJ, 799*(1), 108. https://doi.org/10.1088/0004637X/799/1/108. arXiv: 1407.4730 [astro-ph.CO]

Dolag, K., Bartelmann, M., Perrotta, F., Baccigalupi, C., Moscardini, L., Meneghetti, M., & Tormen, G. (2004). Numerical study of halo concentrations in dark-energy cosmologies. *A & A, 416*, 853–864. https://doi.org/10.1051/00046361:20031757. arXiv: astroph/0309771 [astro-ph]

Duffy A. R., Schaye, J., Kay S. T., & Dalla Vecchia, C. (2008). Dark matter halo concentrations in the Wilkinson Microwave Anisotropy Probe year 5 cosmology. *MNRAS, 390*(1), L64–L68. https://doi.org/10.1111/j.17453933.2008.00537.x. arXiv: 0804.2486 [astro-ph]

Dutton, A. A., & Macciò, A. V. (2014). Cold dark matter haloes in the Planck era: Evolution of structural parameters for Einasto and NFW profiles. *MNRAS, 441*(4), 3359–3374. https://doi.org/10.1093/mnras/stu742. arXiv: 1402.7073 [astro-ph.CO]

Eliasdóttir Á., Limousin, M., Richard, J., Hjorth, J., Kneib, J.-P., Natarajan, P., & Paraficz, D. (2007). Where is the matter in the merging cluster Abell 2218? ArXiv e-prints. arXiv: 0710.5636

Fassnacht, C. D., Gal, R. R., Lubin, L. M., McKean, J. P., Squires, G. K., & Readhead, A. C. S. (2006). Mass along the line of sight to the gravitational lens B1608+656: Galaxy groups and implications for H0. *ApJ, 642*(1), 30–38. https://doi.org/10.1086/500927. arXiv: astroph/0510728 [astro-ph]

Faure, C., Kneib, J.-P., Hilbert, S., Massey R., Covone, G., Finoguenov A., & Koekemoer A. M. (2009). On the contribution of large-scale structure to strong gravitational lensing. *ApJ, 695*(2), 1233–1243. https://doi.org/10.1088/0004637X/695/2/1233. arXiv: 0810.4838 [astro-ph]

Golse, G., & Kneib, J.-P. (2002). Pseudo elliptical lensing mass model: Application to the NFW mass distribution. *A & A, 390*, 821–827. https://doi.org/10.1051/0046361:20020639. arXiv: astro ph/0112138 [astro-ph]

Inoue, K. T. (2016). On the origin of the flux ratio anomaly in quadruple lens systems. *MNRAS, 461*(1), 164–175. https://doi.org/10.1093/mnras/stw1270. arXiv: 1601.04414 [astro-ph.CO]

Kassiola, A., & Kovner I. (1993). Elliptic mass distributions versus elliptic potentials in gravitational lenses. *ApJ, 417*, 450. https://doi.org/10.1086/173325

Kassiola, A., Kovner I., & Fort, B. (1992). Perturbations of cluster cusps by galaxies: The triple arc in CL 0024+1654. *ApJ, 400*, 41. https://doi.org/10.1086/171971

Keeton, C. R. (2001). A catalog of mass models for gravitational lensing. arXiv e-prints astro-ph/0102341. arXiv: astroph/0102341 [astro-ph]

Kochanek, C. S. (1991). The implications of lenses for galaxy structure. *ApJ, 373*, 354. https://doi.org/10.1086/170057

Kormann, R., Schneider P., & Bartelmann, M. (1994). Isothermal elliptical gravitational lens models. *A & A, 284*, 285–299.

Limousin, M., Kneib, J.-P., & Natarajan, P. (2005). Constraining the mass distribution of galaxies using galaxy-galaxy lensing in clusters and in the field. *MNRAS, 356*, 309–322. https://doi.org/10.1111/j.13652966.2004.08449.x. eprint: astroph/0405607

Ludlow A. D., Navarro, J. F., Angulo, R. E., Boylan-Kolchin, M., Springel, V., Frenk, C., & White, S. D. M. (2014). The mass-concentration-redshift relation of cold dark matter haloes. *MNRAS, 441*(1), 378–388. https://doi.org/10.1093/mnras/stu483. arXiv: 1312.0945 [astro-ph.CO]

Mao, S., & Schneider P. (1998). Evidence for substructure in lens galaxies? *MNRAS, 295*(3), 587–594. https://doi.org/10.1046/j.13658711.1998.01319.x. arXiv: astroph/9707187 [astro-ph]

Meneghetti, M., Argazzi, R., Pace, F., Moscardini, L., Dolag, K., Bartelmann, M., Oguri, M. (2007). Arc sensitivity to cluster ellipticity asymmetries, and substructures. *A & A, 461*(1), 25–38. https://doi.org/10.1051/00046361:20065722. arXiv: astroph/0606006 [astro-ph]

Meneghetti, M., Bartelmann, M., & Moscardini, L. (2003). Cluster cross-sections for strong lensing: Analytic and numerical lens models. *MNRAS, 340*(1), 105–114. https://doi.org/10.1046/j.1365-8711.2003.06276.x. arXiv: astroph/0201501 [astro-ph]

Meneghetti, M., Fedeli, C., Pace, F., Gottlöber S., & Yepes, G. (2010). Strong lensing in the MARENOSTRUM UNIVERSE. I. Biases in the cluster lens population. *A & A, 519*, A90. https://doi.org/10.1051/00046361/201014098. arXiv: 1003.4544 [astro-ph.CO]

Meneghetti, M., Natarajan, P., Coe, D., Contini, E., De Lucia, G., Giocoli, C., & Zitrin, A. (2017). The Frontier Fields lens modelling comparison project. *MNRAS, 472*(3), 3177–3216. https://doi.org/10.1093/mnras/stx2064. arXiv: 1606.04548 [astro-ph.CO]

Meneghetti, M., Rasia, E., Vega, J., Merten, J., Postman, M., Yepes, G., & Zitrin, A. (2014). The MUSIC of CLASH: Predictions on the concentration-mass relation. *ApJ, 797*(1), 34. https://doi.org/10.1088/0004637X/797/1/34. arXiv: 1404.1384 [astro-ph.CO]

Metcalf, R. B., & Madau, P. (2001). Compound gravitational lensing as a probe of dark matter substructure within galaxy halos. *ApJ, 563*(1), 9–20. https://doi.org/10.1086/323695. arXiv: astro ph/0108224 [astro-ph]

Metcalf, R. B., & Zhao, H. (2002). Flux ratios as a probe of dark substructures in quadruple-image gravitational lenses. *ApJL, 567*(1), L5–L8. https://doi.org/10.1086/339798. arXiv: astroph/0111427 [astro-ph]

Navarro, J. F., Frenk, C. S., & White, S. D. M. (1997). A Universal density profile from Hierar chical clustering. *ApJ, 490*(2), 493–508. https://doi.org/10.1086/304888. arXiv: astro ph/9611107 [astro-ph]

Oguri, M., Schrabback, T., Jullo, E., Ota, N., Kochanek, C. S., Dai, X., & Fohlmeister J. (2013). The Hidden Fortress: structure and substructure of the complex strong lensing cluster SDSS J1029+2623. *MNRAS, 429*(1), 482–493. https://doi.org/10.1093/mnras/sts351. arXiv: 1209 458 [astro-ph.CO]

Puchwein, E., & Hilbert, S. (2009). Cluster strong lensing in the Millennium simulation: The effect of galaxies and structures along the line-of-sight. *MNRAS, 398*(3), 1298–1308. https://doi.org/10.1111/j.13652966.2009.15227.x. arXiv: 0904.0253 [astro-ph.CO]

Rivera-Thorsen, T. E., Dahle, H., Chisholm, J., Florian, M. K., Gronke, M., Rigby J. R., & Bayliss, M. (2019). Gravitational lensing reveals ionizing ultraviolet photons escaping from a distant galaxy. *Science, 366*(6466), 738–741. https://doi.org/10.1126/science.aaw0978. arXiv: 1904.08186 [astro-ph.GA]

Rusu, C. E., Fassnacht, C. D., Sluse, D., Hilbert, S., Wong, K. C., Huang, K-H., & Koopmans, L. V. E. (2017). H0LiCOW III. Quantifying the effect of mass along the line of sight to the gravitational lens HE 0435-1223 through weighted galaxy counts*. *MNRAS, 467*(4), 4220–4242. https://doi.org/10.1093/mnras/stx285. arXiv: 1607.01047 [astro-ph.GA]

Schneider P., Ehlers, J., & Falco, E. E. (1992). *Gravitational lenses*. https://doi.org/10.1007/978-3-662-03758-4

Tessore, N., & Metcalf, R. B. (2015). The elliptical power law profile lens. *A & A, 580*, A79. https://doi.org/10.1051/00046361/201526773. arXiv: 1507.01819

Torri, E., Meneghetti, M., Bartelmann, M., Moscardini, L., Rasia, E., & Tormen, G. (2004). The impact of cluster mergers on arc statistics. *MNRAS, 349*(2), 476–490. https://doi.org/10.1111/j.1365-2966.2004.07508.x. arXiv: astroph/0310898 [astro-ph]

Wright, C. O., & Brainerd, T. G. (2000). Gravitational lensing by NFW halos. *ApJ, 534*(1), 34–40. https://doi.org/10.1086/308744

Xu, D. D., Mao, S., Cooper A. P., Gao, L., Frenk, C. S., Angulo, R. E., & Helly J. (2012). On the effects of line-of-sight structures on lensing flux-ratio anomalies in a Λ CDM universe. *MNRAS, 421*(3), 2553–2567. https://doi.org/10.1111/j13652966.2012.20484.x. arXiv: 1110.1185 [astro-ph.CO]

Xu, D., Sluse, D., Gao, L., Wang, J., Frenk, C., Mao, S., & Springel, V (2015). How well can cold dark matter substructures account for the observed radio flux-ratio anomalies. *MNRAS, 447*(4), 3189–3206. https://doi.org/10.1093/mnras/stu2673. arXiv: 1410.3282 [astro-ph.CO]

Lensing by Galaxies and Clusters

6

Since the first observations of the lensing phenomena in the extra-galactic sky, which happened in the 1980s (see Chap. 1), gravitational lensing became one of the most powerful tools to investigate the matter distribution within galaxies and galaxy clusters. Several authors developed techniques to produce mass models of the lenses using strong and weak lensing observations. This chapter primarily focuses on mass modeling. We will review the fundamental ideas motivating the most common modeling strategies, and we will illustrate and discuss them through examples. We also discuss the problem of finding strong lenses. Finally, we provide a short review of the possible usages of mass models in several applications ranging from understanding the nature of dark matter to measuring cosmological parameters to using galaxies and clusters as cosmic telescopes to investigate the distant universe.

6.1 Strong Lensing by Galaxies and Galaxy Clusters

6.1.1 Scale of the Lensing Events

As extensively discussed in Chap. 5, strong lensing effects occur near the lens critical lines. These lines separate the multiple images of strongly lensed sources. Gravitational arcs form as the result of merging of multiple images across the critical lines, originated by extended sources overlapping the caustics.

It is natural to use the size of the critical lines to describe the scale of a strong lens. In the case of an axially symmetric lens, we quantify the size of the tangential critical line with the Einstein radius (see Eq. 5.33). However, both galaxies and galaxy clusters are not axially symmetric. Galaxies, in particular the early-type ones, are reasonably well described by elliptical mass distributions. Galaxy clusters are

© Springer Nature Switzerland AG 2021   255
M. Meneghetti, *Introduction to Gravitational Lensing*, Lecture Notes
in Physics 956, https://doi.org/10.1007/978-3-030-73582-1_6

the largest gravitationally bound objects in the universe. Therefore, they are also the youngest cosmic structures, often observed while still forming. Consequently, their mass distributions are often highly asymmetric and multi-modal. The corresponding critical lines are elongated, irregular, and far from being circular.

Regardless of the complexity of the lens mass distribution, we can still use the Einstein radius to quantify the size of a strong lens. In particular, we can define the equivalent Einstein radius as the radius of the circle with the same area A enclosed by the lens critical line,

$$\theta_{E,eq.} = \sqrt{\frac{A}{\pi}} \,. \tag{6.1}$$

For galaxies with masses $\sim 10^{11}$–10^{12} $M_\odot$, assuming typical redshifts of lenses and sources, the size of the Einstein radius is of the order of $\sim 1''$. Galaxy clusters as massive as 10^{14}–10^{15} $M_\odot$ typically have Einstein radii of the order of tens of arcseconds.

6.1.2 Strong Lensing Cross-Section

In the case of microlensing, we defined the cross-section as the area enclosed by the lens Einstein ring (see Sect. 4.1.6). This definition of the cross-section for multiple images would apply to galaxies and galaxy clusters if they were SISs. In that case, multiple images could form if sources were inside the cut, which is the Einstein ring projected onto the source plane. More generally, the strong lensing cross-section for multiple images of galaxies and galaxy clusters is the area enclosed by the lens caustics. This definition is exact in the case of point sources. Extended sources produce multiple images also if they are partially contained within the caustics.

As an order of magnitude, the cross-section for multiple images is up to few sq. arcseconds for galaxies and several hundreds of sq. arcseconds for galaxy clusters. Because of the small cross-section, strong lensing by galaxies is rare. To date, only a few hundreds of strong lensing galaxies are known. Besides, galaxies usually lens only one source, although there are exceptions (see, e.g. Gavazzi et al. 2008). Galaxy clusters are rarer objects than galaxies, but their strong lensing cross-section is significantly larger. Therefore, it is frequent to observe several families of multiple images originated by different sources in the core of massive galaxy clusters.

6.1.3 The Quest for Strong Lensing Galaxies

While the first discoveries of gravitational lenses were serendipitous, starting from the 1990s, several strategies were employed to find strong lenses more systematically. We may divide these search methods into two broad categories, namely the source-oriented and the lens-oriented methods.

In the first case, the targets are sources that are likely lensed. An example of a lens survey that employed this kind of strategy is the "Cosmic Lens All-Sky Survey" (Myers et al. 2003), which searched for gravitationally lensed compact radio sources. The survey focused on over 10^4 radio sources selected from snapshot observations with the Very Large Array (VLA). These sources were chosen because of their flux density above $30\,\mu$Jy at 5 GHz and their flat radio spectra, i.e. with power index $\gtrsim 0.5$ between 1.4 and 5 GHz. These flat-spectrum radio sources should appear point-like in low-resolution observations such as those with the VLA. Instead, if they appear as extended or with multiple components, there is a high chance that they are strongly lensed by some foreground galaxy. In CLASS's procedure, those sources that exhibit these characteristics were re-observed with the Multi-Element Radio Linked Interferometer Network (MERLIN), whose higher spatial resolution allowed resolving the most extended emission permitting to confirm the lensing origin. Those candidates that required even higher spatial resolution were followed-up with the Very Long Baseline Array (VLBA). This instrument allows us to detect substructure in the source components and test the lensing hypothesis by checking if the substructure has similar morphology in all the components. The yield of strong lenses from the CLASS survey consists of 22 lenses. Among them, 12 doubles, nine quadruples, and one sextuple source.

Another example of lens search employing a source-oriented strategy was performed at sub-mm wavelengths by Negrello et al. (2010) (see also Negrello et al. 2017), in the framework of the Herschel Astrophysical Terahertz Large Area Survey (H-ATLAS; Eales et al. 2010). In this case, the so-called magnification bias effect identifies dusty star-forming galaxies (DSFGs) lensed by foreground galaxies. Let us consider a population of sources whose cumulative number density as a function of flux can be approximated with a power-law so that, in the absence of lensing, the cumulative source number density is

$$n_0(> F) = QF^{-\delta} , \tag{6.2}$$

where Q is a proportionality constant, and $\delta > 0$ may depend on flux. Because of lensing, the observed cumulative source number density changes in two ways. On the one hand, lensing magnifies the source fluxes by a factor μ. This effect increases the number density of sources above a given flux, being $n_0(>F/\mu) > n_0(>F)$. On the other hand, lensing also magnifies the space between the sources. Therefore, the observed number density decreases by a factor of μ because of this effect. Accounting for both effects, the observed cumulative number density of sources above the flux F is

$$n(> F) = \frac{Q(F/\mu)^{-\delta}}{\mu} = n_0(> F)\mu^{\delta-1} . \tag{6.3}$$

Thus, which one of the two effects wins depends on the value of δ. Since $\mu \gtrsim 1$, if $\delta > 1$, the source number counts increase. On the contrary, if $\delta < 1$, they decrease. Fig. 6.1 illustrates how this magnification bias effect can be used to find strong lenses. The unlensed number counts of DSFGs is given by the black solid line (see Glenn et al. 2010). The curve steepens with flux. As a result, the number density of DSFGs drops off very rapidly at fluxes above $100\,m$Jy. The purple data-points show the observed number density of such sources in the H-ATLAS survey. These are consistent with the expectations in the case of lensing magnification (orange shaded area). Lensing wins at large fluxes. In particular, all sources with flux $> 100\,m$Jy are likely to be lensed sources. Using this method, Negrello et al. (2017) identified 80 candidate lensed galaxies over 600 squared degrees.

An example of a lens survey that employed a more lens-oriented search strategy is the Sloan Lens ACS survey (SLACS) (Bolton et al. 2006). In this case, the lens candidates are selected from the Sloan Digital Sky Survey (SDSS). They are sources whose spectra show evidence of the presence of more than one galaxy along the

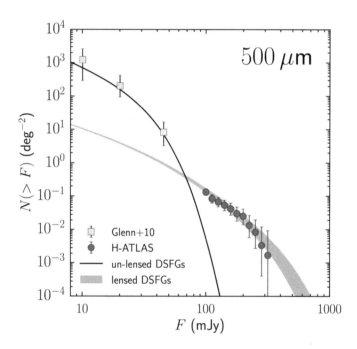

Fig. 6.1 The magnification bias effect on the number density of dusty star-forming galaxies at $500\,\mu$m in the H-ATLAS survey. The black line shows the unlensed cumulative number density as a function of flux. The purple data-points show the H-ATLAS survey measurements, consistent with the lensed cumulative number density function shown by the orange shaded area. Courtesy of M. Negrello

line-of-sight. For example, an Early-Type Galaxy (ETG) should have the typical spectrum of a passive source, without strong emission lines. When these are found in such spectra, they signal a star-forming galaxy along the line-of-sight. If this is at redshift larger than the ETG, it is likely to be a lensed source. The SLACS targets are re-observed with the Advanced Camera for Surveys (ACS) aboard the Hubble Space Telescope. Thanks to the higher spatial resolution of these observations, the presence of multiple images and arcs can be detected, in particular, if multi-band imaging is available. While the passive ETGs have reddish colors, the background star-forming galaxies appear bluer, thus easily identifiable. SLACS has identified nearly 150 lenses and lens candidates (Auger et al. 2009, Brewer et al. 2012, Brownstein et al. 2012, Shu et al. 2017, Treu et al. 2011). These are typically quite massive galaxies, and 80% of them have elliptical morphology. The remainder 20% are equally split between lenticulars and spirals. A subsample of these lenses is shown in Fig. 6.2.

Other methods to find lenses include using arc or ring finders, i.e. software that automatically searches for elongated blue features around red galaxies (e.g. Alard 2006, Cabanac et al. 2007, Gavazzi et al. 2014, Maturi et al. 2014, Seidel & Bartelmann 2007, Sonnenfeld et al. 2018). These methods have been employed successfully in surveys like the Strong Lensing Legacy Survey (Seidel & Bartelmann 2007) or the Survey of Gravitationally lensed Objects in Hyper-Suprime-Cam Imaging (SuGOHI) (Sonnenfeld et al. 2018), delivering several tens of lenses and candidate lenses.

Most recently, the advent of wide galaxy surveys covering thousands of square degrees of the sky at a good depth, like the Kilo-Degree-Survey (KiDS) (de Jong et al. 2017) and the Dark-Energy-Survey (DES) (The Dark Energy Survey Collaboration 2005), propelled the interest for using Artificial Intelligence in the searches of gravitational lenses. In supervised methods, the algorithms are trained (mainly with realistic simulations of gravitational lenses) to recognize the lenses among various galaxy morphologies using specific features. Deep-learning algorithms like Convolutional-Neural-Networks (Lecun et al. 2015) are capable of learning these informative features from the examples provided during the training phase. Examples of the application of these techniques in recent wide-field surveys are given in Canameras et al. 2020, Huang et al. 2020, Jacobs et al. 2019; 2017, Lanusse et al. 2018, Petrillo et al. 2017, Pourrahmani et al. 2018. For a comparison between the algorithms and their performances, we refer the reader to Metcalf et al. 2019. These algorithms will play a crucial role in finding gravitational lenses in future surveys like those which will be operated with the Large-Synoptic-Survey-Telescope (recently renamed the Rubin Observatory LSST Science Collaboration et al. 2009), and with the *Euclid* and *Nancy Grace Roman* space telescopes (Laureijs et al. 2011, Spergel et al. 2015). These surveys are expected to increase the number of known lenses on galaxy scales by several magnitude orders.

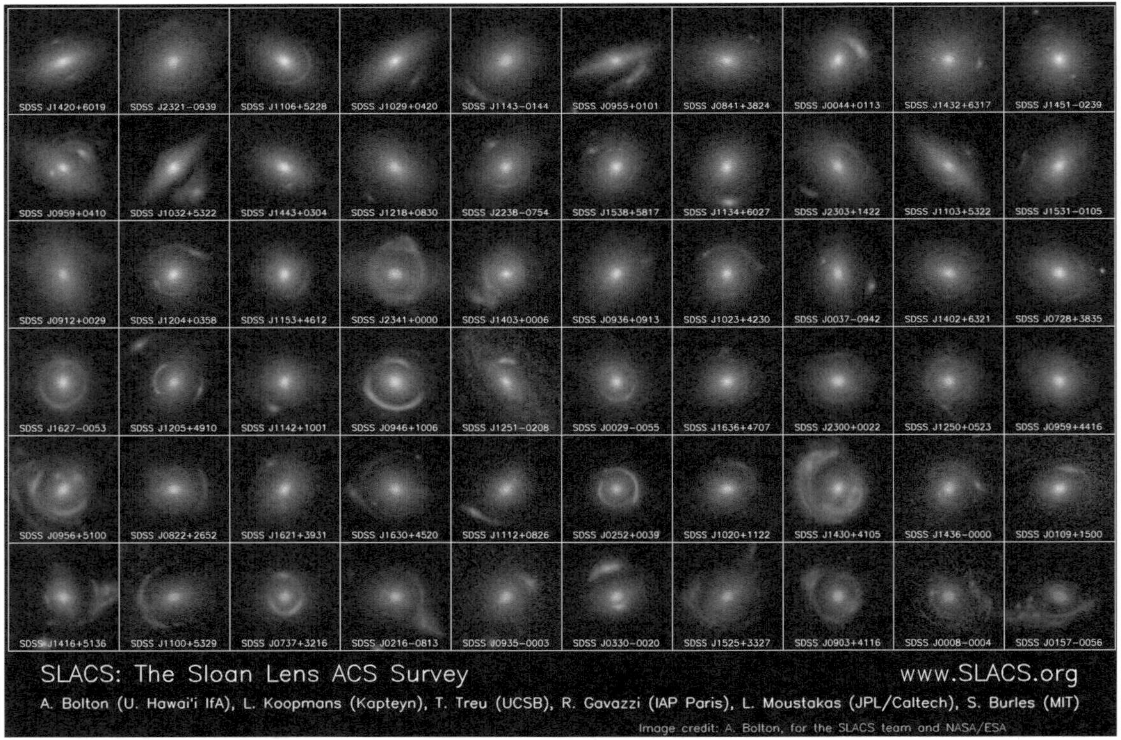

Fig. 6.2 A color-enhanced mosaic of Hubble Space Telescope images of 60 gravitational-lens galaxies discovered by the SLACS survey. In each case, the massive foreground galaxy is seen in yellow to red, and the distorted features of the more distant background galaxy are seen in blue. The images are arranged from the upper left in order of increasing distance of the foreground galaxy from Earth. Credit: A. Bolton (UH/IfA) for SLACS and NASA/ESA

6.1.4 Strong Lensing by Galaxy Clusters

The phenomenology of strong gravitational lensing effects in galaxy clusters is very rich due to their complex mass distributions. As an example, Fig. 6.3 shows the core of the galaxy cluster MACS J1206.2-0847 at $z = 0.439$. This cluster was observed in the framework of the Cluster Lensing and Supernova Survey with Hubble (CLASH, Postman et al. 2012). The most prominent strong lensing features observable are giant tangential and radial arcs, which extend for tens of arcseconds in the sky. A single strong lensing cluster can simultaneously lens many distant sources so that hundreds of multiple images can be discovered in cluster fields. Bergamini et al. (2019) identified and confirmed spectroscopically 82 multiple images of 27 galaxies in the redshift range $z_L \sim 1–7$ in the field of MACSJ1206 (see also Bonamigo et al. 2018, Caminha et al. 2017). They are indicated with circles in Fig. 6.3.

Besides, individual cluster galaxies can act as strong lenses themselves and lens background sources producing multiple images around secondary, smaller critical lines (see, e.g., the bottom-left panel in Fig. 6.3). These galaxy–galaxy strong lensing events are more abundant in clusters than in the field, in part because of the higher background density provided by the cluster environment (Meneghetti et al. 2020).

The image multiplicity of a source depends on its location with respect to the caustics. Lenses consisting of multiple mass components, such as galaxy clusters, can have very complex caustic structures, where the caustics of the individual mass component can overlap or merge. As a result, a single source, or parts of it, can be imaged more than five times (the maximum number of images that an elliptical lens can produce). A very peculiar example of extreme image multiplicity is shown in Fig. 6.4. This cluster at redshift $z = 0.443$, whose name is PSZ1 G311.65-18.48, exhibits a magnificent system of giant tangential arcs, dubbed the *Sunburst arc* (Dahle et al. 2016). These arcs are the images of a source at redshift $z = 2.369$. Rivera-Thorsen et al. (2019) identified a compact star-forming region in the source which is multiply imaged 12 times within the Sunburst arc system. This is due to several cluster galaxies (at least 5) perturbing the smooth cluster halo's critical line. The multiplicity of this bright, compact region is confirmed spectroscopically. Indeed, Rivera-Thorsen et al. (2019) detected a characteristic spectral feature signaling leakage of Lyman-continuum radiation in all these multiple images.

Thus, strong lensing by galaxy clusters can manifest in several ways, implying that it is challenging to automatize the search for these strong lensing effects all at once. Until now, some *arc finders* have been developed to automatically detect elongated features like the gravitational arcs (Alard 2006, Maturi et al. 2014, Seidel and Bartelmann 2007, Stapelberg et al. 2019, Xu et al. 2016). These most apparent features can be the starting point to find more multiple images and arcs.

Fig. 6.3 Color-enhanced image of the galaxy cluster MACS J1206.2-0847, observed with the Hubble Space Telescope. A variety of strong lensing features can be found in this cluster, namely giant tangential and radial arcs, several families of multiple images of distant sources (marked with circles), and even galaxy–galaxy strong lensing events (as shown in the bottom-left panel). Bergamini et al. (2019), Caminha et al. (2017), and Bonamigo et al. (2018) used the multiple images shown here to build the mass model of this galaxy cluster. The labels next to each circle indicate the image family and multiplicity

For example, (Carrasco et al. 2020) proposed to build a rough lens model for the cluster using the detected arcs, assuming that the light of the cluster galaxies traces the cluster mass. Then, the model can be used to guess the position of additional lensed images.

More commonly, however, searching for strong lensing features in the galaxy cluster is not an automatic procedure and involves a significant amount of human interaction. Multiple image candidates are identified gluing together several pieces of information. First of all, when multi-band imaging is available, the image colors' similarity can be used. Second, although perturbations from secondary mass components in the lens can shift the positions of the multiple images by several

Fig. 6.4 A color-enhanced image of the core of galaxy cluster PSZ1 G311.65-18.48, observed with the Hubble Space Telescope. A unique system of gravitational arcs, dubbed the *Sunburst arc*, is visible in the core of the cluster. These arcs are the multiple images of a source at redshift $z = 2.369$. Within the arcs, a compact star-forming region in the source, violet in color, is seen 12 times, as indicated by the white labels. Credit: ESA/Hubble, NASA, Rivera-Thorsen, et al.

arcseconds and even increase the image multiplicity, generally, the geometries of the image systems are grossly consistent with those discussed in Chap. 5 for relatively simple lenses such as the elliptical lenses. Third, as shown in Sect. 5.4.1, multiple images must satisfy parity inversion rules because they form on opposite sides of the lens critical lines. Distant galaxies are often star-forming, irregular, or spiral galaxies. If observed with instruments such as the Hubble Space Telescope, features such as star-forming regions, spiral arms, and others can be easily identified within the multiple images of the same source (as shown earlier for the Sunburst arc), and their relative positions can be used to test the consistency with the parity rules mentioned above.

Once the multiple image candidates have been found, it is essential to confirm the identifications using spectroscopy. Indeed, the multiple images must share the same spectrum of the source. The unambiguous identification of spectral features like spectral lines allows us to measure the source redshift, a crucial ingredient for building the lens's mass model.

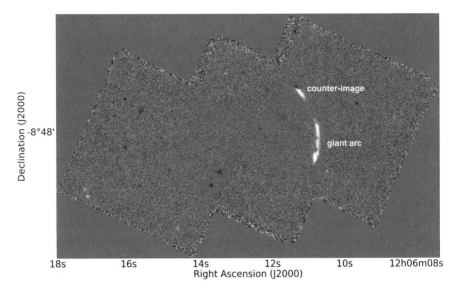

Fig. 6.5 Slice centered at $7593\,\text{Å}$ and with width $5\,\text{Å}$, extracted from the MUSE data-cube of MACS J1206.2-0847. The [OII] emission from the giant arc shown in Fig. 6.3 ($z_S = 1.0369$) is visible and isolated from the other sources. It is easy to recognize the arc counter-image northern of the arc

In recent years, the advent of instruments such as the Multi-Unit-Spectroscopic-Explorer (MUSE) spectrograph of the Very-Large-Telescope (VLT) (Bacon et al. 2010) resulted in being extremely useful for the strong lensing analyses of galaxy clusters. MUSE is an integral-field spectrograph with a field-of-view of 1 sq. arcmin, wide enough to contain the majority of strong lensing clusters' critical lines. Even in larger lenses, MUSE can cover the strong lensing region with a few pointings. A MUSE observation output is a data-cube, consisting of a stack of monochromatic images of the cluster field at different wavelengths between 4750–$9350\,\text{Å}$. Thus, this instrument can be used to extract spectra of all sources (cluster members, foreground and background galaxies, multiple images) in the field-of-view. The instrument is handy to identify the multiple images of line-emitting sources, which can be easily recognized in the continuum subtracted slice of the data-cube centered on the line wavelength. For example, Fig. 6.5 shows a slice centered at $7593\,\text{Å}$ and with width $5\,\text{Å}$, extracted from the MUSE data-cube of MACS J1206.2-0847. The source visible is the one that appears as the giant tangential arc shown in Fig. 6.3. Northern of the arc, a counter-image is also visible at this wavelength, while all the other sources, including the cluster members, do not show up in the continuum subtracted slice. This radiation corresponds to the [OII] emission of a source at $z = 1.0369$. While this arc is also visible in the HST observations, other lensed line-emitting sources are sometimes identified in the MUSE data-cube without any optical counter-part (see e.g. Caminha et al. 2017).

6.1.5   Lens Inversion

One of the primary applications of gravitational lensing is to reconstruct the lens mass distribution. Three types of constraints can be used for this purpose:

1. The positions of the multiple images of lensed source probe the lens's deflection field, i.e. the first derivatives of the lensing potential. These are the easiest constraints to obtain, provided that deep, high-resolution imaging data is available. The current state-of-the-art instrument to observe strong lenses at optical and near-infrared wavelengths is the Hubble Space Telescope, thanks to the sensitivity and resolution of its Advanced Camera for Surveys (ACS) and Wide-Field-Camera3 (WFC3). Alternatively, imaging from the ground via adaptive optics (AO) also delivers the required high-resolution (Chen et al. 2019; 2016). AO is a technology used to correct the wavefront distortions by the atmosphere (Rousset et al. 1990), allowing to obtain images that have nearly diffraction-limited point-spread-function (PSF). Since the distortions mentioned above are evaluated in real time from the images of nearby stars, this technology's usage is currently limited to few galaxy scale systems and is not yet viable on the scale of galaxy clusters. This limit will be eventually overcome soon with the advent of the new generation of *Multi-Conjugate-Adaptive-Optics* systems (Rigaut & Neichel 2020).

2. The fluxes (magnification) and the shapes of the multiple images and gravitational arcs probe the higher-order (primarily second) derivatives of the lensing potential. For this reason, these constraints are also sensitive to the smaller-scale mass components of the lenses. For example, the curvature of the giant tangential arc in Fig. 6.3 reflects the presence of a large dark matter halo centered on the Brightest-Central-Galaxy (BCG) of MACS J1206.2-0847. Besides, the southernmost portion of the arc is also distorted by one of the cluster galaxies. Similarly, substructures in galaxies perturb the images of extended sources, as discussed in Sect. 5.7. Effects of substructures of mass $\sim 10^8$ $M_\odot$, or smaller perhaps, can be detected in very high-resolution observations that are achievable today with interferometry at cm- and mm-wavelengths (Hezaveh et al. 2016, Powell et al. 2020, Spingola et al. 2018). However, this requires follow-up of the lenses with substantial telescope time.

3. The relative time delays between multiple images. As discussed in Sects. 3.6.1 and 5.8, the time delays probe the lensing potential. Unfortunately, time delays are available only for a few tens of lenses because they can only be measured if the lensed sources are intrinsically variable (i.e. Refsdal 1964). These sources are rare: either a quasar or a supernova explosion. Besides, time delays are difficult to measure, as they require dedicated telescope time for continuous monitoring of the sources for long periods and accurate photometry. The source must show noticeable brightness fluctuations on time scales shorter than the monitoring period. Many quasars have little variability, while supernovae explosions are unpredictable. An alert system must be in place to trigger their follow-up after their discovery. Once the light-curves of the individual images are acquired, they

must be cross-correlated to measure the time delay. This is difficult because systematic errors in the photometry or microlensing of individual images can introduce uncorrelated variabilities between the images (Courbin et al. 2011, Eigenbrod et al. 2005, Kochanek et al. 2006, Tewes et al. 2013).

The process of converting the observed strong lensing constraints into matter distributions is called *lens inversion*. There are two general classes of inversion algorithms, namely the so-called *parametric* and *free-form* (a.k.a. non-parametric) algorithms, which are discussed in detail in the following sections.

Parametric Reconstruction Algorithms

In *parametric* lens modeling, the mass distribution is reconstructed by combining one or more clumps of matter, often positioned assuming that the light traces the mass. For example, a galaxy cluster is described by an ensemble of mass clumps attached to each cluster galaxy, in addition to one or more smooth mass halos connected to the BCGs. Each mass clump is characterized by an ensemble of parameters describing its density profile and shape. Some of the most widely used models to describe these mass distributions were presented in Chap. 5. These models' parameter space is explored to find the best combination reproducing the observed positions, shapes, magnitudes, and relative time delays of the multiple images and arcs.

First, we consider a strong lensing event like the one in Fig. 6.6. This simulated image shows an early-type elliptical galaxy, which strongly lenses a background quasar and its host galaxy. We can quickly identify four multiple images of the quasar (indicated with yellow filled circles). Their positions, $\vec{\theta}_i$ with $1 \leq i \leq 4$, can be used to constrain the lens's mass distribution. For this purpose, we assume to know the source and the lens redshifts.

Following the parametric approach, we assume that the lens can be fitted using a single mass clump, described by a model where the lens surface density is a function of n parameters $\vec{p} = [p_1, \ldots, p_n]$. For example, if we decide to fit the lens with a SIE model (see Sect. 5.4.1), the parameters could be the axis ratio, the orientation (i.e. the angle between the major axis of the lens and the vertical axis of the image), and the velocity dispersion σ_v. The lens's center could be fixed at the center of the lens galaxy (thus assuming that the lens light traces the lens mass) or let free of varying.

The procedure to fit the lens consists of finding the best combination of parameters that reproduces the observed positions of the multiple images of the quasar. This can be implemented in several ways. For example, we consider the so-called *lens plane optimization*, which involves the following steps:

1. Given a set of lens parameters, we calculate the lens deflection angle at the observed positions of the quasar images, $\vec{\alpha}(\vec{\theta}_i | \vec{p})$.
2. Using the lens equation, we map these positions on the source plane:

$$\vec{\beta}_i(\vec{p}) = \vec{\theta}_i - \vec{\alpha}(\vec{\theta}_i | \vec{p}) . \tag{6.4}$$

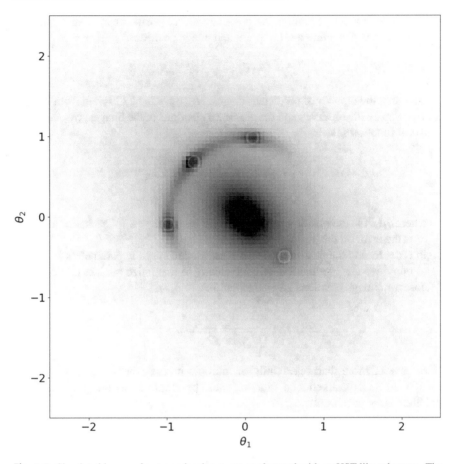

Fig. 6.6 Simulated image of a strong lensing event as observed with an HST-like telescope. The pixel scale is 0.03″, and the image's side-length is 5″. The observing band is not relevant in this example. The event involves an early-type galaxy lens and an elliptical background source hosting a quasar at its center. The simulated observation includes the instrument point-spread-function (see Sect. 6.2.6), here assumed to be of Gaussian shape and with a Full-Width-at-Half-Maximum (FWHM) of 0.1″

3. The resulting points can be used to find the predicted source position. The points $\vec{\beta}_i(\vec{p})$ will be spread over a region of the source plane, so we can estimate the source position by taking their mean position:

$$\vec{\beta}(\vec{p}) = \frac{\sum_{i=1}^{i=N_{ima}} \vec{\beta}_i(\vec{p})}{N_{ima}}, \tag{6.5}$$

where $N_{ima} = 4$ is the number of observed images of the quasar.

4. By solving the lens equation, the predicted source position can be mapped back onto the lens plane into a set of predicted image positions,

$$\vec{\beta}(\vec{p}) \rightarrow \{\vec{\theta}_i(\vec{p}) \; ; i = 1, \ldots, N_{ima,\vec{p}}\} \, . \tag{6.6}$$

Note that the number of predicted images, $N_{ima,\vec{p}}$, can be different from N_{ima}.
5. To compare the observed and the predicted positions of the images, we can define a cost function such as

$$\chi^2(\vec{p}) = \sum_{i=1}^{N_{ima}} \frac{[\vec{\theta}_i - \vec{\theta}_i(\vec{p})]^2}{\sigma_i^2} \, , \tag{6.7}$$

where $\vec{\theta}_i(\vec{p})$ is the position of the predicted image closest to the image at $\vec{\theta}_i$ and σ_i is the error on the position of the $i-$th image.
6. In order to find the best-fit model of lens, we can vary the parameters $\vec{p}$ such as to minimize the χ^2 function, or equivalently to maximize the likelihood for the observed image positions given the parameters $\vec{p}$,

$$\mathcal{L}(\vec{p}) = \frac{1}{\Pi_{i=1}^{N_{ima}} \sigma_i \sqrt{2\pi}} \exp{-\frac{\chi^2(\vec{p})}{2}} \, . \tag{6.8}$$

In case of more than one family of multiple images, the same procedure can be applied, but the likelihood function must be redefined as the product of the likelihoods for each family:

$$\mathcal{L}(\vec{p}) = \Pi_{j=1}^{N_{fam}} \frac{1}{\Pi_{i=1}^{N_{ima}} \sigma_{ji} \sqrt{2\pi}} \exp{-\frac{\chi_j^2(\vec{p})}{2}} \, , \tag{6.9}$$

where χ_j^2 is defined as in Eq. 6.7, but refers to the $j-$th family of multiple images and σ_{ji} is the error on the position of the $i-$th image of the $j-$th family. Note that maximizing this likelihood is equivalent to minimizing the total χ^2 defined as

$$\chi_{tot}^2(\vec{p}) = \sum_{j=1}^{N_{fam}} \chi_j^2(\vec{p}) \, . \tag{6.10}$$

An alternative to the lens plane optimization is given by the *source plane optimization*. Ideally, the multiple images should be mapped to the same position on the source plane, if the model parameters match exactly the lens parameters. Thus, we could define a χ^2 on the source plane as

$$\chi_s^2(\vec{p}) = \sum_{i=1}^{N_{ima}} \frac{[\vec{\beta}_i(\vec{p}) - \vec{\beta}(\vec{p})]^2}{\mu_i(\vec{p})^{-2}\sigma_i^2} \, . \tag{6.11}$$

The major advantage of this approach is that step 4 in the procedure outlined above can be avoided. The solution of the lens equation is possible using numerical methods and is computationally expensive. When many families of multiple images are used as constraints, the lens plane optimization is significantly slower than the source plane optimization. Besides, it is not necessary to match the observed, and the model-predicted images one by one when computing the χ^2. One disadvantage is that the error on each $\beta_i(\vec{p})$ must be estimated using the model estimated magnification $\mu_i(\vec{p}) = \mu(\vec{\theta}_i|\vec{p})$ at the image position. Besides, source plane optimization can lead to biased solutions, favoring models with flat density profile and large ellipticity (Kneib & Natarajan 2011).

The total χ^2 function can be further extended to include additional terms, if other constraints are available. For example, to include the constraints from fluxes (or flux ratios) and from relative time delays between the multiple images, in addition to those from the image positions, we can define the total χ^2 as

$$\chi^2_{tot}(\vec{p}) = \chi^2_{pos}(\vec{p}) + \chi^2_{flux}(\vec{p}) + \chi^2_{time}(\vec{p}) , \tag{6.12}$$

where $\chi^2_{pos}(\vec{p})$ was defined in Eqs. 6.7 and 6.10. The other two terms may be defined in a similar way. For a single source, the contribution to the total χ^2 from fluxes is

$$\chi^2_{flux}(\vec{p}) = \sum_{i=1}^{N_{ima}} \frac{[\mu_i(\vec{p})F_s - F_i]^2}{\sigma^2_{F_i}} , \tag{6.13}$$

where F_i, and σ_{F_i} are the observed flux and the corresponding error of the $i-$th image, respectively. The intrinsic source flux F_s is a free parameter in the fit. Alternatively, we could fit the flux ratios between the images, eliminating this parameter.

The contribution to the total χ^2 from time delays is instead

$$\chi^2_{time}(\vec{p}) = \sum_{ik} \frac{[\Delta t_{ik}(\vec{p}) - \Delta t_{ik}]^2}{\sigma^2_{\Delta t_{ik}}} , \tag{6.14}$$

where $\Delta t_{ik}(\vec{p})$ and Δt_{ik} are the model-predicted and the observed time delays between the $i-$ and the $k-$th images. Obviously, $\sigma^2_{\Delta t_{ik}}$ are the errors on the measured time delays.

Simultaneous Reconstruction of Source and Lens

So far, sources were assumed to be point-like. This approximation is often used even in the presence of extended sources. In the example shown in the previous section, the constraints were derived from the images of a quasar, regardless of its galaxy host's presence. Within very extended arcs, as shown in Fig. 6.4, one can identify bright star-forming regions and knots repeating in the multiple images and use them as point-like constraints.

Nevertheless, in several galaxy–galaxy lensing cases, the full information from extended sources can be used, and source and lens can be reconstructed simultaneously. Consider a source with intrinsic surface brightness $I_s(\vec{\beta})$. The surface brightness of its images is $I(\vec{\theta})$. We can use a set of parameters, $\vec{p}_s$, to model the surface brightness of the source, $I_s(\vec{\beta}|\vec{p}_s)$. Because of the surface brightness conservation, for a set of parameters $\vec{p}$ of the lens, we can derive the model-predicted surface brightness of the image as

$$I(\vec{\theta}|\vec{p}, \vec{p}_s) = I_s[\vec{\theta} - \vec{\alpha}(\vec{\theta}|\vec{p})|\vec{p}_s] \,. \tag{6.15}$$

We may consider the images as a collection of N_{pix} pixels. The i−th pixel provides a measure of the surface brightness, I_i, at the position $\vec{\theta}_i$. These measures can be fit into an observed data vector, $\vec{d}$. Similarly, using Eq. 6.15 we can build the model data vector $\vec{d}_{mod} = [I(\vec{\theta}_i|\vec{p}, \vec{p}_s)]$. Finally, we can define the χ^2 function

$$\chi^2 = [\vec{d} - \vec{d}_{mod}]^T C_I^{-1} [\vec{d} - \vec{d}_{mod}] \,, \tag{6.16}$$

where C_I is the covariance matrix of the pixel surface brightness.

The source can be modeled in different ways. For example, one can use an analytical function to parameterize the surface brightness and assume an elliptical shape. A possible choice could be the Sérsic surface brightness profile (see Eq. 3.100). In this case, the source parameters would be the effective radius, the Sérsic index, the total flux, the center coordinates, the ellipticity, and the position angle.

Alternatively, the source could be modeled as a collection of pixels or expanded into a set of basis functions. In these cases, each pixel or coefficient in the expansion is a free parameter that needs to be determined by minimizing the χ^2 function (Suyu et al. 2006, Vegetti & Koopmans 2009, Warren & Dye 2003) For example, in case of pixelized sources, whose surface brightness is evaluated at finite positions $\vec{\beta}_i$ with $i = [1, \ldots, N_{pix,s}]$, the model data vector can be written as the product of a lensing operator $L(\vec{p})$ of size $N_{pix} \times N_{pix,s}$ and of the source vector $\vec{s} = [I_s(\vec{\beta}_i)]$:

$$\vec{d}_{mod} = L(\vec{p})\vec{s} \,. \tag{6.17}$$

Vegetti and Koopmans (2009) showed that $L(\vec{p})$ can be written as an interpolation operator, i.e. each element $d_k = I(\vec{\theta}_k)$ of $\vec{d}_{mod}$ is obtained by performing a bi-linear interpolation on the source plane using three values of the source surface brightness evaluated at the vertices of a triangle containing the point $\vec{\beta}_k = \vec{\theta}_k - \vec{\alpha}(\vec{\theta}_k|\vec{p})$. For a given set of lens parameters, $\vec{p}$, one can find source vector $\vec{s}$ as

$$\vec{s} = L(\vec{p})^{-1}\vec{d} \,. \tag{6.18}$$

Exploring the lens parameter space, the best combination of parameters $\vec{p}$ and source $\vec{s}$ that matches the data can be found, thus obtaining the simultaneous reconstruction of the source and of the lens.

In describing this algorithm, we have ignored several problems. First of all, the images are noisy. This implies that the source reconstructed through Eq. 6.18 may present severe discontinuities and pixel-to-pixel variations, resulting in being non-physical. A prior on the source parameters must be assumed to force a smooth variation in nearby pixels' value in the source plane. This can be achieved by re-defining the χ^2 by adding a regularization term,

$$\chi^2 \rightarrow \chi^2 + \lambda \vec{s}^T H \vec{s} \,, \tag{6.19}$$

where λ is a regularization constant and H is a regularization matrix (Suyu et al. 2006).

Besides, one needs to account for the smearing effect of the PSF when computing the model data vector to be compared to the data. For this purpose, another operator B, called *blurring operator*, can be introduced, so that Eq. 6.17 is re-written as

$$\vec{d}_{mod} = B L(\vec{p}) \vec{s} \,. \tag{6.20}$$

Complex Parametric Models

Galaxy clusters and groups have complex mass distributions and are often located in dense environments, with filaments of matter departing from their outskirts and connecting to other clusters (Dietrich et al. 2012, Jauzac et al. 2016, Kondo et al. 2020, Limousin et al. 2012, Merten et al. 2011). In the parametric approach, their lens models typically contain many mass components and external perturbations, which are combined as explained in Sect. 5.7. Observations suggest that a smooth dark matter halo dominates the cluster mass budget. When building a lens model, this halo is generally modeled using one or more large-scale mass components with virial radii of order ~ 1 Mpc, typically associated with the BCGs. Multiple large-scale components are used to model merging galaxy clusters and reproduce the characteristic asymmetries and elongations of the mass distributions of clusters and groups. The presence of additional mass components in the cluster surroundings can be modeled using external perturbations (see Sect. 5.6). About 15% of the mass of clusters is in the form of hot, diffuse gas (Ettori et al. 2013, Sarazin 1988), which can also be included as a component of the lens model as shown by Bonamigo et al. (2018). Finally, as discussed in Sect. 6.1.4, the models must include a clumpy component associated with the cluster galaxies.

The lens models comprising multiple mass components and external perturbations have a larger number of free parameters. Thus, more constraints are needed to break the possible model degeneracies and robustly determine the lens mass distributions. Unfortunately, clusters contain many hundreds of galaxies in their cores. When all of them are incorporated in the lens models, the number of free parameters largely exceeds the number of constraints. To mitigate this problem, it is common practice to include only the cluster galaxies brighter than a given luminosity threshold, located within a region around the cluster center, whose size is approximately twice the size of the cluster Einstein radius. For example, Bergamini et al. (2019) only includes galaxies with magnitude in HST band F160W smaller than 24. This magnitude is a good proxy for the stellar mass. Thus, the luminosity selection is motivated by assuming that only the most massive galaxies produce non-negligible lensing effects.

Although using only a subset of cluster galaxies helps, this does not sufficiently reduce the model's number of free parameters. Apart from a small subset of galaxies located near multiple images or arcs, the rest of the galaxies can be treated as a population obeying well-motivated scaling relations.

For example, based on the virial theorem, the total mass of a galaxy can be related to the stellar velocity dispersion and to the effective radius as

$$M \propto \sigma_v^2 r_e \ . \tag{6.21}$$

The galaxy luminosity is related to the mean surface brightness inside the effective radius, I_e, as

$$L \propto I_e r_e^2 \ . \tag{6.22}$$

The mass-to-light ratio is thus

$$\left(\frac{M}{L}\right) \propto \frac{\sigma_v^2}{r_e I_e} \ . \tag{6.23}$$

Assuming that $M/L \propto L^{-\delta}$, we obtain

$$\left(\frac{M}{L}\right) \propto (I_e r_e^2)^{-\delta} \ . \tag{6.24}$$

Combining Eqs. 6.23 and 6.24, we obtain that

$$r_e \propto \sigma_v^{\frac{2}{2-\delta}} I_e^{\frac{\delta-1}{1-2\delta}} \ . \tag{6.25}$$

This relation describes the *fundamental plane of elliptical galaxies* (Binney & Tremaine 2008, Djorgovski & Davis 1987).

It is common practice to model the cluster galaxies using the dPIE model introduced in Sect. 5.5.2 (Brainerd et al. 1996, Eliasdóttir et al. 2007, Kneib & Natarajan 2011, Limousin et al. 2005, Natarajan & Kneib 1997). The shapes are often chosen to be circular or with ellipticities fixed to those of the surface brightness distributions. The parameters of the dPIE density profile are assumed to scale with luminosity as

$$\sigma_v = \sigma_{v,0} \left(\frac{L}{L_0}\right)^\alpha \ ,$$

$$r_{cut} = r_{cut,0} \left(\frac{L}{L_0}\right)^\beta \ ,$$

$$r_{core} = r_{core,0} \left(\frac{L}{L_0}\right)^\gamma \ . \tag{6.26}$$

The core radius of galaxies is known to be very small. For example, no central images are observed in galaxy scale strong lenses, implying that the central density profile is nearly isothermal (Koopmans et al. 2009, Treu 2010a). For this reason, it can be neglected or assumed to be a small constant. The total mass of the dPIE model scales with the cut radius and the velocity dispersion as in Eq. 5.122. Together with Eq. 6.26, this implies that

$$M \propto L^{2\alpha+\beta} . \tag{6.27}$$

Since

$$\frac{M}{L} \propto L^{2\alpha+\beta-1} \propto L^{-\delta} , \tag{6.28}$$

the parameters α and β of the scaling relations in Eq. 6.26 and the parameter δ of the fundamental plane are related as

$$\beta = 1 - \delta - 2\alpha , \tag{6.29}$$

showing that, if we assume that the cluster galaxies lay on the fundamental plane ($\delta \sim 0.2$; Bender et al. 1992, Faber et al. 1987), we get rid of one free parameter. The population of the cluster galaxies is then fully described by only three parameters ($\sigma_{v,0}$, $r_{cut,0}$, and one among α and β).

Free-Form Reconstruction Algorithms

In the *free-form* (a.k.a. non-parametric) modeling approach, the lens is subdivided into a structured or an unstructured mesh onto which the lensing observables are mapped. The mesh is then transformed into a pixelized mass distribution using the relations between the observables and the lens surface density.

This approach has been widely used in the literature with both galaxy- and cluster-scale lenses and implemented in very different ways (Birrer et al. 2015, Blandford et al. 2001, Bradač et al. 2005, Coe et al. 2008, Coles et al. 2014, Diego et al. 2005; 2007, Jee et al. 2007, Koopmans 2005, Liesenborgs et al. 2006, Merten 2016, Merten et al. 2009, Saha & Williams 2004, Sebesta et al. 2016, Suyu & Blandford 2006, Suyu et al. 2009). Here, we discuss a simple example to illustrate the basic principles.

Consider a lens mass distribution corresponding to a lensing potential $\Psi(\vec{\theta})$. A discrete representation of this potential is given by a vector of potential values $\vec{\Psi} = [\Psi_k]$, $k = 1, \ldots, N_{pix}$ at the positions $'\vec{\theta}_k$. The elements of this vector are the unknowns to be determined.

Consider now a family of multiple images with positions $\vec{\theta}_i$ with $i = [1, \ldots, N_{ima}]$. The corresponding positions on the source plane are

$$\vec{\beta}_i = \vec{\theta}_i - \vec{\alpha}(\vec{\theta}_i) . \tag{6.30}$$

The deflection angle is the gradient of the lensing potential (Eq. 3.9). Given that the lensing potential is defined on a grid, the components of its gradient can be estimated from a linear combination of the neighboring grid-points, via a finite-difference scheme. In matrix notation, the components of the deflection angle are

$$\alpha_1(\vec{\theta}_i) = D_{jk}^{(1)} \Psi_k \,,$$

$$\alpha_2(\vec{\theta}_i) = D_{jk}^{(2)} \Psi_k \,. \tag{6.31}$$

where D_{jk} are sparse band matrices encoding the information on the finite-difference scheme. These matrices are perfectly known (Fornberg 1988).

From now on, we can proceed as in the source plane optimization for the parametric approach. Starting from an initial guess of the Ψ_k, we can compute the mean source position for the family of multiple images as a function of the potential values Ψ_k as done in Eq. 6.5 and define the χ_s^2 as done in Eq. 6.11. The pixelized potential of the lens can then be determined by solving

$$\frac{d\chi_s^2}{d\Psi_k} = 0 \,. \tag{6.32}$$

Given that the deflection angle components are linear functions of the potential values, Eq. 6.32 reduces to a system of linear equations which can be solved to find the Ψ_k. This is done iteratively: at each iteration step, new values of Ψ_k are determined, used to compute a new mean source position, which is used to update the values of Ψ_k until convergence. In the case of many families of multiple images, the χ_s^2 can be computed by summing each family's contribution.

The advantage of the free-form approach is that no assumptions on the lensing potential are necessary. However, the number of multiple images is typically smaller than the potential values to be determined. For this reason, the introduction of a regularization term in the χ_s^2 meant to disfavor small-scale fluctuations in the potential is necessary.

Another advantage is that other complementary constraints can be easily combined with the multiple images' positions by adding other terms to the χ^2 function. For example, Merten et al. (2009) use pairs of multiple images to trace the lens critical lines. Similarly, weak lensing measurements can also be combined with strong lensing to constrain the outer region of lenses such as galaxy clusters, as it will be shown in the following sections.

6.2   Weak Lensing by Galaxy Clusters

6.2.1   The Principle

Far from the critical lines of galaxy clusters, the deflection field's spatial variations are nearly linear over the typical angular scales of background galaxies. In this limit, convergence and shear are constant, and the lensing Jacobian fully describes the lens mapping.

As discussed in Sect. 3.3, lensing causes circular sources to become ellipses with their major axes pointing in the direction of the eigenvector of the shear tensor Γ with eigenvalue γ. The semi-major and semi-minor axes of the ellipses are related to the local values of convergence and shear as in Eqs. 3.52. Defining the source ellipticity as

$$\epsilon = \frac{a-b}{a+b} \, , \tag{6.33}$$

Equation 3.53 states that

$$\epsilon = g \, , \tag{6.34}$$

where g is the *reduced shear*,

$$g \equiv \frac{\gamma}{1-\kappa} \, . \tag{6.35}$$

Being the shear Γ a 2×2 tensor, by analogy we define the reduced shear tensor $\tilde{g}$:

$$\begin{aligned}
\tilde{g} &\equiv \begin{pmatrix} g_1 & g_2 \\ g_2 & -g_1 \end{pmatrix} \\
&= \frac{1}{1-\kappa} \begin{pmatrix} \gamma_1 & \gamma_2 \\ \gamma_2 & -\gamma_1 \end{pmatrix} \, .
\end{aligned} \tag{6.36}$$

Thus, the eigenvectors of Γ are also eigenvectors of $\tilde{g}$, and the components of the reduced shear can be written as

$$g_1 = g \cos(2\varphi) \, , \tag{6.37}$$

$$g_2 = g \sin(2\varphi) \, . \tag{6.38}$$

Using the complex notation introduced in Sect. 3.5.1, we can also define the complex reduced shear and ellipticity:

$$g = g_1 + i g_2 \, , \tag{6.39}$$

$$\epsilon = \epsilon_1 + i \epsilon_2 \, . \tag{6.40}$$

Figure 6.7 illustrates how a regular grid of circular sources (shown in the left panel) is lensed by a foreground elliptical lens placed at the center of the field-of-view. The solid blue curves show the lens critical lines. Some tangential and radial

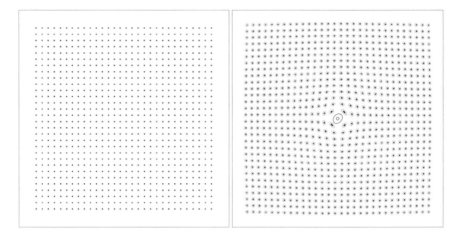

Fig. 6.7 Left panel: Regular grid of 30×30 circular sources, each modeled with a Sérsic surface brightness profile with $n = 1$ and an effective radius $r_e = 1''$, distributed over a region of side-length $600'' \times 600''$ on a plane at $z_S = 1$. Right panel: the source grid is lensed by a foreground elliptical lens, modeled as a dPIE with $\sigma_v = 1200$ km/s and $r_{core} = 2''$ at redshift $z_L = 0.5$. The lens is at the center of the field-of-view. The lens critical lines are shown in blue. The best-fit ellipses of all the sources are shown with red dotted lines enclosing the lensed images

arcs are easily recognizable around them and are the images of the sources closest to the lens center. We can see that as the distance from the lens center increases, the shapes of the lensed images become more similar to ellipses, as the limit discussed above is reached. We notice that there is a coherent alignment of the ellipses around the lens and that the ellipticity decreases as a function of the distance, reflecting the lens shear's behavior, which decreases as a function of distance from the lens.

This example and the equations above illustrate the basic idea of weak lensing. By measuring the ellipticity of the lensed images of background sources, we can infer the lenses' mass distribution. Indeed, the ellipticity measures a combination of shear and convergence, which are related to the second derivatives of the lensing potential.

6.2.2 Ellipticity Measurements

The ellipticity of the images must be measured from their surface brightness distributions. Let us consider an elliptical image. We begin by defining the image centroid, which is given by the first moment of the surface brightness $I(\vec{\theta})$:

$$\vec{\theta}_0 = \frac{\int I(\vec{\theta})\vec{\theta}d^2\theta}{\int I(\vec{\theta})d^2\theta} \ . \tag{6.41}$$

Instead, the shape of the image is determined by the quadrupole (or second) moments,

$$Q_{ij} = \frac{\int I(\vec{\theta})(\theta_i - \theta_{0i})(\theta_j - \theta_{0j})d^2\theta}{\int I(\vec{\theta})d^2\theta} \tag{6.42}$$

where $(i, j) \in \{1, 2\}$.

Note that Q_{ij} are the elements of a 2×2 symmetric tensor. Indeed, $Q_{12} = Q_{21}$. This tensor can be diagonalized and its eigenvectors define the principal axes of the surface brightness distribution. Thus, the ellipse describing the image has major and minor axes oriented as the principal axes of Q. The lengths of its major and minor semi-axes, a and b, are the reciprocals of the square root of the eigenvalues of Q, given by

$$\lambda_+ = \frac{1}{2}\left(Q_{11} + Q_{22} + \sqrt{(Q_{11} - Q_{22})^2 + 4Q_{12}^2}\right) = \frac{1}{a^2},$$

$$\lambda_- = \frac{1}{2}\left(Q_{11} + Q_{22} - \sqrt{(Q_{11} - Q_{22})^2 + 4Q_{12}^2}\right) = \frac{1}{b^2}. \tag{6.43}$$

For an image with circular isophotes, $Q_{11} = Q_{22}$ and $Q_{12} = Q_{21} = 0$.

Thus, the image ellipticity is given, in terms of the elements of the Q tensor, by

$$|\epsilon| = \frac{\sqrt{(Q_{11} - Q_{22})^2 + 4Q_{12}^2}}{Q_{11} + Q_{22} + 2(Q_{11}Q_{22} - Q_{12})^{1/2}}. \tag{6.44}$$

Being $|\epsilon| = \sqrt{\epsilon_1^2 + \epsilon_2^2}$, the components of the ellipticity tensor are

$$\epsilon_1 = \frac{Q_{11} - Q_{22}}{Q_{11} + Q_{22} + 2(Q_{11}Q_{22} - Q_{12})^{1/2}},$$

$$\epsilon_2 = \frac{2Q_{12}}{Q_{11} + Q_{22} + 2(Q_{11}Q_{22} - Q_{12})^{1/2}}. \tag{6.45}$$

These are also the real and the imaginary parts of the complex ellipticity ϵ.

The position angle of the ellipse gives the shear direction φ,

$$\tan(2\varphi) = \frac{\epsilon_2}{\epsilon_1} = \frac{2Q_{12}}{Q_{11} - Q_{22}}. \tag{6.46}$$

In practice, shape measurements are performed on pixelized images, so that the integrals in Eqs. 6.41 and 6.42 must be substituted by sums over the image pixels. In addition, the extent of the image is defined by some brightness threshold, I_{th}. We may then multiply the surface brightness $I(\vec{\theta})$ in Eqs. 6.41 and 6.42 by a Heaviside step function $q_I = H(I - I_{th})$,

$$H(I - I_{th}) = \begin{cases} 1 & I \geq I_{th} \\ 0 & I < I_{th} \end{cases} . \tag{6.47}$$

Then, only the pixels above the brightness level I_{th} will concur to the computation of the image centroid and shape. This is particularly important because the noise in real astronomical images compromises the shape measurements at low surface brightness levels.

Applying the equations above to the images in Fig. 6.7, we obtain ellipticity measurements, as shown by the dotted red contours.

6.2.3   Tangential and Cross Component of the Shear

Because of the scalar nature of the lensing potential, gravitational lensing by a single lens can only induce tangential and radial distortions of the images (see e.g. Fig. 3.4). It is often convenient to redefine the shear components in a rotated frame where the axis θ_1' passes through the image centroid and the lens center. Obviously, the axis θ_2' is perpendicular to θ_1'. Let the angle between θ_1 and θ_1' be ϕ. The transformation between the coordinates (θ_1, θ_2) and (θ_1', θ_2') is then given by

$$R(\phi) = \begin{pmatrix} \cos(\phi) & -\sin(\phi) \\ \sin(\phi) & \cos(\phi) \end{pmatrix} . \tag{6.48}$$

The shear tensor in the rotated frame is then

$$\Gamma' = R^T(\phi)\Gamma R(\phi) , \tag{6.49}$$

and its components are

$$\gamma_1' = \gamma_1 \cos(2\phi) + \gamma_2 \sin(2\phi) , \tag{6.50}$$

$$\gamma_2' = \gamma_1 \sin(2\phi) - \gamma_2 \cos(2\phi)) . \tag{6.51}$$

As we can see from Fig. 3.4, any distortion in the radial or in the tangential directions with respect to the lens is caused by a shear tensor with $\gamma_2' = 0$. Images are tangentially elongated if $\gamma_1' < 0$. On the contrary, images are radially elongated if $\gamma_1' > 0$.

The quantities $\gamma_t = -\gamma_1'$ and $\gamma_\times = -\gamma_2'$ are called *tangential* and *cross* components of the shear. The first encodes the pure lensing signal, while the second should be zero unless some systematics affect the measurements.

Analogously, we can define the tangential and the cross components of the reduced shear $(g_t, g_\times)$ and the ellipticity $(\epsilon_t, \epsilon_\times)$.

Figure 6.8 shows that the tangential component of the ellipticity of the sources in the right panel of Fig. 6.7 decreases as a function of the distance from the lens center. The orange circles indicate the measurements made for each source. The blue dots show the input values of the reduced tangential shear at the image centroids. They match the measured ellipticities very well except at the center of the field-of-view, close to the lens critical lines. In this region, the lens mapping cannot be described adequately by a linear mapping, as higher-order distortions need to be accounted for. The cross component of the ellipticity is consistent with zero at all radii, thus validating the measurements against systematic errors.

For a given lens model and set of parameters $\vec{p}$, we can now compute the expected ellipticities of all N_{ima} lensed images, $\epsilon_i(\vec{p}) = g_{t,i}(\vec{p})$, where $i \in N_{ima}$, and compare then to the measured ones to assess how well the model matches the data.

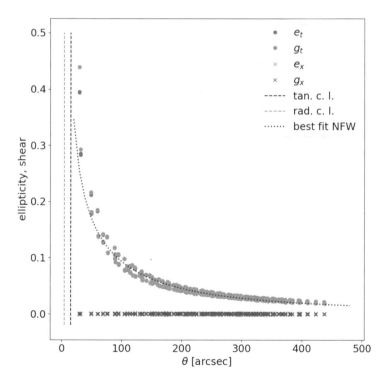

Fig. 6.8 Tangential and cross components of the true reduced shear and the ellipticity of the sources in the simulation illustrated in Fig. 6.7. The vertical dashed lines show the mean positions of the lens critical lines. The true tangential shear values are indicated with orange circles. They are almost identical to the values estimated from the tangential component of the image ellipticities, shown by the blue circles, particularly far from the lens critical lines. The red and green crosses show the cross components of the true shear and image ellipticity, respectively. They are consistent with zero, as expected. The dotted blue line shows the best-fit NFW model to the data (see Sect. 6.4.2)

We define the χ^2

$$\chi_{WL}^2(\vec{p}) = \sum_{i=1}^{N_{ima}} \frac{[\epsilon_{t,i} - g_{t,i}(\vec{p})]^2}{\sigma_i^2} , \qquad (6.52)$$

where σ_i indicates the error associated with the i-th ellipticity measurement. By minimizing χ_{WL}^2 with respect to $\vec{p}$, we find the best-fit parameters describing the lens.

6.2.4 Aperture Mass Densitometry

Using the Divergence Theorem, we can obtain a relation between the mass enclosed within radius θ and the tangential shear:

$$m(\theta) = \frac{1}{\pi} \int \kappa(\vec{\theta}) d^2\theta = \frac{\theta}{2\pi} \oint \frac{\partial\hat{\Psi}}{\partial\theta} d\phi , \qquad (6.53)$$

where the first and the second integrals are performed onto the surface and the perimeter of the circle of radius θ, respectively. We use the lower-case m because this mass is dimensionless.

By taking the derivative with respect to θ, we find that

$$\frac{dm(\theta)}{d\theta} = \frac{m(\theta)}{\theta} + \frac{\theta}{2\pi} \oint \frac{\partial^2\hat{\Psi}}{\partial\theta^2} d\phi . \qquad (6.54)$$

Being $\partial^2\hat{\Psi}/\partial\theta^2 = \kappa - \gamma_t$, the above integral leads to

$$\frac{dm(\theta)}{d\theta} = \frac{m(\theta)}{\theta} + \theta[\langle\kappa\rangle(\theta) - \langle\gamma_t\rangle(\theta)] , \qquad (6.55)$$

where $\langle\kappa\rangle(\theta)$ and $\langle\gamma_t\rangle(\theta)$ denote the mean convergence and shear along the circle of radius θ. Since $dm(\theta)/d\theta = 2\theta\langle\kappa\rangle(\theta)$, we obtain that

$$\langle\gamma_t\rangle = -\frac{1}{2\pi\theta} \left[\frac{dm(\theta)}{d\theta} - 2\frac{m(\theta)}{\theta} \right] . \qquad (6.56)$$

Note that this relation is formally identical to that in Eq. 5.21 for axially symmetric lenses. Indeed,

$$m(\theta) = \theta^2\bar{\kappa}(<\theta) = 2\int_0^\theta \langle\kappa\rangle(\theta)\theta \, d\theta , \qquad (6.57)$$

and therefore

$$\langle \gamma_t \rangle(\theta) = \overline{\kappa}(< \theta) - \langle \kappa \rangle(\theta) \, . \tag{6.58}$$

This shows that the tangential shear is related to the differential surface density $\Delta\Sigma(\theta) = \overline{\Sigma}(< \theta) - \Sigma(\theta)$:

$$\Delta\Sigma(\theta) = \langle \gamma_t \rangle(\theta)\Sigma_{cr} \, . \tag{6.59}$$

In addition, we can easily see that

$$\frac{\partial \overline{\kappa}(< \theta)}{\partial \ln \theta} = \frac{1}{\pi\theta}\left[\frac{dm(\theta)}{d\theta} - 2\frac{m(\theta)}{\theta}\right] = -2\langle \gamma_t \rangle(\theta) \, . \tag{6.60}$$

Thus, we can express the difference between the mean convergences within two radii, θ_1 and θ_2 as

$$\overline{\kappa}(< \theta_1) - \overline{\kappa}(< \theta_2) = 2\int_{\theta_1}^{\theta_2}\langle \gamma_t \rangle(\theta)d\ln\theta \, . \tag{6.61}$$

In addition, the following relation holds between $\overline{\kappa}(< \theta_1)$, $\overline{\kappa}(< \theta_2)$ and the mean convergence in the annulus defined by the two radii θ_1 and θ_2:

$$\theta_2^2\overline{\kappa}(< \theta_2) = \theta_1^2\overline{\kappa}(< \theta_1) + (\theta_2^2 - \theta_1^2)\overline{\kappa}(\theta_1 < \theta < \theta_2) \, . \tag{6.62}$$

Combining Eqs. 6.61 and 6.62, we obtain

$$\overline{\kappa}(< \theta_1) - \overline{\kappa}(\theta_1 < \theta < \theta_2) = \frac{2}{1 - \theta_1^2/\theta_2^2}\int_{\theta_1}^{\theta_2}\langle \gamma_t \rangle(\theta)d\ln\theta \, . \tag{6.63}$$

This equation shows that we can determine the mean convergence within the radius θ by integrating the azimuthally averaged tangential shear profile within an annulus external to that radius. Unfortunately, this is only a lower limit to the real mean convergence of the lens because of the unknown term $\overline{\kappa}(\theta_1 < \theta < \theta_2)$. Clowe et al. (1998) introduced a modified version of the formula above, showing that

$$\overline{\kappa}(< \theta_1) - \overline{\kappa}(\theta_2 < \theta < \theta_{max}) = 2\int_{\theta_1}^{\theta_2}\langle \gamma_t \rangle d\ln\theta + \frac{2}{1 - \theta_2^2/\theta_{max}^2}\int_{\theta_2}^{\theta_{max}}\langle \gamma_t \rangle(\theta)d\ln\theta \, . \tag{6.64}$$

In case of very wide observations, the radii θ_2 and θ_{max} can be chosen to be large so that $\overline{\kappa}(\theta_2 < \theta < \theta_{max})$ is small.

This method to obtain the convergence of the lens is called *aperture mass densitometry*. It belongs to the class of the so-called *free-form* methods to measure the mass distribution of the lenses because no assumption is made about the shape of the lens mass profile, in contrast to the techniques discussed earlier, where a set of parameters described the lens mass distribution.

6.2.5   The Kaiser and Squires Inversion Algorithm

Both the convergence and the shear components are linear combinations of the second derivatives of the lensing potential. Thus, the relations between these quantities in Fourier space are linear. As a result, we can express the convergence as a convolution of the shear with an appropriate kernel function Kaiser and Squires (1993).

The Fourier transforms of the convergence and of the shear are given by

$$\tilde{\kappa} = -\frac{1}{2}(k_1^2 + k_2^2)\tilde{\Psi} \ , \tag{6.65}$$

$$\tilde{\gamma}_1 = -\frac{1}{2}(k_1^2 - k_2^2)\tilde{\Psi} \ , \tag{6.66}$$

$$\tilde{\gamma}_2 = -k_1 k_2 \tilde{\Psi} \ , \tag{6.67}$$

where (k_1, k_2) are the elements of the wave vector and the symbol $\sim$ denotes the Fourier transforms.

We can now eliminate the potential from the equations above, finding the relations between the Fourier transforms of the shear components and of the convergence:

$$\begin{pmatrix} \tilde{\gamma}_1 \\ \tilde{\gamma}_2 \end{pmatrix} = k^{-2}\begin{pmatrix} k_1^2 - k_2^2 \\ 2k_1 k_2 \end{pmatrix}\tilde{\kappa} \ . \tag{6.68}$$

Using

$$\left[k^{-2}\begin{pmatrix} k_1^2 - k_2^2 \\ 2k_1 k_2 \end{pmatrix} \right]\left[k^{-2}\begin{pmatrix} k_1^2 - k_2^2 \\ 2k_1 k_2 \end{pmatrix} \right]^T = 1 \ , \tag{6.69}$$

Equation 6.68 can be inverted to obtain

$$\tilde{\kappa} = \left[k^{-2}\begin{pmatrix} k_1^2 - k_2^2 \\ 2k_1 k_2 \end{pmatrix} \right]^T \begin{pmatrix} \tilde{\gamma}_1 \\ \tilde{\gamma}_2 \end{pmatrix} \ . \tag{6.70}$$

Taking the inverse transform of Eq. 6.70, we obtain

$$\kappa(\vec{\theta}) = \frac{1}{\pi}\int\left[D_1(\vec{\theta} - \vec{\theta}')\gamma_1(\vec{\theta}') + D_2(\vec{\theta} - \vec{\theta}')\gamma_2(\vec{\theta}') \right]d^2\theta' \ , \tag{6.71}$$

where

$$D_1(\theta_1, \theta_2) = \frac{\theta_2^2 - \theta_1^2}{\theta^4} \, , \tag{6.72}$$

$$D_2(\theta_1, \theta_2) = \frac{2\theta_1\theta_2}{\theta^4} \, . \tag{6.73}$$

The Kaiser and Squires algorithm also belongs to the class of the free-form methods. Differently from the previous methods based on the measurement of the tangential shear profile, it allows to obtain a two-dimensional model of the mass distribution of lens.

6.2.6   Challenges in Shear Measurements

Intrinsic Source Ellipticity

So far, we have neglected several complications that affect the real measurements of galaxy shapes. For example, we have assumed that the sources were circular. In reality, galaxies do not have circular shapes. Typical sources used in weak lensing measurements are faint and irregular galaxies. At first approximation, we may assume that they have their own intrinsic ellipticities. These can be described by the quadrupole moments of the intrinsic surface brightness, $I_s(\vec{\beta})$,

$$Q_{s,ij} = \frac{\int I_s(\vec{\beta})(\beta_i - \beta_{0i})(\beta_j - \beta_{0j})d^2\beta}{\int I_s(\vec{\beta})d^2\beta} \, . \tag{6.74}$$

In the former equation, β_{0i} are the coordinates of the source centroid in analogy with the definition in Eq. 6.41.

Using Eqs. 3.4 and 3.30, after some easy calculations, we find that

$$Q_{s,ij} = \sum_k \sum_l A_{ik} Q_{kl} A_{jl} \, , \tag{6.75}$$

which shows that the relation between the intrinsic and the observed quadrupole moment tensors is

$$Q_s = AQA^T = AQA \, . \tag{6.76}$$

Schneider and Seitz (1995) showed that, combining Eqs. 6.75 and 6.45, the relation between intrinsic and observed complex ellipticities is

$$\epsilon_s = \begin{cases} \frac{\epsilon - g}{1 - g^*\epsilon} & \text{if } |g| \leq 1 \, , \\ \frac{1 - g\epsilon^*}{\epsilon^* - g^*} & \text{if } |g| > 1 \, . \end{cases} \tag{6.77}$$

The inverse transformation is obtained by interchanging the source and the image ellipticity and making the substitution $g \rightarrow -g$.

If we make the assumption that the source galaxies are randomly oriented, so that $\langle \epsilon_s \rangle = 0$, when averaging over a sufficiently large number of them, then Eq. 6.77 tells us that

$$\langle \epsilon \rangle = \begin{cases} g & \text{if } |g| \leq 1 , \\ 1/g^* & \text{if } |g| > 1 , \end{cases} \tag{6.78}$$

i.e. the expectation value of the image ellipticity is the reduced shear (or the inverse of the complex conjugate of g).

Thus, even if sources are not intrinsically circular, a relation identical to Eq. 6.34 still holds between the image ellipticity and the reduced shear, which enables us to derive the mass of the lens exactly as described in Sect. 6.2.3. Indeed, each image ellipticity is an unbiased estimator of the local reduced shear. Unfortunately, this estimator is affected by a large noise, which is determined by the dispersion of the intrinsic ellipticities:

$$\sigma_\epsilon = \sqrt{\langle \epsilon_s \epsilon_s^* \rangle} . \tag{6.79}$$

This means that, by averaging over N_g images at whose positions the reduced shear is g, the 1-σ deviation of their ellipticity from the true g is $\sigma_\epsilon / \sqrt{N_g}$ (see e.g. Schneider 2006).

For a centrally concentrated mass distribution, like the lens in the example outlined in the previous sections, the shear can be assumed to be nearly the same for all galaxies at a given distance from the lens center. Thus, by averaging over a sufficiently high number of galaxies in concentric rings of increasing size, we can obtain an estimate of the tangential shear profile of the lens. Then we can fit it with a parametric model to get the lens mass distribution. Similarly, averaging over sources in apertures around a given position, the shear's local value can be estimated and used to obtain the lens convergence using, e.g. the Kaiser and Squires algorithm.

Effects of the Point-Spread-Function

All those instrumental and non-instrumental effects that alter the shape of the sources are critical for shear measurements. The PSF $P(\vec{\theta})$ describes how a point source is smeared by the telescope optics, the telescope guiding, the several steps involved in the data reduction process, and the atmospheric turbulence (in the case of ground-based observations). Of course, the same smearing also applies to extended images of distant galaxies. Their observed surface brightness becomes

$$I_{\text{obs}}(\vec{\theta}) = \int I(\vec{\theta}') P(\vec{\theta} - \vec{\theta}') d^2\theta' . \tag{6.80}$$

In ground-based observations, atmospheric blurring is the major contributor to the PSF, called seeing. In case of good seeing conditions, its Full-Width-at-Half-

Maximum (FWHM) is $\sim$0.6–0.8, which is comparable or larger than the size of typical distant galaxies used in the weak lensing analyses. The main effect of seeing is to reduce the sources' ellipticity, thus making them rounder and letting to under-estimate the shear. On the other hand, optics distortions and other detector effects can contribute to the PSF by introducing anisotropies that can mimic a lensing signal. Thus it is essential to correct for all these effects when carrying out a lensing analysis.

The PSF can be measured using the stars within the field-of-view. Their appear-ance on the astronomical image shows the PSF at the stars' locations. However, the PSF may have spatial variations. To describe them, it is common practice to interpolate the PSF between the star positions using low order polynomials.

Once the PSF has been determined, several methods have been proposed to correct the observed image ellipticity and reduce the systematic errors on the reduced shear measurements. Some of them, including the so-called KSB method (Hoekstra et al. 1998, Kaiser et al. 1995, Luppino & Kaiser 1997), consist of correcting the moments of the surface brightness distribution analytically, using tensors to describe how the image ellipticity responds to PSF anisotropies and shear in the presence of seeing. Other methods attempt to fit the observed images with some models of the surface brightness, convolved with the measured PSF (e.g. Kuijken 1999, Miller et al. 2013, Refregier & Bacon 2003). Most recently, also methods that rely on artificial intelligence have been proposed (e.g. Ribli et al. 2019).

6.2.7 Redshift Dependence of the Signal

We have yet not considered that the weak lensing signal, i.e. the induced ellipticity of the images, depends of the source redshift. Both convergence and shear scale with the distances as $\nu(z_S) = D_{LS}/D_S$. We may write them in terms of their value for sources at infinite redshift. Then, $\gamma = \gamma_\infty \nu(z_S)/\nu_\infty$ and $\kappa = \kappa_\infty \nu(z_S)/\nu_\infty$ Thus, the reduced shear varies with the source redshift as

$$g(z_S) = \frac{\gamma_\infty \nu(z_S)/\nu_\infty}{1 - \kappa_\infty \nu(z_S)/\nu_\infty} . \tag{6.81}$$

In observations, the ellipticity of many sources distributed over a range of redshifts is measured. Their redshifts are necessary to convert the measured lensing signal into an estimate of the lens mass.

The redshift of each galaxy used to perform a shear measurement is generally impossible to determine. Typical sources used in weak lensing analyses are faint distant galaxies that are difficult to target in spectroscopic observations. Photometric redshifts are an alternative solution, but they require multi-band observations, and their accuracy becomes smaller for fainter sources.

In many cases, an effective value of $\langle \nu(z_S) \rangle$ is used to describe the entire population of sources behind a lens. It is determined from catalogs of photometric

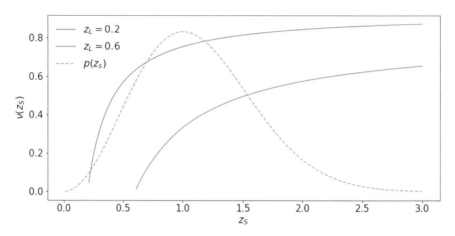

Fig. 6.9 The blue and the orange solid lines show the function $\nu(z_S)$ for two lenses at redshifts $z_L = 0.2$ and $z_L = 0.6$. The dashed green lines show a source redshift distribution peaking at $z_S = 1$

redshifts in the literature (e.g. the COSMOS survey Ilbert et al. 2009) by matching them to the depth of the observations. Unfortunately, this approach can introduce systematic errors. The solid curves in Fig. 6.9 show $\nu(z_S)$ for two lenses at redshifts $z_L = 0.2$ and $z_L = 0.6$. Both curves grow rapidly and then reach an asymptotic value higher for lower redshift lenses. Depending on the observations' depth, the redshift distribution of the sources in current lensing surveys peaks at $z_S \sim 0.6$–1. The dashed curve shows a tentative redshift distribution peaking at $z_s = 1$. For a lens at $z_L = 0.2$, most sources are at distances characterized by similar values of $\nu(z_S)$. On the contrary, for the lens at $z_L = 0.6$ the evolution of $\nu(z_S)$ within the redshift range covered by the majority of the sources is strong. Thus, especially in the case of high-redshift lenses, estimating the effective $\nu(z_S)$ requires a good knowledge of the true source redshift distribution.

It is worth noting that since the dispersion of a shear estimate from averaging over N_g galaxy ellipticities is σ_ϵ^2/N_g, the shear signal-to-noise is

$$\frac{S}{N} = \frac{\gamma}{\sigma_\epsilon} N_g^{1/2} . \tag{6.82}$$

Since higher redshift lenses have fewer detectable sources in their background at a fixed depth, the shear measurement is much more difficult for these galaxy clusters.

Let us consider a SIS lens. Within an annulus of inner and outer radii θ_1 and θ_2 centered on the lens, the number of lensed sources is

$$N = n_{gal}\pi(\theta_2^2 - \theta_1^2) , \tag{6.83}$$

where n_{gal} is the number density of galaxies. The mean shear within the annulus is

$$\overline{\gamma}(\theta_1 < \theta < \theta_2) = \frac{\theta_E}{\theta_1 + \theta_2} \tag{6.84}$$

Plugging Eqs. 6.83 and 6.84 into Eq. 6.82, and remembering that $\theta_E = 4\pi(\sigma_v/c)^2 D_{LS}/D_S$, we obtain that the signal-to-noise ratio is

$$\frac{S}{N} = \frac{\theta_E}{\sigma_\epsilon}\sqrt{n_{gal}\pi}\sqrt{\frac{\theta_2/\theta_1 - 1}{\theta_2/\theta_1 + 1}} = \frac{4\pi^{3/2}}{c^2}\sigma_v^2\sqrt{n_{gal}}\sigma_\epsilon^{-1}\sqrt{\frac{\theta_2/\theta_1 - 1}{\theta_2/\theta_1 + 1}}\frac{D_{LS}}{D_S}. \tag{6.85}$$

This formula shows that clusters at intermediate redshifts with masses $M_{vir} \sim 10^{15} M_\odot$ corresponding to a velocity dispersion $\sigma_v \sim 1000 \text{ km s}^{-1}$ are detectable in deep ground-based observations ($n_{gal} \sim 20\text{--}30 \text{ arcmin}^{-2}$) at $S/N \sim 10$, assuming $\theta_2/\theta_1 \sim 10$ and $\sigma_\epsilon \sim 0.2$. Hence, massive galaxy clusters can be studied individually. For lower mass or higher redshift clusters of groups of galaxies, the S/N becomes too small and it is necessary to increase the number density of lensed sources using deeper observations with the Hubble Space Telescope. Alternatively, clusters with similar masses can be stacked to measure their average mass distributions (Umetsu 2020)

6.2.8 Limitations of the Methods

Unfortunately, both Eqs. 6.64 and 6.71 relate the convergence to the shear, not to the reduced shear, which we measure from the data. One method to mitigate the problem is to adopt an iterative approach. For example, we can start from the assumption that $\kappa = 0$. Then, the sources' measured ellipticities provide an estimate of the shear, and Eq. 6.71 allows us to estimate $\kappa(\vec{\theta})$. In the next iteration, we use this convergence to convert the measured reduced shear into the shear, and we repeat the calculation to find a new estimate of $\kappa(\vec{\theta})$. After a few iterations, we usually find a stable solution.

The Kaiser and Squires method also assumes that the integral in Eq. 6.71 extends to infinity, while observed fields are finite. Thus, as seen for the aperture mass densitometry, wide-field imaging helps to mitigate this problem.

The methods to recover the lens convergence based on the tangential shear profile measurement suffer from another critical limitation. They assume that there is a unique lens center. As discussed by Meneghetti et al. (2010), this causes systematic errors if the lens is not isolated or massive substructures are present because the assumption that the shear is tangential to a unique mass clump is not correct.

All methods are affected by biases related to the possible contamination by foreground or cluster galaxies. If we include these accidentally among the sources used to measure the shear, they dilute the signal, thus underestimating the convergence and, eventually, the lens mass. Methods to identify the cluster background sources are generally based on photometric redshift measurements or, when these are not possible, on color selection (Formicola et al. 2016, Masters et al. 2015, Medezinski et al. 2018).

6.3   Applications of Lensing by Galaxies and Galaxy Clusters

This chapter focuses on how the gravitational lensing effects can map the distribution of matter in galaxies and galaxy clusters. The resulting mass models are used in many applications, aiming at answering outstanding questions in cosmology and astrophysics. Several review articles do a great job illustrating these applications in detail (Kneib & Natarajan 2011, Limousin et al. 2013, Meneghetti et al. 2013, Treu 2010b, Treu & Marshall 2016, Umetsu 2020). In this section, we provide a very short and concise summary of some of them.

6.3.1   The Nature of Dark Matter

In the standard cosmological model, the matter content of the universe is dominated by cold dark matter (CDM), collision-less particles that interact with ordinary matter (baryons) only through gravity. Gravitationally bound dark matter halos form hierarchically, with the most massive systems forming through mergers of smaller ones. As structure assembles in this fashion, large dark matter halos contain smaller-scale substructure in the form of embedded sub-halos. Theoretically, it is expected that the mass function of these sub-halos follows a power-law behavior,

$$\frac{dn}{d \ln m} \propto m^\delta \, , \tag{6.86}$$

where m is the sub-halo mass and $\delta \sim 0.8$ (Despali and Vegetti 2017, Giocoli et al. 2010).

Besides, several numerical simulations of structure formation show that CDM halos develop a characteristic density profile, similar to the NFW density profile introduced in Sect. 5.5.1. Recent theoretical results indicate that this profile should be universal over 20 orders of magnitude in halo mass (Wang et al. 2020). Also, halo mass and concentration are tightly related in a way that depends on cosmology.

Gravitational lensing is a potentially powerful tool to investigate the predictions of the standard cosmological model. As discussed, substructures in the lens produce their characteristic lensing effects, such as brightness fluctuations, image shifts, or even additional images of the background sources. Interestingly, substructures can also be detected if they are dark, i.e. if they do not host stars emitting light. A common practice to catch them in strong lensing galaxies is to fit extended images using smooth mass distributions. Substructures are detected as residuals in the source-subtracted images. This technique is called "gravitational imaging" (Vegetti & Koopmans 2009, Vegetti et al. 2010). The mass of the substructure can be estimated by adding perturbers in the lens model. Their positions and masses are adjusted to minimize the residuals (Birrer et al. 2015, Hezaveh et al. 2016, Vegetti et al. 2012).

A great effort is being made to be able to identify the smallest substructures in galaxies because alternative models to cold dark matter predict that these are suppressed below a characteristic scale that depends on the assumed physics of the dark matter particles (see, e.g. Lovell et al. 2012). The perturbations that such small dark matter clumps can produce are tiny. Detecting them in the images of extended sources requires to achieve a very high spatial resolution, currently achievable only with interferometry techniques in the sub-mm and radio wavelengths (Hezaveh et al. 2016, Powell et al. 2020). Alternatively, substructures' presence is inferred from the anomalous image fluxes of multiply imaged point sources (Nierenberg et al. 2020). Observations did not show evidence of a discrepancy with the CDM paradigm so far.

Galaxies are expected to trace the most massive substructures in galaxy clusters. Their strong lensing effects occur on scales of the order of $\sim$1 arcsec. They enable to study the dark matter in the galactic halos, measuring their masses and spatial distribution within the clusters (Grillo et al. 2015, Limousin et al. 2007, Natarajan et al. 2017; 2007, Natarajan & Kneib 1997, Natarajan et al. 2002, Natarajan & Springel 2004). Interestingly, recent studies recently showed that the dark matter halos of cluster galaxies are more concentrated and closer to the cluster centers than expected in the CDM model (Meneghetti et al. 2020). For these reasons, the probability of galaxy–galaxy strong lensing events in galaxy clusters is relatively high.

The inner density profiles of galaxies can be constrained using strong lensing combined with stellar kinematics, measurements obtained from spectroscopy. Indeed, strong lensing alone is insufficient to measure the density profile's inner slope, as it is sensitive to the total mass within the Einstein radius. Thus, independent measurements of the mass within radii smaller than those probed by lensing, i.e. in regions of few kpcs where stars dominate the galaxy mass budget, are necessary to measure the slope and break the degeneracy between dark and baryonic mass. Generally, stars and dark matter are found to "conspire" to create a density profile that is close to isothermal (e.g. Auger et al. 2009, Koopmans et al. 2009).

The outer profile of cluster-size dark matter halos can be constrained using weak lensing (Umetsu 2020). The combination of weak and strong lensing allows us to measure the dark matter distribution in the largest cosmic structures from the inner region, down to where radial arcs are detected, out to the virial radius. Recent results from the Cluster Lensing And Supernova Survey with Hubble (CLASH) show that the NFW profile describes well the mass distribution in massive galaxy clusters and that the relationship between concentration and mass is consistent with expectations in the standard cosmological model (Meneghetti et al. 2014, Merten et al. 2015, Umetsu et al. 2014). Weak lensing has been used to also characterize the outer regions of dark matter halos for statistical samples of early-type galaxies out to intermediate redshifts (e.g. Hoekstra et al. 2004).

Constraints from stellar kinematics in the BCGs can break the degeneracy between stellar and dark matter and measure the very inner slope of the cluster density profile. Some results suggest that the dark matter profiles at the center of some galaxy clusters may be shallower than expected. (Newman et al. 2015; 2013, Sand et al. 2008). It is not yet clear if this feature is a hint of some tension with the CDM model or if it is the result of the interplay between dark matter and baryons.

In the standard CDM scenario, dark matter particles are assumed to be collision-less, meaning that they have negligible self-interactions. Stars are also effectively collision-less. Instead, the hot intracluster gas is a collisional fluid. It is subject to pressure forces and dissipation, and shocks and turbulence are developed during collisions. Two colliding clumps of collision-less dark matter (and the galaxies therein) pass through each other and are only slowed down by their gravitational interaction. On the contrary, the gas clouds stick on each other due to the much stronger interactions between their particles. Thus, during violent events such as head-on mergers of galaxy clusters, the different nature of dark matter, gas, and stars manifests itself through displacements of their respective mass distributions. By measuring these displacements, it is possible to place constraints on the possible collisional (or self-interacting) cross-section of dark matter particles. Mass mapping through weak lensing allows us to determine the distribution of the dark matter. The cluster gas distribution is observable through its X-ray emission, instead. The displacements measured in a few merging clusters support that dark matter is effectively collision-less like galaxies, placing a robust upper limit on the self-interacting cross-section of $\sigma_{DM}/m \lesssim 1.25\,cm^2/g$ (Bradač et al. 2008, Markevitch et al. 2004, Randall et al. 2008). Strong lensing also puts constraints on the self-interaction cross-section, as shown by Meneghetti et al. (2001), Miralda-Escudé (2002), Natarajan et al. (2002)

6.3.2 The Interplay Between Dark Matter and Baryons

Observations of strong and weak gravitational lensing can be compared to complementary probes of the matter distribution in galaxies and galaxy clusters to accurately measure the distribution of dark matter and baryons in the cosmic structures. These include: stellar or galaxy kinematics, observations of the Intracluster medium in the X-ray or radio wavelengths, etc. This comparison provides important hints on what is the interplay between dark and ordinary matter. Baryons condense inside halos to form stars. Star formation is regulated by cooling and heating mechanisms that depend, among others, on many physical processes such as energy feedback from active-galactic-nuclei and supernovae explosions, galactic winds, metal enrichment of the interstellar medium, mergers, etc. Strong lensing can help us understand these processes by providing precise measurements of the fraction of total mass in the form of dark matter within the Einstein radius and the mass-to-light ratios in galaxies and clusters (e.g. Jiang & Kochanek 2007). Dark matter, particularly in the central regions of galaxies and galaxy clusters, is potentially affected by baryon physics due to the gravitational interactions between its particles and ordinary matter. Thus, it is interesting to measure how the shape of the density profiles changes under the effects of baryons across a range of lens masses. For example, (Newman et al. 2013) find hints for a systematic decreasing trend of the inner slope of the density profiles as a function of mass.

6.3.3 Cosmic Telescopes

Thanks to their magnification, galaxies and galaxy clusters are powerful cosmic telescopes. Sources near the lens caustics are so highly magnified that details as small as globular clusters (i.e. with intrinsic sizes of $\lesssim 200$ parsec) can be resolved in their images Vanzella et al. (2017a;b). Besides, some of the most distant sources in the universe were observed behind galaxy clusters (e.g. Coe et al. 2013, Zheng et al. 2012).

More generally, gravitational lenses offer the possibility to study in detail source populations that would be out of reach even with the current most powerful and highest resolution telescopes. Some of these source populations are particularly interesting because they can help understand the early phases of star and galaxy formation and evolution. For example, the most intense star formation in the universe occurs in dust-rich galaxies at $z \gtrsim 1$, where the star formation rate amounts to $> 100 - 1000 M_\odot$ yr^{-1} (Casey et al. 2014). The UV radiation emitted by massive young stars is largely reprocessed by the interstellar dust and re-emitted at far-infrared and sub-mm wavelengths. As seen in Sect. 6.1.3, dusty star-forming galaxies (DSFGs) lensed by other galaxies along the line-of-sight are identified as the brightest sources in the infrared surveys thanks to the magnification bias (Negrello et al. 2010, Vieira et al. 2010). Once detected, they can be observed at higher resolution using interferometers like the Atacama Large Millimeter Array (ALMA) (Hezaveh et al. 2016, Spilker et al. 2016). A careful lens modeling allows to de-lens the sources and to study their intrinsic properties. Reproducing a realistic population of DSFGs has long been a challenge for theoretical models of galaxy evolution (Narayanan et al. 2015) and lensing offers a way to study them in detail.

An important reason for studying the distant universe is to witness the stars and galaxies in the first 1 billion years of cosmic history. Unlensed sources in this epoch are extremely faint (>28 mag). They might have played an essential role in the process of re-ionizing the gas in the universe between redshift 15 and 8. Before that epoch, in the so-called dark ages, the universe did not yet contain stars and galaxies, and this gas was neutral (Ciardi & Ferrara 2005, Dijkstra 2014, Gunn & Peterson 1965, Madau et al. 1997, Planck Collaboration et al. 2020). Then, the re-ionization occurred likely due to the UV and X-ray photons escaping from some of the first objects to form. It is yet unclear which sources might have contributed most to the cosmic re-ionization. Several candidates have been proposed: the first population of stars (the so-called Pop III stars), a population of faint dwarf galaxies, nuclear black holes, X-ray binaries, etc. Zaroubi (2013). While the James-Webb-Space-Telescope (JWST) will reveal this epoch in detail, we can get a jump-start by combining HST's resolution with the gravitational lensing magnification of massive clusters. Indeed, extensive surveys of galaxy clusters with the HST, including the Cluster Lensing And Supernova survey with Hubble and the Frontier Fields, have demonstrated the power of strong gravitational lensing to deliver large samples of high-redshift galaxies efficiently (Atek et al. 2015, Bradley et al. 2014). Combining imaging data from the HST and deep integral-field spectroscopic data from MUSE at the VLT, Vanzella et al. (2020) also found a candidate complex Pop. III stars at redshift 6.63, enormously magnified by the galaxy clusters MACSJ0416.

6.3.4 Cosmological Applications

Finally, it is impossible not to mention the applications of lensing by galaxies and galaxy clusters in cosmology. There are at least three important reasons that make gravitational lensing an important cosmological tool. First, as we have discussed extensively, the intensity of the gravitational lensing effects is determined by the distances of lenses and sources from the observer. Thus, lensing is strongly dependent on the universe's geometry, which in turn depends on the values of the cosmological parameters. Second, the ability of galaxies and galaxy clusters to produce lensing effects depends on their internal structure. For example, at fixed mass, lenses that are more concentrated are more efficient at producing strong lensing effects, as their central surface density can more easily exceed Σ_{cr}. It is well known that the concentration of dark matter halos depends on cosmology (e.g. Bullock et al. 2001, Dolag et al. 2004, Eke et al. 2001, Navarro et al. 1997). For example, in cosmological models where they form earlier, when the universe's mean background density is higher, galaxies and galaxy clusters are more concentrated. Third, the abundance of lenses in the universe depends on the epoch of their formation. If galaxies and galaxy clusters form earlier, a larger number of them exists at redshifts between $z \sim 0.2$ and $z \sim 0.6$, where their lensing strength is highest for a population of sources whose redshift distribution peaks at $z \sim 1$. As the formation and the evolution of the cosmic structures are regulated by the content of matter and energy of the universe, which determines how it expands across the cosmic time, the frequency of gravitational lensing events can be used to constrain cosmological parameters such as the contributions to the density parameter by dark matter and dark energy.

We briefly mention some of the cosmological tests based on gravitational lensing by galaxies and galaxy clusters:

- **Time delay cosmography:** Time delays can be measured in gravitational lensing events where variable sources (e.g. quasars or supernovae) are multiply imaged by foreground galaxies or galaxy clusters (Refsdal 1964, Treu et al. 2016, Treu & Marshall 2016). As seen in Sect. 5.8, the total time delay is proportional to the time delay distance, $D_{\Delta t}$, which depends on cosmological parameters (see Sect. 9.5). In particular, it is inversely proportional to the present value of the Hubble constant, H_0, which measures the relative rate of expansion of the universe (Sect. 9.4). To sample the light-curves of the individual images accurately, compare them, and measure the time delays, continuous monitoring of the multiple images on time scales of years is required. Projects like the

COSmological MOnitoring of GRAvItational Lenses (COSMOGRAIL) are currently carrying out these measurements from optical light-curves obtained with small but (almost) dedicated telescopes in the northern and southern hemispheres (Bonvin et al. 2019, Eigenbrod et al. 2005, Millon et al. 2020). The goal is to measure individual time delays with an accuracy below 3%. The time delays and the other lensing observables (i.e. the positions of the multiple images, etc.) are fitted simultaneously to measure the lens mass distribution and the cosmological parameters. The success of the technique depends on the availability and size of a suitable sample of lensed quasars or supernovae, precise measurements of the time delays, accurate modeling of the gravitational potential of the main deflector, and the ability to characterize the distribution of mass along the line-of-sight to the source. Indeed, as seen in Sect. 5.9, in presence of an external convergence κ_{ext}, the time delay changes by a factor $\lambda = (1 - \kappa_{ext})$.

Measurement of H_0 with lensing is highly complementary to other cosmological probes, such as the observations of the Cosmic Microwave Background made by the ESA's Planck mission (Planck Collaboration et al. 2020) or of type Ia supernovae (SNae) calibrated via the distance ladder (Freedman et al. 2012, Riess et al. 2019; 2018; 2016, Sandage et al. 2006). The latest results from these two kinds of probes are in strong tension. The Supernovae, H0, for the Equation of State of Dark Energy (SH0ES, Riess et al. 2016) collaboration found a higher value of H_0 ($H_0 = 74.03 \pm 1.42$ km s^{-1} Mpc^{-1}) compared to Planck ($H_0 = 67.4 \pm 0.5$ km s^{-1} Mpc^{-1}). The H_0 Lenses in COSMOGRAIL's Wellspring (H0LiCOW) collaboration recently measured $H_0 = 73.3^{+1.7}_{-1.8}$ km s^{-1} Mpc^{-1} in a ΛCDM cosmology, based on the analysis of six multiply imaged quasars (Wong et al. 2020). Such value is consistent with the measurement of H_0 from the SH0ES collaboration, confirming a tension between the measurements of H_0 from observations of the universe at early and late times.

- **Strong lensing cosmography:** Strong gravitational lensing can be used to constrain the cosmological parameters through the dependence of deflection angle on the distance ratio D_{LS}/D_S. To illustrate this technique, we consider a simple SIS lens. The Einstein radius, i.e. the angular separation of two multiple images, scales with the velocity dispersion and with the distances between the lens and with the angular diameter distances D_S and D_{LS} as $\theta_E \propto D_{LS}/D_S \sigma_0^2$. Measuring the Einstein radius from the positions of the multiple images of a source at a given redshift, z_S, provides a measurement of the distance ratio. Unfortunately, this is degenerate with the lens velocity dispersion (and thus with its mass), so that we cannot learn much about cosmology. However, if the multiple images of a second source at a different redshift z_S' are used to obtain a second estimate of the Einstein radius, then the ratio between the two Einstein radii only depends on angular distances, and thus on cosmology. More precisely, such a ratio provides a measurement of the so-called family ratio,

$$\Xi(z_{LS}, z_S, z_S', \vec{\Pi}) = \frac{D_{LS}}{D_S} \frac{D_S'}{D_{LS}'} , \qquad (6.87)$$

where $\vec{\Pi}$ denotes the ensemble of cosmological parameters. In lenses with complex mass distributions and density profiles that are not isothermal, the family ratio remains degenerate with the mass distribution. Still, the degeneracy can be alleviated if several sources at different redshifts are lensed simultaneously, as it often happens in massive galaxy clusters. For example, Jullo et al. (2010) used a parametric lens inversion method to simultaneously constrain the mass distribution of the cluster Abell 1689 and cosmological parameters such as the contribution of matter to the density parameter, Ω_m, and the dark energy equation of state w_{DE}, finding $\Omega_{m,0} = 0.25 \pm 0.05$ and $w_{DE} = -0.97 \pm 0.07$. More recently, Caminha et al. (2016), applying the same method to the galaxy cluster RXCJ2248.7-4431 and assuming a flat cosmology, measured $\Omega_{m,0} = 0.2^{+0.13}_{-0.16}$ and $w_{DE} = -1.07^{+0.16}_{-0.42}$. These values are in agreement with other cosmological probes, (e.g. Planck Collaboration et al. 2020). D'Aloisio and Natarajan (2011) showed that unaccounted structures along the line-of-sight to the lensed sources can introduce biases in the estimate of the cosmological parameters and suggested to combine observations of several clusters to mitigate their effects.

The Supernova "Refsdal," exploded in a galaxy multiply imaged by galaxy cluster MACS J1149.6+2223 (Kelly et al. 2016; 2015), offered the possibility to obtain time delay measurements in a galaxy cluster. Time delays are generally measured in galaxy scale lenses. Several families of multiple images of distant galaxies are detected in the deep observations of MACSJ1149 with the HST. Thus, for this cluster, the combined dependence of the multiple image positions and time delays on the cosmological parameters allowed to infer the values of H_0 and $\Omega_{m,0}$ with a relative (1σ) statistical errors of, respectively, 6% and 31% in flat cosmological models (Grillo et al. 2018).

- **Mass calibration for cluster cosmology:** Galaxy clusters represent the high-mass tail of hierarchical structure formation, which is exponentially sensitive to the growth of cosmic structure. For this reason, the normalization and the redshift evolution of the mass function are very sensitive to cosmological parameters like $\Omega_{m,0}$, σ_8, and the equation of state of dark energy w_{DE} (Rosati et al. 2002). Galaxy clusters became a key cosmological probe in several past, ongoing, and future surveys (Abbott et al. 2020, Ivezić et al. 2019, Laureijs et al. 2011, Merloni et al. 2012, Planck Collaboration et al. 2014, Vikhlinin et al. 2009). Some of these surveys, conducted at very different wavelengths, will scan large fractions of the whole sky, delivering observations of thousands of galaxy clusters. One example is the Planck mission, whose PSZ2 catalog contains 1653 cluster candidates identified through their Sunyaev–Zel'dovich effect (Planck Collaboration et al. 2014). The mission eRosita will detect 50-100 thousand galaxy clusters and groups of galaxies through the X-ray emission from their intracluster-medium (ICM) (Pillepich et al. 2012). Masses for thousands of clusters cannot be measured by modeling each of them individually because this would require a huge workload. Instead, the masses are estimated through scaling relations between mass and observables that can more easily be accessed, like the temperature, the pressure, the luminosity of the X-ray emitting ICM Vikhlinin et al. (2009), the Sunyaev–Zel'dovich effect signal (quantified by the integrated Compton-y parameter, which is proportional to the line-of-sight integral of the temperature weighted thermal electron density of the ICM) (Planck Collaboration et al. 2014), or the cluster richness (a measure of the number of galaxies associated with the cluster) (Johnston et al. 2007, McClintock et al. 2019, Rozo et al. 2010). To obtain unbiased estimates of the cosmological parameters, it is crucial to calibrate these scaling relations using accurate mass measurements like those that (strong and weak) gravitational lensing can provide. Meneghetti et al. (2010) showed that the mass derived from weak and strong lensing is unbiased on average, although affected by scatter due to projection effects (see also Becker & Kravtsov 2011, Giocoli et al. 2014). We remind that lensing measures the integrated mass along the line-of-sight. Galaxy clusters are probably triaxial objects (Bonamigo et al. 2015, Despali et al. 2017, Jing & Suto 2002, Limousin et al. 2013), whose projected mass in a given aperture can vary significantly along different lines-of-sight. On the contrary, other measures of the cluster mass based on the observation of the X-ray emission of the ICM, the SZE effect, or even the cluster galaxies' kinematics may be less sensitive to projection effects. Still, they could be affected by significant biases because they rely on the assumption that the ICM or the galaxies are in some sort of equilibrium (hydrostatic or virial) within the cluster potential. Several recent studies compared weak lensing and X-ray or SZE mass estimates to quantify the amplitude of these biases (Hilton et al. 2018, Hoekstra et al. 2015, Miyatake et al. 2019, Penna-Lima et al. 2017, Sereno et al. 2017, Smith et al. 2016, von der Linden et al. 2014).

6.4   Python Applications

6.4.1   Parametric Strong Lensing Mass Reconstruction

In this example, we will use the SIE model to fit some strong lensing observables. The goal is to illustrate how the parametric approach, discussed in Sect. 6.1.5, works.

Simulating a Lens

Let us start by mocking the observation of a lens. For this purpose, we will use the python package LENSTRONOMY[1] (Birrer & Amara 2018). This package is rich in features and well documented. In the following, we took inspiration from one of the jupyter notebooks distributed with the `lenstronomy_extensions`[2].

---

[1] https://lenstronomy.readthedocs.io/en/latest/index.html

[2] https://github.com/sibirrer/lenstronomy_extensions

We begin by importing the necessary modules:

```
import lenstronomy.Util.simulation_util as sim_util
import lenstronomy.Util.image_util as image_util
from lenstronomy.Util import param_util
from lenstronomy.ImSim.image_model import ImageModel
from lenstronomy.PointSource.point_source import PointSource
from lenstronomy.LensModel.lens_model import LensModel
from lenstronomy.LensModel.Solver.lens_equation_solver import
LensEquationSolver
from lenstronomy.LightModel.light_model import LightModel
from lenstronomy.Sampling.parameters import Param
from lenstronomy.Data.imaging_data import ImageData
from lenstronomy.Data.psf import PSF
```

To set up the simulation, we need to specify a few parameters to characterize the mock observation, for example, the noise per pixel, the exposure time, and the size of the output image. In this example, we create a cutout 100×100 pixels centered on the lens. The pixel scale is 0.05″. We assume a Gaussian PSF with FWHM 0.1″:

```
# observation parameters:
background_rms = 0.5  # background noise per pixel
exp_time = 100   # exposure time (arbitrary units)
numPix = 100   # number of pixels
deltaPix = 0.05  # pixel size in arcsec

# PSF specification
fwhm = 0.1  #  PSF FWHM
kwargs_data = sim_util.data_configure_simple(numPix, deltaPix,
                                             exp_time,
                                             background_rms)
data_class = ImageData(**kwargs_data)
kwargs_psf = {'psf_type': 'GAUSSIAN',
              'fwhm': fwhm,
              'pixel_size': deltaPix,
              'truncation': 5}
psf_class = PSF(**kwargs_psf)
```

The lens mass distribution is modeled using a SIE model. The lens is characterized by a set of parameters, namely the velocity dispersion σ_v, the axis ratio f and the position angle φ. The center of the lens is at the origin of the lens plane, i.e. at position $(0, 0)$. For the input lens parameters, we set $\sigma_v = 200$ km/s, $f = 0.7$, and $\varphi = 45$ degrees. The lens redshift is $z_L = 0.3$. LENSTRONOMY has its own implementation of the SIE model and uses a different definition of ellipticity compared to that we used e.g. in Sect. 5.11.1. Indeed, the eccentricity components are related to the parameter f as

$$e_1 = \frac{1-f}{1+f} \cos(2\varphi)$$

$$e_2 = \frac{1-f}{1+f} \sin(2\varphi). \tag{6.88}$$

In LENSTRONOMY, the size of the Einstein radius is used to define the lens scale. To compute it using Eq. 5.50, we set the source redshift to be $z_S = 1.5$. We further assume a flat ΛCDM cosmological model with $\Omega_{m,0} = 0.3$ to compute the angular diameter distances:

```
# lens parameters
f=0.7
sigmav=200.
pa=np.pi/4.0 # position angle in radians
zl=0.3 # lens redshift
zs=1.5 # source redshift

# lens Einstein radius
from astropy.cosmology import FlatLambdaCDM
co = FlatLambdaCDM(H0=70, Om0=0.3)
from astropy.constants import c, G
dl=co.angular_diameter_distance(zl)
ds=co.angular_diameter_distance(zs)
dls=co.angular_diameter_distance_z1z2(zl,zs)

# compute the Einstein radius
thetaE=1e6*(4.0*np.pi*sigmav**2/c**2*dls/ds*180.0/np.pi*3600.0).value

# eccentricity computation
e1,e2=(1-f)/(1+f)*np.cos(-2*pa),(1-f)/(1+f)*np.sin(-2*pa)
lens_model_list = ['SIE']
kwargs_sie = {'theta_E': thetaE,
              'center_x': 0,
              'center_y': 0,
              'e1': e1,
              'e2': e2}
kwargs_lens = [kwargs_sie]
lens_model_class = LensModel(lens_model_list=lens_model_list)
```

We model the lens galaxy's surface brightness distribution using a Sérsic elliptical model, implemented in the LightModel class of LENSTRONOMY. The ellipticity and center of the surface brightness distribution are assumed to be consistent with the mass distribution. The effective radius and the Sérsic index are chosen to be $r_e = 2''$ and $n = 4$, respectively.

We assume the source to be a galaxy hosting a quasar at the center. Its position on the source plane is $(-0.1\theta_E, \theta_E)$. The host galaxy is an extended elliptical source and is modeled using a Sérsic profile with $r_e = 0.1''$ and $n = 3$. The eccentricity components are $(0.1, 0.01)$. Instead, the quasar is a point source, modeled using the functions provided by the module PointSource.

The piece of code below uses LENSTRONOMY to perform the following operations:

1. It solves the lens equation for the provided source position and lens model. In this step, the source is treated as a point source.
2. It computes the lensed fluxes of the quasar and also adds some fluctuations mimicking microlensing in the lens galaxy.
3. It creates the images of lens and source, adding noises that affect observations and including the effects of the PSF.
4. It produces a simulated observation of the lensing event.

```
# create the light model for the lens (SERSIC_ELLIPSE)
lens_light_model_list = ['SERSIC_ELLIPSE']
kwargs_sersic = {'amp': 3500, # flux of the lens (arbitrary units)
                 'R_sersic': 2., # effective radius
                 'n_sersic': 4, # sersic index
                 'center_x': 0, # x-coordinate
                 'center_y': 0, # y-coordinate
                 'e1': e1,
                 'e2': e2}
kwargs_lens_light = [kwargs_sersic]
lens_light_model_class = LightModel(light_model_list=lens_light_model_list)

# create the light model for the source (SERSIC_ELLIPSE)
source_model_list = ['SERSIC_ELLIPSE']

# set the position of the source
ra_source, dec_source = -0.1*thetaE, thetaE

kwargs_sersic_ellipse = {'amp': 4000.,
                         'R_sersic': .1,
                         'n_sersic': 3,
                         'center_x': ra_source,
                         'center_y': dec_source,
                         'e1': 0.1,
                         'e2': 0.01}
kwargs_source = [kwargs_sersic_ellipse]
source_model_class = LightModel(light_model_list=source_model_list)

# solve the lens equation and find the image positions
# using the LensEquationSolver class of Lenstronomy.
lensEquationSolver = LensEquationSolver(lens_model_class)
x_image, y_image = lensEquationSolver.image_position_from_source(ra_source,
                   dec_source,
                   kwargs_lens,
                   min_distance=deltaPix,
                   search_window=search_window=numPix * deltaPix,
                   precision_limit=1e-10, num_iter_max=100,
                   arrival_time_sort=True,
```

```
                    initial_guess_cut=True,
                    verbose=False,
                    x_center=0,
                    y_center=0,
                    num_random=0,
                    non_linear=False,
                    magnification_limit=None)

# compute lensing magnification at image positions
mag = lens_model_class.magnification(x_image, y_image,
kwargs=kwargs_lens)
mag = np.abs(mag)  # ignore the sign of the magnification

# perturb observed magnification due to e.g. micro-lensing
# the noise is generated from a normal distribution
# with mean 'mag' and standard deviation 0.5
mag_pert = np.random.normal(mag, 0.5, len(mag))

# quasar position in the lens plane
kwargs_ps = [{'ra_image': x_image,
              'dec_image': y_image,
              'point_amp': point_amp}]

point_source_list = ['LENSED_POSITION']
point_source_class =
PointSource(point_source_type_list=point_source_list,
            fixed_magnification_list=[False])

# create the simulated observation of lens and (lensed)
# source
kwargs_numerics = {'supersampling_factor': 1,
                   'supersampling_convolution': False}
# imageModel includes the details of the instrument, psf, lens,
# and source models
imageModel = ImageModel(data_class, psf_class, lens_model_class,
source_model_class,lens_light_model_class,point_source_class,
kwargs_numerics=kwargs_numerics)
# now, the simulated image is saved in image_sim
image_sim = imageModel.image(kwargs_lens, kwargs_source,
kwargs_lens_light, kwargs_ps)

# add noise and background
poisson = image_util.add_poisson(image_sim, exp_time=exp_time)
bkg = image_util.add_background(image_sim, sigma_bkd=background_rms)
image_sim = image_sim + bkg + poisson
```

The resulting simulated image is shown in Fig. 6.6. The host galaxy is distorted to form a broken Einstein ring. The quasar has four images, whose positions are indicated with yellow filled circles. We will use these images as constraints to determine the mass distribution of the lens.

Lens Modeling

To begin, we only use the positions of the multiple images of the quasar as
constraints to model the lens galaxy. In a realistic situation, these positions,
measured on the astronomical image, are affected by some uncertainty. We mimic a
positional error by adding a small scatter (0.015″) to the true image positions stored
in the variables x_image and y_image:

```
mu, sigma = 0, 0.015 # mean and standard deviation
s1 = np.random.normal(mu, sigma, len(x_image))
s2 = np.random.normal(mu, sigma, len(y_image))
x1_ima=x_image_+s1
x2_ima=y_image_+s2
```

Now, we can start the fitting process and attempt to recover the input parameters
of the lens. Instead of using the SIE model implementation of LENSTRONOMY, we
use the sie_lens class introduced in Sect. 5.11.1.

We make a guess about the lens parameters ($\sigma_v = 180$ km/s, $f = 0.3$, and
$\varphi = 40$ degrees) and use the lens equation to map the images onto the source plane:

```
# function that calculates the source positions of an ensemble
# of point images
def guess_source(sie_m,x1_ima,x2_ima):
    # calculate the deflection angle at the image positions
    phi=np.arctan2(x2_ima,x1_ima)
    a1_ima_,a2_ima_=sie_m.alpha(phi-sie_m.pa)
    # apply rotation by the lens position angle
    a1_ima=a1_ima_*np.cos(sie_m.pa)-a2_ima_*np.sin(sie_m.pa)
    a2_ima=a1_ima_*np.sin(sie_m.pa)+a2_ima_*np.cos(sie_m.pa)
    # use lens equation to find source positions
    y1_ima=x1_ima-a1_ima
    y2_ima=x2_ima-a2_ima
    return y1_ima, y2_ima

# first guess for the lens model: assuming sigmav=180 km/s,
# f=0.3, and pa=40 degrees
sie_m=sie_lens(co,sigmav=180.0,zl=0.3,zs=2.0,f=0.3,pa=40.0)

# calculate the source positions using the model
y1_ima,y2_ima = guess_source(sie_m,
                             x1_ima/sie_m.theta0,
                             x2_ima/sie_m.theta0)
```

We find four different source positions, one for each quasar image, because the
lens model is not the correct one. The orange stars indicate these in the left panel of
Fig. 6.10. The tangential caustic and the lens model cut are marked with solid and
dashed lines, respectively.

We may assume that the best guess for the quasar unlensed position, given this lens model, is the mean position of these four sources, indicated by the blue star. We can now map this source back to the lens plane by solving the lens equation:

```
x_m,phi_m=sie_m.phi_ima(y1_m,y2_m,checkplot=False,verbose=False)
x1_ima_m=x_m*np.cos(phi_m)
x2_ima_m=x_m*np.sin(phi_m)
```

In the central panel of Fig. 6.10, the model-predicted image positions of the quasar are shown as red dots. The green dots indicate the observed positions of the quasar images, instead. As discussed in Sect. 6.1.5, to perform the image-plane optimization, we need to construct a cost function to compare the observed and the model-predicted image positions. We proceed as follows. For each observed image of the quasar, we find the closest model-predicted image. Then, we measure the distance between the two positions. Such distances are visualized as blue sticks in the central panel of Fig. 6.10.

The next step consists of exploring the parameter space looking for the combination of σ_v, f, and φ that minimizes these distances. As done in Sect. 4.9.2, we use the package lmfit for this purpose. The cost function implemented below returns an array of differences between data and model. This is inconsistent with the formula in Eq. 6.7. However, note the sum of the squares of the array will be sent to the fitting function. This is how the minimize function works.

```
import lmfit

# parameter set for the initial guesses. For each parameter
# we indicate also the range of values where the solution will
# be searched (e.g. sigma_v is initially set to 130 km/s and
# the range is between 50 and 300 km/s)
p = lmfit.Parameters()
p.add_many(('sigmav', 130., True, 50, 300),
           ('f', 0.8, True, 0.2, 1.0),
           ('pa', 45.0, True, 20., 60.))

# implementation of the cost function:
# returns the distances between observed and model predicted
# image positions.
def cost_function(p,x1_ima,x2_ima,sigma_ima):
    sie_m=sie_lens(co,sigmav=p['sigmav'],
                   zl=0.3,zs=2.0,
                   f=p['f'],pa=p['pa'])
    y1_ima,y2_ima =
    guess_source(sie_m,x1_ima/sie_m.theta0,x2_ima/sie_m.theta0)
    y1_m=y1_ima.mean()
    y2_m=y2_ima.mean()
    x_m,phi_m=sie_m.phi_ima(y1_m,y2_m,checkplot=False,
                            verbose=False)
    x1_ima_m=x_m*np.cos(phi_m)*sie_m.theta0
    x2_ima_m=x_m*np.sin(phi_m)*sie_m.theta0
    imod=[]
    for i in range(len(x1_ima)):
```

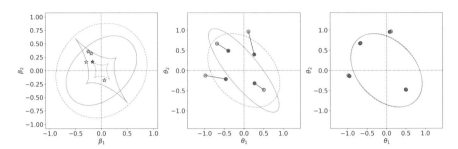

Fig. 6.10 Visualization of the several steps involved in the parametric reconstruction of the lens in Fig. 6.6. The algorithm implements the image-plane optimization procedure. Left panel: using a guess model, the multiple images of the quasar are mapped onto the source plane (yellow stars). The mean of their positions (blue star) is used as a guess for the source position. The solid gray curves indicate the caustic and the cut of the lens model, respectively. The dashed curves give true caustic and cut. Central panel: solving the lens equation, the blue star images are found (red circles) and compared to the observed image positions of the quasar (orange circles). The blue sticks show the distances between each quasar image and the closest model-predicted image. Right panel: by minimizing the distances between the model-predicted and the observed quasar positions, the best combination of lens parameters is found. The solid gray curves in the central and in the right panels are the lens model's critical lines before and after optimization, respectively. The dashed gray line gives the true lens critical line in both panels

```
d=(x1_ima[i]-x1_ima_m)**2+(x2_ima[i]-x2_ima_m)**2
imod.append(np.argmin(d))

res1=(x1_ima_m[imod]-x1_ima)/sigma_ima
res2=(x2_ima_m[imod]-x2_ima)/sigma_ima

return res1, res2

# minimize the cost function (here using the 'powell' method)
mi = lmfit.minimize(cost_function, p,
                    method='powell',
                    args=(x1_ima,x2_ima,sigma_ima))
```

We assume that the error on each image position (`sigma_ima`) is 0.015″. The minimization returns the following values for the best-fit model parameters: $\sigma_v = 200.52$ km/s, $f = 0.69$, $\varphi = 46.15$ degrees. With these parameters, the model-predicted image positions match very well the observed image positions, as shown in the right panel of Fig. 6.10. The reduced χ^2 of the fit is 0.98.

Finally, using the `emcee` package, we calculate the posterior probability distributions for the parameters and estimate the errors. We can visualize them using the corner package (Foreman-Mackey 2016), as shown in Fig. 6.11. The medians of the probability distributions and a 1σ quantile, estimated as half the difference between the 16th and 84th percentiles, are printed above the 1D-histograms showing the marginalized distributions. Note that when using the `minimize` function with the emcee method, the cost function has been redefined such as to return a float value,

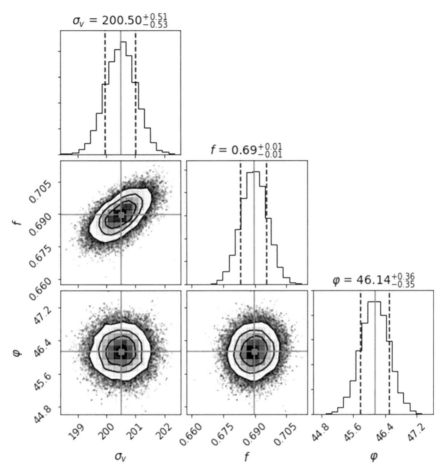

Fig. 6.11 Corner plot showing the posterior probability distributions of the model parameters based on the positions of the multiple images of the lensed quasar shown in Fig. 6.6. The medians and the 16th and 84th percentiles of the marginalized distributions for the velocity dispersion (σ_v), axis ratio (f), and position angle (φ) are quoted and shown as solid and dashed vertical lines in each histogram

i.e. the χ^2, as specified by setting `float_behavior='chi2'` in the function call:

```
# redefine the cost function such that it returns the chi2
def chi2(p,x1_ima,x2_ima):
    d1,d2=cost_function(p,x1_ima,x2_ima)
    return np.sqrt(d1**2+d2**2)

# call the minimize function using the method 'emcee' and
# the cost function chi2. We choose to run 1000 steps of MCMC
#and to set the 'burn-in' steps to 300
```

```
res = lmfit.minimize(chi2, method='emcee',
                     nan_policy='omit', burn=300, steps=1000,
                     params=mi.params,
                     float_behavior='chi2',
                     progress=True,args=(x1_ima,x2_ima))

# show corner plot (confidence limits, parameter
# distributions, correlations)
import corner
figure = corner.corner(res.flatchain,
                labels=[r"$\sigma_v$", r"$f$",
                        r"$\varphi$"],
                truths=list(res.params.valuesdict().values()),
                quantiles=[0.16, 0.84],
                show_titles=True,
                title_kwargs={"fontsize": 14},
                label_kwargs={"fontsize": 14})
for ax in figure.get_axes():
    ax.tick_params(axis='both', labelsize=12)
```

Using More Constraints

If additional constraints are available, they can be included in the cost function. For example, we may use the information derived from the fluxes of the multiple images. While creating the mock lens, the magnification factors of the quasar images (with some perturbations) were stored in the variable `mag_pert`. Instead of fitting the fluxes (which would imply adding another free parameter, namely the intrinsic quasar flux), we opt here for fitting the flux ratios between the images. In particular, we choose to compute the fluxes relative to the first image. Note that, when we used the `LensEquationSolver` class of LENSTRONOMY, we requested to sort the multiple images by their arrival time by setting the keyword `arrival_time_sort=True`. The flux ratios are then computed as

```
fr_ima=(mag_pert/mag_pert[0])
```

The cost function which also accounts for the measured flux ratios can be written as follows:

```
def cost_with_flux_ratios(p,x1_ima,x2_ima,fr_ima,sigma_ima,sigma_fr):
    sie_m=sie_lens(co,sigmav=p['sigmav'],zl=0.3,zs=2.0,
                   f=p['f'],pa=p['pa'])
    y1_ima,y2_ima = guess_source(sie_m,
                                 x1_ima/sie_m.theta0,
                                 x2_ima/sie_m.theta0)
    y1_m=y1_ima.mean()
    y2_m=y2_ima.mean()
    x_m,phi_m=sie_m.phi_ima(y1_m,y2_m,checkplot=False,
                            verbose=False)
    x1_ima_m=x_m*np.cos(phi_m)*sie_m.theta0
    x2_ima_m=x_m*np.sin(phi_m)*sie_m.theta0
    mu_ima_m=sie_m.mu(x_m,phi_m)
    imod=[]
```

```
    for i in range(len(x1_ima)):
        d=(x1_ima[i]-x1_ima_m)**2+(x2_ima[i]-x2_ima_m)**2
        imod.append(np.argmin(d))
    # compute the flux ratios
    m_ord=np.abs(mu_ima_m[imod])
    fr_ima_m=m_ord/m_ord[0]

    # compute differences between model and data
    res1=(x1_ima_m[imod]-x1_ima)/sigma_ima
    res2=(x2_ima_m[imod]-x2_ima)/sigma_ima
    res3=(fr_ima_m-fr_ima)/sigma_fr

    return res1, res2, res3
```

where `sigma_fr` are the errors on the flux ratio measurements. Repeating the fitting procedure outlined above using this cost function slightly improves the accuracy and precision of the fit. The estimated parameters are $\sigma_v = 200.40^{+0.49}_{-0.51}$, $f = 0.72^{+0.01}_{-0.01}$, and $\varphi = 44.91^{+0.34}_{-0.35}$.

We can also combine constraints from other families of multiple images. For example, the blue circles in Fig. 6.12 show the two images of a second source at $z_S = 4$, which can be used in combination with the four images of the quasar indicated again with the orange circles. Let us assume that the observed positions of these additional multiple images are stored in the arrays `x1_ima2`, `x2_ima2`. We can proceed as follows:

```
# pack positions and errors of all images of the first and the
# second source into lists
x1_allima = [x1_ima,x1_ima2]
x2_allima = [x2_ima,x2_ima2]
sigma_allima = [sigma_ima,sigma_ima2]

# here is the new cost function: iterate over families of
# multiple images to compute the differences between the
# data and model

def cost_with_morefamilies(p,x1_allima,x2_allima,
                           sigma_allima,zs_fam):

    res1=[]
    res2=[]
    # loop over image families
    for j in range(len(x1_allima)):
        sie_m=sie_lens(co,sigmav=p['sigmav'],
                    zl=0.3,zs=zs_fam[j],
                    f=p['f'],pa=p['pa'])
        x1_ima=x1_allima[j]
        x2_ima=x2_allima[j]
        sigma_ima=sigma_allima[j]
        # compute the source position
        y1_ima,y2_ima = guess_source(sie_m,
                        x1_ima/sie_m.theta0,
```

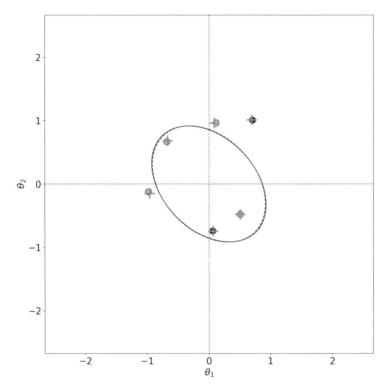

Fig. 6.12 Representation of a so-called *compound lens*. In this simulation, we used the same lens shown in Fig. 6.6. In addition to the four images of a quasar at redshift $z_S = 2$ (orange circles), the lens produces two images of another source at redshift $z_{S,2} = 4$. These two families of multiple images are used to perform the lens inversion. After optimizing in the image plane, the model-predicted image positions are shown with gray crosses

```
                                        x2_ima/sie_m.theta0)
y1_m=y1_ima.mean()
y2_m=y2_ima.mean()
# compute the images
x_m,phi_m=sie_m.phi_ima(y1_m,y2_m,
                        checkplot=False,
                        verbose=False)
x1_ima_m=x_m*np.cos(phi_m)*sie_m.theta0
x2_ima_m=x_m*np.sin(phi_m)*sie_m.theta0
# associate the predicted images to the observed ones by
using the distance
imod=[]
for i in range(len(x1_ima)):
    d=(x1_ima[i]-x1_ima_m)**2+(x2_ima[i]-x2_ima_m)**2
    res1.append((x1_ima_m[np.argmin(d)]-x1_ima[i])
    /sigma_ima[i])
    res2.append((x2_ima_m[np.argmin(d)]-x2_ima[i])
```

```
              /sigma_ima[i])
    return np.array(res1), np.array(res2)

# we specify the redshift of the two families of multiple
# images
zs_fam=[2.0,4.0]

# minimize the cost function using lmfit.minimize
mi = lmfit.minimize(cost_with_morefamilies, p,
                    method='powell',
                    args=(x1_allima,x2_allima,
                    sigma_allima,zs_fam))
```

The gray crosses in Fig. 6.12 show the model-predicted positions of the multiple images after optimization in the image plane. The model and the lens's true critical lines are given by the solid and dashed curves, respectively. The agreement between the model and the truth seems very good.

Optimization in the Source Plane

The optimization in the image plane is often computationally demanding because it requires to find the solutions of the lens equation at each iteration and for each family of multiple images. An alternative and faster approach consists of finding the best combination of model parameters that minimize the scatter between the predicted source positions obtained by de-lensing each family of multiple images. As said in Sect. 6.1.5, this process is called optimization in the source plane and it is illustrated in Fig. 6.13. The figure is similar to the left panel of Fig. 6.10. The yellow stars are the positions of the quasar's multiple images mapped onto the source plane using the lens model. The red star is the average of the positions of the four yellow stars. The blue sticks show the distances between the yellow and red stars. After optimization, the four yellow stars are brought very close to each other and the red star.

We can implement the optimization in the source plane very easily by writing the appropriate cost function (e.g. see Eq. 6.11). For example,

```
def cost_sp(p,x1_allima,x2_allima):
    sie_m=sie_lens(co,sigmav=p['sigmav'],zl=0.3,zs=2.0,f=p['f'],pa=p['pa'])
    res1=[]
    res2=[]
    # loop over image families
    for j in range(len(x1_allima)):
        x1_ima=x1_allima[j]
        x2_ima=x2_allima[j]
        # compute the source position
        y1_ima,y2_ima = guess_source(sie_m,
                                     x1_ima/sie_m.theta0,
                                     x2_ima/sie_m.theta0)
        y1_m=y1_ima.mean()
```

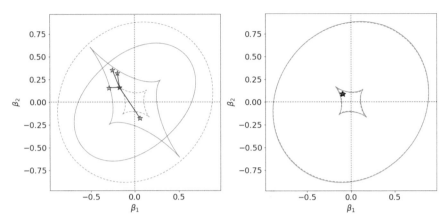

Fig. 6.13 Optimization in the source plane. Left panel: as in the left panel of Fig. 6.10, the positions of the four images of the quasar, mapped onto the source plane using the guessing model, are given by the yellow stars. The mean of these positions is given by the red star instead. We compute the distances between the yellow and the red stars, here indicated by the blue sticks. Right panel: by minimizing these distances, we find the best combination of parameters fitting the data. After optimization, the four yellow stars are brought very close to the red star. The model-predicted and the true lens caustics are shown with solid and dashed curves in both the left and the right panels

```
        y2_m=y2_ima.mean()
        # compute distances and pack them into two arrays
        for i in range(len(x1_ima)):
            res1.append(y1_ima[i]-y1_m)
            res2.append(y2_ima[i]-y2_m)
    return np.array(res1), np.array(res2)
```

Note that the returned distances `res1` and `res2` are not weighted by the errors on the image positions on the source plane. The estimate of these errors is tricky, because it involves the calculation of the magnification factor of each image. The magnification diverges near the critical lines. We encourage the interested readers to modify the cost function above to include the errors.

6.4.2   Parametric Weak Lensing Mass Measurement

This example shows how the mass of a galaxy cluster can be estimated by fitting with analytical models the profiles of the tangential ellipticities obtained from the weak lensing analysis.

Weak Lensing Measurements
We consider the very unrealistic situation represented in Fig. 6.7. The sources are circular and distributed on a regular grid on a single plane at $z_S = 1$. They all have Sérsic brightness profiles with parameters $n = 1$ and $r_e = 1''$.

To lens each source, we perform a full ray-tracing simulation, as shown in the example in Sect. 3.7.6. However, in this case, we use a dPIE potential for the lens. This lens model can be easily implemented using the equations in Sect. 5.5.2. The lens velocity dispersion is $\sigma_v = 1200$ km/s. The core radius is $r_{core} = 2''$ and the cut radius is $r_{cut} = 1000''$. The lens major axis points at 45 degrees from the θ_1 axis. Finally, the lens axis ratio is $q = 0.8$.

We begin by measuring the ellipticity of each source. The methods discussed in Sect. 6.2.2 are implemented in the `image_fit` class. An instance of the class is created using an image in the form of a `numpy.array` and a threshold value I_{th} which sets the source boundary (see Eq. 6.47). The centroid and the quadrupole moments of the surface brightness are measured using discrete versions of Eqs. 6.41 and 6.42. The components of the ellipticity are then obtained from Eqs. 6.45:

```python
class image_fit(object):
    """
    Measure image ellipticity and centroid from its
    surface brightness distribution
    """

    def __init__(self,img,ith=0):
        """
        Input:
        - img = image to be analyzed
        - ith = threshold defining the source boundaries
        """
        self.img=img
        self.ith=ith
        # use only pixels above threshold
        isel= img > ith
        self.img_sel=self.img[isel].flatten()
        # compute array of pixel coordinates
        theta1 = np.linspace(0, img.shape[0] - 1 ,img.shape[0])
        theta2 = np.linspace(0, img.shape[0] - 1, img.shape[0])
        x_,y_=np.meshgrid(theta1,theta2)
        self.r = np.stack((x_[isel].flatten(), y_[isel].flatten()), axis=1)
        # compute total flux, centroid, and quadrupole moments
        self.flux=self.img_sel.sum()
        self.c=self.centroid()
        self.Q=self.quadrupoles()

    def centroid(self):
        """
        Centroid calculation
        """
        c=[np.sum(self.img_sel*self.r[:,i])/np.sum(self.img_sel)
            for i in range(2)]
        return c

    def quadrupoles(self):
        """
        Quadrupole moments
```

```
    """
    Q=np.zeros((2,2))
    for i in range(2):
        for j in range(2):
            Q[i,j]=np.sum(self.img_sel*(self.r[:,i]-self.c[i])
            *(self.r[:,j]-self.c[j]))
    Q/=self.flux
    return Q

def ellipticity(self):
    """
    Ellipticity from quadrupole moments
    """
    a,b=self.axes()
    Q=self.Q
    e1=(Q[0,0]-Q[1,1])/(Q[0,0]+Q[1,1]+2*np.sqrt(Q[0,0]*Q[1,1]-Q[0,1]))
    e2=2*Q[0,1]/(Q[0,0]+Q[1,1]+2*np.sqrt(Q[0,0]*Q[1,1]-Q[0,1]))
    if (a>b):
        return e1,e2
    else:
        return -e1, -e2

def axes(self):
    """
    Axes of the ellipse
    """
    l1=0.5*(self.Q[0,0]+self.Q[1,1]+np.sqrt((self.Q[0,0]-
        self.Q[1,1])**2+4.0*self.Q[0,1]**2))
    l2=0.5*(self.Q[0,0]+self.Q[1,1]-np.sqrt((self.Q[0,0]-
        self.Q[1,1])**2+4.0*self.Q[0,1]**2))
    return 1.0/sqrt(l1),1.0/sqrt(l2)
```

Let us assume that the images of each lensed source are stored in a list `image_list` of images of size (N_{pix}, N_{pix}). In each image, a single source is present. We process all images to obtain the measurements of the ellipticity components (ϵ_1, ϵ_2), as well as the centroid coordinates of each lensed source:

```
e1_=[]
e2_=[]
c1_=[]
c2_=[]
phi_=[]
f = 0.05 # custom threshold
for image in image_list:
    imafit=image_fit(image,ith=f)
    if (len(imafit.img_sel)>0):
        # if the source has some pixel above the threshold,
        # measure the ellipticity
        e1, e2 = imafit.ellipticity()
        e1_.append(e1)
        e2_.append(e2)
        # the centroid:
        center = imafit.c
```

```
c1_.append(center[0])
c2_.append(center[1])
```

The ellipticity components (ϵ_1, ϵ_2), stored in the lists `e1_` and `e2_`, can be converted into the tangential and the cross components of the ellipticity using Eqs. 6.50 and 6.51:

```
varphi=np.arctan2(c2_,c1_)
et_=-(e1_*np.cos(2*varphi)+e2_*np.sin(2*varphi))
ex_=(e2_*np.cos(2*varphi)-e1_*np.sin(2*varphi))
```

These are shown as blue circles and green crosses in Fig. 6.8, where they are also compared to the true values of the tangential and cross components of the reduced shear at the position of each lensed source. The values are plotted as a function of the cluster-centric distance θ. In this example, we assume that the error on each ellipticity measurement `sigma_et` is constant. In more realistic cases, the errors on the measurements of the ellipticity components ϵ_1 and ϵ_2 can be expressed in terms of the errors on the quadrupole moments Q_{ij}, which depend on the sky brightness and on the shot-noise due to the source (see e.g. Appendix A of Hoekstra et al. 2000).

Fit of the Tangential Shear Profile

To estimate the lens's mass, we can fit the tangential shear profile using some parametric model. For example, if we assume that an NFW density profile well describes the mass distribution of the cluster, then, under the assumption of circular symmetry, we can use Eqs. 5.21, 5.117, and 5.116 to write a fitting function for the reduced shear. The following code shows the implementation of the class `nfwcirc`:

```
class nfwcirc(object):
    """
    class circular NFW lens
    """

    def __init__(self,co,zl=0.3,zs=2.0,c200=4.0,m200=1e15):
        """
        Initialize the nfwcirc object
        """
        self.co = co # cosmological model
        self.zl = zl # lens redshift
        self.zs = zs # source redshift
        self.m200 = m200 # Mass of the lens
        self.c200 = c200 # concentration of the lens

        # compute rhos
        self.rhos = 200./3.*
                    (self.co.critical_density(self.zl).to('Msun/Mpc3'))*
                    self.c200**3/
                    (np.log(1.+self.c200)-self.c200/
                    (1.0+self.c200))
```

```python
    # compute r200 and rs
    f200=4./3.*np.pi*200.*
        self.co.critical_density(self.zl).to('Msun/Mpc3').value
    self.r200 = (self.m200/f200)**(1./3.)
    self.rs=self.r200/self.c200

    # compute the angular diameter distances:
    self.dl=self.co.angular_diameter_distance(self.zl)
    self.ds=self.co.angular_diameter_distance(self.zs)
    self.dls=self.co.angular_diameter_distance_z1z2(self.zl,self.zs)

    # surface critical density:
    self.sc = self.sigma_crit()

    # convergence scale:
    self.ks = self.rhos.value * self.rs / self.sc

def sigma_crit(self):
    """
    Function to compute the critical surface density
    """
    c2G = (const.c ** 2 / const.G).to(units.Msun / units.Mpc)
    factor = c2G / (4 * np.pi)
    return (factor * (self.ds / (self.dl * self.dls))).value

def kappap(self,r):
    """
    Convergence at radius r [Mpc].
    """
    x = r/self.rs
    fx = np.piecewise(x, [x > 1., x < 1., x==1.],
                      [lambda x: (1 - (2.0 / np.sqrt(x * x - 1.) *
                                      np.arctan(np.sqrt((x - 1.) /
                                      (x + 1.))))))/
                                      (x**2 - 1),
                       lambda x: (1 - (2.0 / np.sqrt(1. - x * x) *
                                      np.arctanh(np.sqrt((1.- x) /
                                      (1. + x))))))/
                                      (x**2 - 1),
                       0.])
    kappa = 2.0 * self.ks * fx
    return kappa

def massp(self,r):
    """
    Dimensionless mass at radius r [Mpc]
    """
    x = r/self.rs
    fx = np.piecewise(x, [x > 1., x < 1., x==1.],
                      [lambda x:  (2.0 / np.sqrt(x * x - 1.) *
                                      np.arctan(np.sqrt((x - 1.) /
                                      (x + 1.)))),
                       lambda x:  (2.0 / np.sqrt(1. - x * x) *
```

```
                                      np.arctanh(np.sqrt((1.- x) /
                                      (1. + x))))),
                          0])
    massp = 4.0 * self.ks * (np.log(x / 2.) + fx)
    return massp

def shearp(self,r):
    """
    Shear at radius r [Mpc]
    """
    kp = self.kappap(r)
    mp = self.massp(r)
    x = r/self.rs
    gammap = -(kp-mp/x**2)
    return gammap

def redshearp(self,r):
    """
    Reduced shear at radius r [Mpc]
    """
    redgammap = np.abs(self.shearp(r) / (1. - self.kappap(r)))
    return redgammap
```

The function `redshearp` was used to fit the blue data-points in Fig. 6.8 (employing the same method used in the previous examples). The input parameters are the logarithm of the mass M_{200} (i.e. the mass within a sphere enclosing a mean over-density equal to 200 times the critical density of the universe) and the concentration c_{200}.

```
def residuals(p,co,zl,zs,r_as,e,err):
    """
    The cost function to be minimized: it returns the differences
    between the model predicted and the measured source
    ellipticities
    """
    # given a set of parameters (c200 and logm200),
    # create an instance of the nfwcirc class
    nc=nfwcirc(co,zl=zl,zs=zs,c200=p['c200'],m200=10**p['logm200'])
    # convert arcseconds to Mpc
    r = r_as/180.0/3600.0*np.pi*nc.dl.value
    # compute the reduced shear at the distances of each
    # source from the lens center
    gmodel = nc.redshearp(r)
    # calculate the difference between data and model
    res = (e - gmodel)/err
    return res

import lmfit

# initial guesses
p = lmfit.Parameters()
p.add_many(('c200', 4.0,True, 0.01,30.0), ('logm200', 15, True, 14, 16))
```

```
# redshifts of lens and source
zl = 0.5
zs = 1.0
# find best fit: the data are stored in three arrays:
# r: distances of the lensed sources from the cluster center
# et: tangential component of the ellipticity
# sigma_et: error on the ellipticity measurement
mi = lmfit.minimize(residuals,p,args=(co,zl,zs,r,et,sigma_et))
```

The dotted blue line gives the best-fit reduced shear profile in Fig. 6.8. The best-fit parameters are $\log_{10}(M_{200}) \sim 15$ and $c_{200} \sim 5.2$. For comparison, the true mass of the lens is $\log_{10}(M_{200,true}) = 15.04$. It is not surprising that the fit is not perfect. Indeed, we used a dPIE lens to simulate the lensing distortions, not an NFW lens. Note that, even in realistic cases, the true shape of the lens density profile is unknown! Besides, the lens is not circular, which is why the data-points are not perfectly aligned.

The data to be fit is much noisier in more realistic cases than shown in Fig. 6.8. Indeed, as discussed in Sects. 6.2.6, galaxies are not intrinsically circular and not at the same redshift. They are typically faint objects, with irregular morphologies, immersed in a noisy background. Besides, their images are smeared and distorted by the PSF.

6.4.3   The Kaiser-Squires Inversion Algorithm

In this example, we discuss the implementation of the Kaiser and Squires (1993) inversion algorithm discussed in Sect. 6.2.5.

We consider a more complex lens compared to the previous examples. It is composed of two mass clumps described by dPIE models. The first represents a massive galaxy cluster with velocity dispersion $\sigma_v = 1200$ km/s positioned at the center of the field-of-view; the second is $\sigma_v = 800$ km/s companion, which is located approximately 300 arcsec away from the first lens. Their redshift is $z_L = 0.5$. The maps of the convergence and the two shear components for source redshift $z_S = 1$ are shown in Fig. 6.14. The maps are 128×128 pixel and cover a field-of-view of 1800×1800 arcsec.

The algorithm is as follows:

• We compute the Fourier transforms of the two components of the shear, $\tilde{\gamma}_1$ and $\tilde{\gamma}_2$. We use zero-padding to limit the artifacts due to boundary periodic conditions.

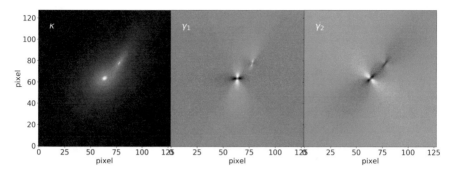

Fig. 6.14 Convergence and shear maps of the lens used in the example in Sect. 6.4.3. The field-of-view is 1800×1800 arc s

- We multiply $\tilde{\gamma}_1$ and $\tilde{\gamma}_2$ by the functions

$$\tilde{D}_1 = \frac{k_1^2 - k_2^2}{k^2} \tag{6.89}$$

$$\tilde{D}_2 = \frac{2k_1 k_2}{k^2} , \tag{6.90}$$

as shown in Eq. 6.70. We obtain the Fourier transform of the convergence, $\tilde{\kappa}$.
- Finally, we compute the inverse Fourier transform of $\tilde{\kappa}$ to obtain the convergence map κ.

The Fourier transforms are computed using the `scipy.fftpack` package. The functions employed for zero-padding and cropping the maps were introduced in Sect. 2.5.2:

```python
import scipy.fftpack as fftengine
def KS93(g1,g2):
    """
    Implementation of the KS93 algorithm:
    :parameters: g1, g2 - maps of the shear
                 components
    :returns: kappa - convergence map
    """
    # zero-padding of the shear maps
    g1_pad=gpad(g1)
    g2_pad=gpad(g2)

    # computation of the kernel on the
    # grid
    D1,D2=kernel(g1_pad.shape[0])

    # Fourier transforms of the shear
    # components
    g1ft = fftengine.fftn(g1_pad)
```

```
    g2ft = fftengine.fftn(g2_pad)

    # computation of the FT of the
    # convergence
    kappaft=D1*g1ft+D2*g2ft

    # inverse FT to obtain the convergence
    # map
    kappa=fftengine.ifftn(kappaft)

    # return the map cropping out the zero
    # padded region
    return mapcrop(kappa.real,g1.shape[0])

def kernel(n):
    """
    implement the kernel function D:
    :parameters: n - size of the grid
                 (assumed to be a square
                  grid of size n x n)
    :returns: D1, D2
    """
    kx,ky = np.meshgrid(fftengine.fftfreq(n),fftengine.fftfreq(n))
    norm=(kx**2+ky**2+1e-12)
    D1=(kx**2-ky**2)/norm
    D2=2*kx*ky/norm
    return(D1,D2)

def gpad(gmap):
    """
    Zero pad the input map

    :return: zero-padded convergence map
    """
    def padwithzeros(vector, pad_width, iaxis, kwargs):
        vector[:pad_width[0]] = 0
        vector[-pad_width[1]:] = 0
        return vector
    return np.lib.pad(gmap, 2*gmap.shape[0],padwithzeros)

def mapcrop(inmap,n):
    """
    Crop the map by removing the region added for zero padding
    :param inmap: input map to be cropped (e.g. the lensing potential
    :return: outmap - cropped map
    """
    xmin=int(inmap.shape[0]/2-n/2)
    ymin=int(inmap.shape[1]/2-n/2)
    xmax=int(xmin+n)
    ymax=int(ymin+n)
    outmap=inmap[xmin:xmax,ymin:ymax]
    return(outmap)
```

To compute the convergence map, we call the function KS93, passing the shear components γ_1 and γ_2:

```
kappa=KS93(gamma1,gamma2)
```

Unfortunately, it is not that simple. For instance, we cannot measure γ_1 and γ_2. Instead, we can use the galaxy ellipticities to estimate the reduced shear components, g_1 and g_2.

Kaiser (1995) and Seitz and Schneider (1996) generalized the Kaiser and Squires inversion to the non-linear regime, by solving the integral equation obtained from Eq. 6.71 by replacing γ by $(1 - \kappa)g$:

$$\kappa(\vec{\theta}) = \frac{1}{\pi} \int \left[D_1(\vec{\theta} - \vec{\theta}')g_1(\vec{\theta}')[1 - \kappa(\vec{\theta}')] + D_2(\vec{\theta} - \vec{\theta}')g_2(\vec{\theta}')[1 - \kappa(\vec{\theta}')] \right] d^2\theta' .$$

(6.91)

The integral can then be solved through an iterative approach. We begin with setting $\kappa = 0$ inside the integral and obtain a first convergence map. In the next iteration, this map is inserted in the integral, and a new solution is found. After a few iterations, the result becomes stable, and the procedure returns the final convergence map:

```
# compute the reduced shear maps to be fed into KS93:
g1=gamma1/(1.0-kappa)
g2=gamma2/(1.0-kappa)

# initial guess for the convergence:
# kappa=0
kiter=np.zeros((npix,npix))

# we display the maps at each iteration in
# a row of figures
fig,ax=plt.subplots(1,5,figsize=(20,10),sharey=True,
                    gridspec_kw={'wspace': 0})

# we make 5 iterations
for iteray in range(5):
    k_new=KS93(g1*(1.-kiter),g2*(1.-kiter))
    # we assume that k_new > 0
    kiter=k_new-k_new.min()

    # visualize the map
    ax[iteray].imshow(np.sqrt(kiter),origin='low',vmax=kiter.max()*0.7,
                    cmap='cubehelix')

    # figure cosmetics
    ax[iteray].set_xlabel('pixel',fontsize=20)
    if iteray==0:
        ax[iteray].set_ylabel('pixel',fontsize=20)
    ax[iteray].text(20,110,'Iter.'+str(iteray+1),color='yellow',
                    fontsize=20)
    ax[iteray].xaxis.set_tick_params(labelsize=20)
    ax[iteray].yaxis.set_tick_params(labelsize=20)
```

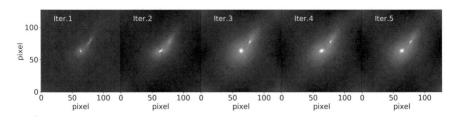

Fig. 6.15 Convergence maps produced by solving iteratively Eq. 6.91. Five iterations are performed. The maps at each iteration 1–5 are shown in a sequence of images from the left to the right

The results of after five iterations are shown in Fig. 6.15. From the left to the right, we can notice how the reconstruction improves until we obtain a map that looks very similar to that in the left panel of Fig. 6.14. Note that the convergence map returned by the function KS93 is re-scaled by shifting its minimum to zero at each iteration. This avoids non-physical negative convergence values. Given that the convergence decreases toward the boundaries of the field-of-view, this is equivalent to assuming that the convergence is zero at large distance from the lens center. This is possible way to break the mass-sheet degeneracy.

References

Abbott, T. M. C., Aguena, M., Alarcon, A., Allam, S., Allen, S., Annis, J., . . . DES Collaboration. (2020). Dark energy survey year 1 results: Cosmological constraints from cluster abundances and weak lensing. *Physics Review, 102*(2), 023509. https://doi.org/10.1103/PhysRevD.102. 023509. arXiv: 2002.11124 [astro-ph.CO]

Alard, C. (2006). Automated detection of gravitational arcs. arXiv e-prints, astro-ph/0606757. arXiv: astro-ph/0606757 [astro-ph]

Atek, H., Richard, J., Jauzac, M., Kneib, J.-P., Natarajan, P., Limousin, M., . . . Clement, B. (2015). Are ultra-faint galaxies at z = 6–8 responsible for cosmic reionization? Combined constraints from the hubble frontier fields clusters and parallels. *ApJ, 814*(1), 69. https://doi.org/10.1088/ 0004-637X/814/1/69. arXiv: 1509.06764 [astro-ph.GA]

Auger, M. W., Treu, T., Bolton, A. S., Gavazzi, R., Koopmans, L. V. E., Marshall, P. J., . . . Moustakas, L. A. (2009). The Sloan lens ACS survey. IX. Colors, lensing, and stellar masses of early-type galaxies. *ApJ, 705*(2), 1099–1115. https://doi.org/10.1088/0004-637X/705/2/1099. arXiv: 0911.2471 [astro-ph.CO]

Bacon, R., Accardo, M., Adjali, L., Anwand, H., Bauer, S., Biswas, I., . . . Yerle, N. (2010). The MUSE second-generation VLT instrument. In *Ground-based and airborne instrumentation for astronomy III* (Vol. 7735, pp. 773508). Society of Photo-Optical Instrumentation Engineers (SPIE) Conference Series. https://doi.org/10.1117/12.856027

Becker, M. R., & Kravtsov, A. V. (2011). On the accuracy of weak-lensing cluster mass reconstructions. *ApJ, 740*(1), 25. https://doi.org/10.1088/0004-637X/740/1/25. arXiv: 1011.1681 [astro-ph.CO]

Bender, R., Burstein, D., & Faber, S. M. (1992). Dynamically hot galaxies. I. Structural properties. *ApJ, 399*, 462. https://doi.org/10.1086/171940

Bergamini, P., Rosati, P., Mercurio, A., Grillo, C., Caminha, G. B., Meneghetti, M., ... Vanzella, E. (2019). Enhanced cluster lensing models with measured galaxy kinematics. *A&A, 631*, A130. https://doi.org/10.1051/0004-6361/201935974

Binney, J., & Tremaine, S. (2008). *Galactic dynamics* (2d ed.).

Birrer, S., & Amara, A. (2018). Lenstronomy: Multi-purpose gravitational lens modelling software package. *Physics of the Dark Universe, 22*, 189–201. https://doi.org/10.1016/j.dark.2018.11. 002. arXiv: 1803.09746 [astro-ph.CO]

Birrer, S., Amara, A., & Refregier, A. (2015). Gravitational lens modeling with basis sets. *ApJ, 813*(2), 102. https://doi.org/10.1088/0004-637X/813/2/102. arXiv: 1504.07629 [astro-ph.CO]

Blandford, R. D., Surpi, G., & Kundić, T. (2001). Modeling galaxy lenses. In T. G. Brainerd & C. S. Kochanek (Eds.), *Gravitational lensing: Recent progress and future go* (Vol. 237, p. 65). Astronomical Society of the Pacific Conference Series. arXiv: astro-ph/0001496 [astro-ph]

Bolton, A. S., Burles, S., Koopmans, L. V. E., Treu, T., & Moustakas, L. A. (2006). The Sloan lens ACS survey. I. A large spectroscopically selected sample of massive early-type lens galaxies. *ApJ, 638*(2), 703–724. https://doi.org/10.1086/498884. arXiv: astro-ph/0511453 [astro-ph]

Bonamigo, M., Despali, G., Limousin, M., Angulo, R., Giocoli, C., & Soucail, G. (2015). Universality of dark matter haloes shape over six decades in mass: Insights from the Millennium XXL and SBARBINE simulations. *MNRAS, 449*(3), 3171–3182. https://doi.org/10.1093/mnras/ stv417. arXiv: 1410.0015 [astro-ph.CO]

Bonamigo, M., Grillo, C., Ettori, S., Caminha, G. B., Rosati, P., Mercurio, A., ... Lombardi, M. (2018). Dissection of the collisional and collisionless mass components in a mini sample of CLASH and HFF massive galaxy clusters at z ≈ 0.4. *ApJ, 864*, 98. https://doi.org/10.3847/ 1538-4357/aad4a7. arXiv: 1807.10286 [astro-ph.GA]

Bonvin, V., Millon, M., Chan, J. H.-H., Courbin, F., Rusu, C. E., Sluse, D., ... Meylan, G. (2019). COSMOGRAIL. XVIII. Time delays of the quadruply lensed quasar WFI2033-4723. *A & A, 629*, A97. https://doi.org/10.1051/0004-6361/201935921. arXiv: 1905.08260 [astro-ph.CO]

Bradač, M., Allen, S. W., Treu, T., Ebeling, H., Massey, R., Morris, R. G., ... Applegate, D. (2008). Revealing the properties of dark matter in the merging cluster MACS J0025.4-1222. *ApJ, 687*(2), 959–967. https://doi.org/10.1086/591246. arXiv: 0806.2320 [astro-ph]

Bradač, M., Erben, T., Schneider, P., Hildebrand t, H., Lombardi, M., Schirmer, M., ... Schindler, S. (2005). Strong and weak lensing united. II. The cluster mass distribution of the most X-ray luminous cluster RX J1347.5-1145. *A & A, 437*(1), 49–60. https://doi.org/10.1051/0004-6361: 20042234

Bradley, L. D., Zitrin, A., Coe, D., Bouwens, R., Postman, M., Balestra, I., ... Molino, A. (2014). CLASH: A census of magnified star-forming galaxies at z ∼ 6–8. *ApJ, 792*(1), 76. https://doi. org/10.1088/0004-637X/792/1/76. arXiv: 1308.1692 [astro-ph.CO]

Brainerd, T. G., Blandford, R. D., & Smail, I. (1996). Weak gravitational lensing by galaxies. *ApJ, 466*, 623. https://doi.org/10.1086/177537. arXiv: astro-ph/9503073 [astro-ph]

Brewer, B. J., Dutton, A. A., Treu, T., Auger, M. W., Marshall, P. J., Barnabè, M., ... Koopmans, L. V. E. (2012). The SWELLS survey—III. Disfavouring 'heavy' initial mass functions for spiral lens galaxies. *MNRAS, 422*(4), 3574–3590. https://doi.org/10.1111/j.1365-2966.2012.20870.x. arXiv: 1201.1677 [astro-ph.CO]

Brownstein, J. R., Bolton, A. S., Schlegel, D. J., Eisenstein, D. J., Kochanek, C. S., Connolly, N., ... Weaver, B. A. (2012). The BOSS emission-line lens survey (BELLS). I. A large spectroscopically selected sample of lens galaxies at redshift ∼0.5. *ApJ, 744*(1), 41. https:// doi.org/10.1088/0004-637X/744/1/41. arXiv: 1112.3683 [astro-ph.CO]

Bullock, J. S., Kolatt, T. S., Sigad, Y., Somerville, R. S., Kravtsov, A. V., Klypin, A. A., ... Dekel, A. (2001). Profiles of dark haloes: evolution, scatter and environment. *MNRAS, 321*(3), 559–575. https://doi.org/10.1046/j.1365-8711.2001.04068.x. arXiv: astro-ph/9908159 [astro-ph]

Cabanac, R. A., Alard, C., Dantel-Fort, M., Fort, B., Gavazzi, R., Gomez, P., ... Valls-Gabaud, D. (2007). The CFHTLS strong lensing legacy survey. I. Survey overview and T0002 release sample. *A&A, 461*(3), 813–821. https://doi.org/10.1051/0004-6361:20065810. arXiv: astro- ph/0610362 [astro-ph]

Caminha, G. B., Grillo, C., Rosati, P., Balestra, I., Karman,W., Lombardi, M., ... Ziegler, B. (2016). CLASH-VLT: A highly precise strong lensing model of the galaxy cluster RXC J2248.7-4431 (Abell S1063) and prospects for cosmography. *A & A, 587*, A80. https://doi.org/10.1051/0004-6361/201527670. arXiv: 1512.04555 [astro-ph.CO]

Caminha, G. B., Grillo, C., Rosati, P., Meneghetti, M., Mercurio, A., Ettori, S., ... Zitrin, A. (2017). Mass distribution in the core of MACS J1206. Robust modeling from an exceptionally large sample of central multiple images. *A&A, 607*, A93. https://doi.org/10.1051/0004-6361/201731498. arXiv: 1707.00690

Canameras, R., Schuldt, S., Suyu, S. H., Taubenberger, S., Meinhardt, T., Leal-Taixe, L., ... Savary, E. (2020). HOLISMOKES—II. Identifying galaxy-scale strong gravitational lenses in Pan-STARRS using convolutional neural networks. arXiv e-prints, arXiv:2004.13048. arXiv: 2004.13048 [astro-ph.GA]

Carrasco, M., Zitrin, A., & Seidel, G. (2020). MIFAL: Fully automated multiple-image finder algorithm for strong-lens modelling—proof of concept. *MNRAS, 491*(3), 3778–3792. https://doi.org/10.1093/mnras/stz3040. arXiv: 1905.09802 [astro-ph.CO]

Casey, C. M., Narayanan, D., & Cooray, A. (2014). Dusty star-forming galaxies at high redshift. *Biophysical Reports, 541*(2), 45–161. https://doi.org/10.1016/j.physrep.2014.02.009. arXiv: 1402.1456 [astro-ph.CO]

Chen, G. C.-F., Fassnacht, C. D., Suyu, S. H., Rusu, C. E., Chan, J. H. H., Wong, K. C., ... Treu, T. (2019). A SHARP view of H0LiCOW: H0 from three time-delay gravitational lens systems with adaptive optics imaging. *MNRAS, 490*(2), 1743–1773. https://doi.org/10.1093/mnras/stz2547. arXiv: 1907.02533 [astro-ph.CO]

Chen, G. C.-F., Suyu, S. H., Wong, K. C., Fassnacht, C. D., Chiueh, T., Halkola, A., ... Vegetti, S. (2016). SHARP - III. First use of adaptive-optics imaging to constrain cosmology with gravitational lens time delays. *MNRAS, 462*(4), 3457–3475. https://doi.org/10.1093/mnras/stw991. arXiv: 1601.01321 [astro-ph.CO]

Ciardi, B., & Ferrara, A. (2005). The first cosmic structures and their effects. *Space Science Reviews, 16*(3–4), 625–705. https://doi.org/10.1007/s11214-005-3592-0. arXiv: astro-ph/0409018 [astro-ph]

Clowe, D., Luppino, G. A., Kaiser, N., Henry, J. P., & Gioia, I. M. (1998). Weak lensing by two z approximately 0.8 clusters of galaxies. *ApJL, 497*(2), L61–L64. https://doi.org/10.1086/311285

Coe, D., Fuselier, E., Benítez, N., Broadhurst, T., Frye, B., & Ford, H. (2008). LensPerfect: Gravitational lens mass map reconstructions yielding exact reproduction of all multiple images. *ApJ, 681*(2), 814–830. https://doi.org/10.1086/588250. arXiv: 0803.1199 [astro-ph]

Coe, D., Zitrin, A., Carrasco, M., Shu, X., Zheng, W., Postman, M., ... Rosati, P. (2013). CLASH: Three strongly lensed images of a candidate z ≈ 11 galaxy. *ApJ, 762*(1), 32. https://doi.org/10.1088/0004-637X/762/1/32. arXiv: 1211.3663 [astro-ph.CO]

Coles, J. P., Read, J. I., & Saha, P. (2014). Gravitational lens recovery with GLASS: Measuring the mass profile and shape of a lens. *MNRAS, 445*(3), 2181–2197. https://doi.org/10.1093/mnras/stu1781. arXiv: 1401.7990 [astro-ph.CO]

Courbin, F., Chantry, V., Revaz, Y., Sluse, D., Faure, C., Tewes, M., ... Meylan, G. (2011). COSMOGRAIL: The cosmological monitoring of gravitational lenses. IX. Time delays, lens dynamics and Baryonic fraction in HE 0435-1223. *A & A, 536*, A53. https://doi.org/10.1051/0004-6361/201015709. arXiv: 1009.1473 [astro-ph.CO]

D'Aloisio, A., & Natarajan, P. (2011). Cosmography with cluster strong lenses: The influence of substructure and line-of-sight haloes. *MNRAS, 411*(3), 1628–1640. https://doi.org/10.1111/j.1365-2966.2010.17795.x. arXiv: 1010.0004 [astro-ph.CO]

Dahle, H., Aghanim, N., Guennou, L., Hudelot, P., Kneissl, R., Pointecouteau, E., ... Sunyaev, R. (2016). Discovery of an exceptionally bright giant arc at z = 2.369, gravitationally lensed by the Planck cluster PSZ1 G311.65-18.48. *A & A, 590*, L4. https://doi.org/10.1051/0004-6361/201628297

de Jong, J. T. A., Verdoes Kleijn, G. A., Erben, T., Hildebrandt, H., Kuijken, K., Sikkema, G., ... Viola, M. (2017). The third data release of the Kilo-Degree Survey and associated data products. *A & A, 604*, A134. https://doi.org/10.1051/0004-6361/201730747. arXiv: 1703.02991 [astro-ph.GA]

Despali, G., Giocoli, C., Bonamigo, M., Limousin, M., & Tormen, G. (2017). A look into the inside of haloes: A characterization of the halo shape as a function of overdensity in the Planck cosmology. *MNRAS, 466*(1), 181–193. https://doi.org/10.1093/mnras/stw3129. arXiv: 1605.04319 [astro-ph.CO]

Despali, G., & Vegetti, S. (2017). The impact of baryonic physics on the subhalo mass function and implications for gravitational lensing. *MNRAS, 469*(2), 1997–2010. https://doi.org/10.1093/mnras/stx966. arXiv: 1608.06938 [astro-ph.GA]

Diego, J. M., Protopapas, P., Sandvik, H. B., & Tegmark, M. (2005). Non-parametric inversion of strong lensing systems. *MNRAS, 360*(2), 477–491. https://doi.org/10.1111/j.1365-2966.2005.09021.x. arXiv: astro-ph/0408418 [astro-ph]

Diego, J. M., Tegmark, M., Protopapas, P., & Sand vik, H. B. (2007). Combined reconstruction of weak and strong lensing data with WSLAP. *MNRAS, 375*(3), 958–970. https://doi.org/10.1111/j.1365-2966.2007.11380.x. arXiv: astro-ph/0509103 [astro-ph]

Dietrich, J. P., Werner, N., Clowe, D., Finoguenov, A., Kitching, T., Miller, L., & Simionescu, A. (2012). A filament of dark matter between two clusters of galaxies. *Nature, 487*(7406), 202–204. https://doi.org/10.1038/nature11224. arXiv: 1207.0809 [astro-ph.CO]

Dijkstra, M. (2014). Lya Emitting Galaxies as a Probe of Reionisation. *PASA, 31*, e040. https://doi.org/10.1017/pasa.2014.33. arXiv: 1406.7292 [astro-ph.CO]

Djorgovski, S., & Davis, M. (1987). Fundamental Properties of Elliptical Galaxies. *ApJ, 313*, 59. https://doi.org/10.1086/164948

Dolag, K., Bartelmann, M., Perrotta, F., Baccigalupi, C., Moscardini, L., Meneghetti, M., & Tormen, G. (2004). Numerical study of halo concentrations in dark-energy cosmologies. *A & A, 416*, 853–864. https://doi.org/10.1051/0004-6361:20031757. arXiv: astro-ph/0309771 [astro-ph]

Eales, S., Dunne, L., Clements, D., Cooray, A., De Zotti, G., Dye, S., … White, G. J. (2010). The Herschel ATLAS. *Proc. ASP, 122*(891), 499. https://doi.org/10.1086/653086. arXiv: 0910.4279. [astro-ph.CO]

Eigenbrod, A., Courbin, F., Vuissoz, C., Meylan, G., Saha, P., & Dye, S. (2005). COSMOGRAIL: The cosmological monitoring of gravitational lenses. I. How to sample the light curves of gravitationally lensed quasars to measure accurate time delays. *A & A, 436*(1), 25–35. https://doi.org/10.1051/0004-6361:20042422. arXiv: astro-ph/0503019 [astro-ph]

Eke, V. R., Navarro, J. F., & Steinmetz, M. (2001). The power spectrum dependence of dark matter halo concentrations. *ApJ, 554*(1), 114–125. https://doi.org/10.1086/321345. arXiv: astro-ph/0012337 [astro-ph]

Eliasdóttir, À., Limousin, M., Richard, J., Hjorth, J., Kneib, J.-P., Natarajan, P., … Paraficz, D. (2007). Where is the matter in the Merging Cluster Abell 2218? ArXiv e-prints. arXiv: 0710.5636

Ettori, S., Donnarumma, A., Pointecouteau, E., Reiprich, T. H., Giodini, S., Lovisari, L., & Schmidt, R. W. (2013). Mass profiles of galaxy clusters from X-ray analysis. *SpaceScienceReview,177*(1–4), 119–154. https://doi.org/10.1007/s11214-013-9976-7. arXiv: 1303.3530 [astro-ph.CO]

Faber, S. M., Dressler, A., Davies, R. L., Burstein, D., Lynden Bell, D., Terlevich, R., & Wegner, G. (1987). Global scaling relations for elliptical galaxies and implications for formation. In S. M. Faber (Ed.), *Nearly normal galaxies. From the Planck time to the present* (p. 175).

Foreman-Mackey, D. (2016). Corner.py: Scatterplot matrices in python. *The Journal of Open Source Software, 1*(2), 24. https://doi.org/10.21105/joss.00024

Formicola, I., Radovich, M., Meneghetti, M., Mazzotta, P., Grado, A., & Giocoli, C. (2016). Selecting background galaxies in weak-lensing analysis of galaxy clusters. *MNRAS, 458*(3), 2776–2792. https://doi.org/10.1093/mnras/stw493. arXiv: 1603.05690 [astro-ph.CO]

Fornberg, B. (1988). Generation of finite difference formulas on arbitrarily spaced grids. *Mathematics of Computation, 51*(184), 699–699. https://doi.org/10.1090/S0025-5718-1988-0935077-0

Freedman, W. L., Madore, B. F., Scowcroft, V., Burns, C., Monson, A., Persson, S. E., . . . Rigby, J. (2012). Carnegie hubble program: A mid-infrared calibration of the hubble constant. *ApJ, 58*(1), 24. https://doi.org/10.1088/0004-637X/758/1/24. arXiv: 1208.3281 [astro-ph.CO]

Gavazzi, R., Marshall, P. J., Treu, T., & Sonnenfeld, A. (2014). RINGFINDER: Automated detection of galaxy-scale gravitational lenses in ground-based multi-filter imaging data. *ApJ, 785*(2), 144. https://doi.org/10.1088/0004-637X/785/2/144. arXiv: 1403.1041 [astro-ph.CO]

Gavazzi, R., Treu, T., Koopmans, L. V. E., Bolton, A. S., Moustakas, L. A., Burles, S., & Marshall, P. J. (2008). The sloan lens ACS survey. VI. Discovery and analysis of a double Einstein ring. *ApJ, 677*(2), 1046–1059. https://doi.org/10.1086/529541. arXiv: 0801.1555 [astro-ph]

Giocoli, C., Meneghetti, M., Metcalf, R. B., Ettori, S., & Moscardini, L. (2014). Mass and concentration estimates from weak and strong gravitational lensing: A systematic study. *MNRAS, 440*(2), 1899–1915. https://doi.org/10.1093/mnras/stu303. arXiv: 1311.1205 [astro-ph.CO]

Giocoli, C., Tormen, G., Sheth, R. K., & van den Bosch, F. C. (2010). The substructure hierarchy in dark matter haloes. *MNRAS, 404*(1), 502–517. https://doi.org/10.1111/j.1365-2966.2010.16311.x. arXiv: 0911.0436 [astro-ph.CO]

Glenn, J., Conley, A., Béthermin, M., Altieri, B., Amblard, A., Arumugam, V., . . . Zemcov, M. (2010). HerMES: Deep galaxy number counts from a P(D) fluctuation analysis of SPIRE Science Demonstration Phase observations. *MNRAS, 409*(1), 109–121. https://doi.org/10.1111/j.1365-2966.2010.17781.x. arXiv: 1009.5675 [astro-ph.CO]

Grillo, C., Rosati, P., Suyu, S. H., Balestra, I., Caminha, G. B., Halkola, A., . . . Treu, T. (2018). Measuring the value of the hubble constant "à la Refsdal". *ApJ, 860*(2), 94. https://doi.org/10.3847/1538-4357/aac2c9. arXiv: 1802.01584 [astro-ph.CO]

Grillo, C., Suyu, S. H., Rosati, P., Mercurio, A., Balestra, I., Munari, E., . . . Frye, B. (2015). CLASHVLT: Insights on the mass substructures in the frontier fields cluster MACS J0416.1-2403 through accurate strong lens modeling. *ApJ, 800*(1), 38. https://doi.org/10.1088/0004-637X/800/1/38. arXiv: 1407.7866 [astro-ph.CO]

Gunn, J. E., & Peterson, B. A. (1965). On the density of neutral hydrogen in intergalactic space. *ApJ, 142*, 1633–1636. https://doi.org/10.1086/148444

Hezaveh, Y. D., Dalal, N., Marrone, D. P., Mao, Y.-Y., Morningstar, W., Wen, D., . . . Wechsler, R. H. (2016). Detection of lensing substructure using ALMA observations of the dusty galaxy SDP.81. *ApJ, 823*(1), 37. https://doi.org/10.3847/0004-637X/823/1/37. arXiv: 1601.01388 [astro-ph.CO]

Hilton, M., Hasselfield, M., Sifón, C., Battaglia, N., Aiola, S., Bharadwaj, V., . . . Wollack, E. J. (2018). The Atacama Cosmology Telescope: The Two-season ACTPol Sunyaev-Zel'dovich effect selected cluster catalog. *ApJS, 235*(1), 20. https://doi.org/10.3847/1538-4365/aaa6cb. arXiv: 1709.05600 [astro-ph.CO]

Hoekstra, H., Franx, M., & Kuijken, K. (2000). Hubble space telescope weak-lensing study of the z=0.83 cluster MS 1054-03. *ApJ, 532*(1), 88–108. https://doi.org/10.1086/308556. arXiv: astro-ph/9910487 [astro-ph]

Hoekstra, H., Franx, M., Kuijken, K., & Squires, G. (1998). Weak lensing analysis of CL 1358+62 using hubble space telescope observations. *ApJ, 504*(2), 636–660. https://doi.org/10.1086/306102

Hoekstra, H., Yee, H. K. C., & Gladders, M. D. (2004). Properties of galaxy dark matter halos from weak lensing. *ApJ, 606*(1), 67–77. https://doi.org/10.1086/382726. arXiv: astro-ph/0306515 [astro-ph]

Hoekstra, H., Herbonnet, R., Muzzin, A., Babul, A., Mahdavi, A., Viola, M., & Cacciato, M. (2015). The Canadian Cluster Comparison Project: Detailed study of systematics and updated weak lensing masses. *MNRAS, 449*(1), 685–714. https://doi.org/10.1093/mnras/stv275. arXiv: 1502.01883 [astro-ph.CO]

Huang, X., Storfer, C., Ravi, V., Pilon, A., Domingo, M., Schlegel, D. J., . . . Yèche, C. (2020). Finding strong gravitational lenses in the DESI DECam legacy survey. *ApJ, 894*(1), 78. https://doi.org/10.3847/1538-4357/ab7ffb. arXiv: 1906.00970 [astro-ph.GA]

Ilbert, O., Capak, P., Salvato, M., Aussel, H., McCracken, H. J., Sanders, D. B., ... Zucca, E. (2009). Cosmos photometric redshifts with 30-bands for 2-deg². *ApJ, 690*(2), 1236–1249. https://doi.org/10.1088/0004-637X/690/2/1236. arXiv: 0809.2101 [astro-ph]

Ivezić, Ž., Kahn, S. M., Tyson, J. A., Abel, B., Acosta, E., Allsman, R., ... et al. (2019). LSST: From science drivers to reference design and anticipated data products. *ApJ, 873*, 111. https://doi.org/10.3847/1538-4357/ab042c. arXiv: 0805.2366

Jacobs, C., Collett, T., Glazebrook, K., McCarthy, C., Qin, A. K., Abbott, T. M. C., ... DES Collaboration. (2019). Finding high-redshift strong lenses in DES using convolutional neural networks. *MNRAS, 484*(4), 5330–5349. https://doi.org/10.1093/mnras/stz272. arXiv: 1811.03786 [astro-ph.GA]

Jacobs, C., Glazebrook, K., Collett, T., More, A., & McCarthy, C. (2017). Finding strong lenses in CFHTLS using convolutional neural networks. *MNRAS, 471*(1), 167–181. https://doi.org/10.1093/mnras/stx1492. arXiv: 1704.02744 [astro-ph.IM]

Jauzac, M., Eckert, D., Schwinn, J., Harvey, D., Baugh, C. M., Robertson, A., ... Tchernin, C. (2016). The extraordinary amount of substructure in the Hubble Frontier Fields cluster Abell 2744. *MNRAS, 463*(4), 3876–3893. https://doi.org/10.1093/mnras/stw2251. arXiv: 1606.04527 [astro-ph.CO]

Jee, M. J., Ford, H. C., Illingworth, G. D., White, R. L., Broadhurst, T. J., Coe, D. A., ... Mei, S. (2007). Discovery of a ringlike dark matter structure in the core of the galaxy cluster Cl 0024+17. *ApJ, 661*(2), 728–749. https://doi.org/10.1086/517498. arXiv: 0705.2171 [astro-ph]

Jiang, G., & Kochanek, C. S. (2007). The Baryon fractions and mass-to-light ratios of early-type galaxies. *ApJ, 671*(2), 1568–1578. https://doi.org/10.1086/522580. arXiv: 0705.3647 [astro-ph]

Jing, Y. P., & Suto, Y. (2002). Triaxial modeling of halo density profiles with high-resolution N-body simulations. *ApJ, 574*(2), 538–553. https://doi.org/10.1086/341065. arXiv: astro-ph/0202064 [astro-ph]

Johnston, D. E., Sheldon, E. S., Wechsler, R. H., Rozo, E., Koester, B. P., Frieman, J. A., ... Annis, J. (2007). Cross-correlation weak lensing of SDSS galaxy Clusters II: Cluster density profiles and the mass-richness relation. arXiv e-prints, arXiv:0709.1159. arXiv: 0709.1159 [astro-ph]

Jullo, E., Natarajan, P., Kneib, J.-P., D'Aloisio, A., Limousin, M., Richard, J., & Schimd, C. (2010). Cosmological constraints from strong gravitational lensing in clusters of galaxies. *Science, 329*, 924–927. https://doi.org/10.1126/science.1185759. arXiv: 1008.4802 [astro-ph.CO]

Kaiser, N. (1995). Nonlinear cluster lens reconstruction. *ApJL, 439*, L1. https://doi.org/10.1086/187730. arXiv: astro-ph/9408092 [astro-ph]

Kaiser, N., & Squires, G. (1993). Mapping the dark matter with weak gravitational lensing. *ApJ, 404*, 441. https://doi.org/10.1086/172297

Kaiser, N., Squires, G., & Broadhurst, T. (1995). A method for weak lensing observations. *ApJ, 449*, 460. https://doi.org/10.1086/176071. arXiv: astro-ph/9411005 [astro-ph]

Kelly, P. L., Rodney, S. A., Treu, T., Strolger, L.-G., Foley, R. J., Jha, S. W., ... Zitrin, A. (2016). Deja Vu all over again: The reappearance of supernova Refsdal. *ApJL, 819*(1), L8. https://doi.org/10.3847/2041-8205/819/1/L8. arXiv: 1512.04654 [astro-ph.CO]

Kelly, P. L., Rodney, S. A., Treu, T., Foley, R. J., Brammer, G., Schmidt, K. B., ... Tucker, B. E. (2015). Multiple images of a highly magnified supernova formed by an early-type cluster galaxy lens. *Science, 347*(6226), 1123–1126. https://doi.org/10.1126/science.aaa3350. arXiv: 1411.6009 [astro-ph.CO]

Kneib, J.-P., & Natarajan, P. (2011). Cluster lenses. *The Astronomy and Astrophysics Review, 19*, 47. https://doi.org/10.1007/s00159-011-0047-3. arXiv: 1202.0185 [astro-ph.CO]

Kochanek, C. S., Morgan, N. D., Falco, E. E., McLeod, B. A., Winn, J. N., Dembicky, J., & Ketzeback, B. (2006). The time delays of gravitational lens HE 0435-1223: An early-type galaxy with a rising rotation curve. *ApJ, 640*(1), 47–61. https://doi.org/10.1086/499766. arXiv: astro-ph/0508070 [astro-ph]

Kondo, H., Miyatake, H., Shirasaki, M., Sugiyama, N., & Nishizawa, A. J. (2020). Weak lensing measurement of filamentary structure with the SDSS BOSS and Subaru Hyper Suprime-Cam data. *MNRAS, 495*(4), 3695–3704. https://doi.org/10.1093/mnras/staa1390. arXiv: 1905.08991 [astro-ph.CO]

Koopmans, L. V. E. (2005). Gravitational imaging of cold dark matter substructures. *MNRAS, 363*(4), 1136–1144. https://doi.org/10.1111/j.1365-2966.2005.09523.x. arXiv: astro-ph/0501324 [astro-ph](4), 1136

Koopmans, L. V. E., Bolton, A., Treu, T., Czoske, O., Auger, M. W., Barnabè, M., . . . Burles, S. (2009). The structure and dynamics of massive early-type galaxies: On homology, isothermality, and isotropy inside one effective radius. *ApJl, 703*, L51–L54. https://doi.org/10.1088/0004-637X/703/1/L51. arXiv: 0906.1349 [astro-ph.CO]

Kuijken, K. (1999). Weak weak lensing: correcting weak shear measurements accurately for PSF anisotropy. *A & A, 352*, 355–362. arXiv: astro-ph/9904418 [astro-ph]

Lanusse, F., Ma, Q., Li, N., Collett, T. E., Li, C.-L., Ravanbakhsh, S., . . . Póczos, B. (2018). CMU DeepLens: Deep learning for automatic image-based galaxy-galaxy strong lens finding. *MNRAS, 473*(3), 3895–3906. https://doi.org/10.1093/mnras/stx1665. arXiv: 1703.02642 [astro-ph.IM]

Laureijs, R., Amiaux, J., Arduini, S., Auguères, J.-L., Brinchmann, J., Cole, R., . . . Zucca, E. (2011). Euclid definition study report. arXiv e-prints, arXiv:1110.3193. arXiv: 1110.3193 [astro-ph.CO]

Lecun, Y., Bengio, Y., & Hinton, G. (2015). Deep learning. *Nature, 521*(7553), 436–444. https://doi.org/10.1038/nature14539

Liesenborgs, J., De Rijcke, S., & Dejonghe, H. (2006). A genetic algorithm for the non-parametric inversion of strong lensing systems. *MNRAS, 367*(3), 1209–1216. https://doi.org/10.1111/j.1365-2966.2006.10040.x. arXiv: astro-ph/0601124 [astro-ph]

Limousin, M., Ebeling, H., Richard, J., Swinbank, A. M., Smith, G. P., Jauzac, M., . . . Kneib, J.-P. (2012). Strong lensing by a node of the cosmic web. The core of MACS J0717.5+3745 at z = 0.55. *A & A, 544*, A71. https://doi.org/10.1051/0004-6361/201117921. arXiv: 1109.3301 [astro-ph.CO]

Limousin, M., Kneib, J.-P., Bardeau, S., Natarajan, P., Czoske, O., Smail, I., . . . Smith, G. P. (2007). Truncation of galaxy dark matter halos in high density environments. *A & A, 461*(3), 881–891. https://doi.org/10.1051/0004-6361:20065543. arXiv: astro-ph/0609782 [astro-ph]

Limousin, M., Kneib, J.-P., & Natarajan, P. (2005). Constraining the mass distribution of galaxies using galaxy-galaxy lensing in clusters and in the field. *MNRAS, 356*, 309–322. https://doi.org/10.1111/j.1365-2966.2004.08449.x. eprint: astro-ph/0405607

Limousin, M., Morandi, A., Sereno, M., Meneghetti, M., Ettori, S., Bartelmann, M., & Verdugo, T. (2013). The three-dimensional shapes of galaxy clusters. *Space Science Reviews, 177*(1–4), 155–194. https://doi.org/10.1007/s11214-013-9980-y. arXiv: 1210.3067 [astro-ph.CO]

Lovell, M. R., Eke, V., Frenk, C. S., Gao, L., Jenkins, A., Theuns, T., . . . Ruchayskiy, O. (2012). The haloes of bright satellite galaxies in a warm dark matter universe. *MNRAS, 420*(3), 2318–2324. https://doi.org/10.1111/j.1365-2966.2011.20200.x. arXiv: 1104.2929 [astro-ph.CO]

LSST Science Collaboration, Abell, P. A., Allison, J., Anderson, S. F., Andrew, J. R., Angel, J. R. P., . . . Zhan, H. (2009). LSST science book, version 2.0. arXiv e-prints, arXiv:0912.0201. arXiv: 0912.0201 [astro-ph.IM]

Luppino, G. A., & Kaiser, N. (1997). Detection of Weak Lensing by a Cluster of Galaxies at z = 0.83. *ApJ, 475*(1), 20–28. https://doi.org/10.1086/303508. arXiv: astro-ph/9601194 [astro-ph]

Madau, P., Meiksin, A., & Rees, M. J. (1997). 21 Centimeter Tomography of the Intergalactic Medium at High Redshift. *ApJ, 475*(2), 429–444. doi:10.1086/303549. arXiv: astro-ph/9608010 [astro-ph]

Markevitch, M., Gonzalez, A. H., Clowe, D., Vikhlinin, A., Forman, W., Jones, C., . . . Tucker, W. (2004). Direct constraints on the dark matter self-interaction cross section from the merging galaxy cluster 1E 0657-56. *ApJ, 606*(2), 819–824. https://doi.org/10.1086/383178. arXiv: astro-ph/0309303 [astro-ph]

Masters, D., Capak, P., Stern, D., Ilbert, O., Salvato, M., Schmidt, S., . . . Cavuoti, S. (2015). Mapping the galaxy color-redshift relation: Optimal photometric redshift calibration strategies for cosmology surveys. *ApJ, 813*(1), 53. https://doi.org/10.1088/0004-637X/813/1/53. arXiv: 1509.03318 [astro-ph.CO]

Maturi, M., Mizera, S., & Seidel, G. (2014). Multi-colour detection of gravitational arcs. *A & A, 567*, A111. https://doi.org/10.1051/0004-6361/201321634. arXiv: 1305.3608 [astro-ph.CO]

McClintock, T., Varga, T. N., Gruen, D., Rozo, E., Rykoff, E. S., Shin, T., ... DES Collaboration. (2019). Dark energy survey year 1 results: Weak lensing mass calibration of redMaPPer galaxy clusters. *MNRAS, 482*(1), 1352–1378. https://doi.org/10.1093/mnras/sty2711. arXiv: 1805.00039 [astro-ph.CO]

Medezinski, E., Oguri, M., Nishizawa, A. J., Speagle, J. S., Miyatake, H., Umetsu, K., ... Komiyama, Y. (2018). Source selection for cluster weak lensing measurements in the Hyper Suprime-Cam survey. *Publications of the Astronomical Society of Japan, 70*(2), 30. https://doi.org/10.1093/pasj/psy009. arXiv: 1706.00427 [astro-ph.CO]

Meneghetti, M., Bartelmann, M., Dahle, H., & Limousin, M. (2013). Arc Statistics. *Space Science Reviews, 177*(1–4), 31–74. https://doi.org/10.1007/s11214-013-9981-x. arXiv: 1303.3363 [astro-ph.CO]

Meneghetti, M., Davoli, G., Bergamini, P., Rosati, P., Natarajan, P., Giocoli, C., ... Vanzella, E. (2020). An excess of small-scale gravitational lenses observed in galaxy clusters. arXiv e-prints, arXiv:2009.04471. arXiv: 2009.04471 [astro-ph.GA]

Meneghetti, M., Rasia, E., Merten, J., Bellagamba, F., Ettori, S., Mazzotta, P., ... Marri, S. (2010). Weighing simulated galaxy clusters using lensing and X-ray. *A & A, 514*, A93. https://doi.org/10.1051/0004-6361/200913222. arXiv: 0912.1343

Meneghetti, M., Rasia, E., Vega, J., Merten, J., Postman, M., Yepes, G., ... Zitrin, A. (2014). The MUSIC of CLASH: Predictions on the concentration-mass relation. *ApJ, 797*(1), 34. https://doi.org/10.1088/0004-637X/797/1/34. arXiv: 1404.1384 [astro-ph.CO]

Meneghetti, M., Yoshida, N., Bartelmann, M., Moscardini, L., Springel, V., Tormen, G., & White, S. D. M. (2001). Giant cluster arcs as a constraint on the scattering cross-section of dark matter. *MNRAS, 325*(1), 435–442. https://doi.org/10.1046/j.1365-8711.2001.04477.x. arXiv: astroph/0011405 [astro-ph]

Merloni, A., Predehl, P., Becker, W., Böhringer, H., Boller, T., Brunner, H., ... German eROSITA Consortium, t. (2012). eROSITA science book: Mapping the structure of the energetic universe. arXiv e-prints, arXiv:1209.3114. arXiv: 1209.3114 [astro-ph.HE]

Merten, J. (2016). Mesh-free free-form lensing—I. Methodology and application to mass reconstruction. *MNRAS, 461*(3), 2328–2345. https://doi.org/10.1093/mnras/stw1413. arXiv: 1412.5186 [astro-ph.CO]

Merten, J., Cacciato, M., Meneghetti, M., Mignone, C., & Bartelmann, M. (2009). Combining weak and strong cluster lensing: Applications to simulations and MS 2137. *A & A, 500*(2), 681–691. https://doi.org/10.1051/0004-6361/200810372. arXiv: 0806.1967 [astro-ph]

Merten, J., Coe, D., Dupke, R., Massey, R., Zitrin, A., Cypriano, E. S., ... Bregman, J. N. (2011). Creation of cosmic structure in the complex galaxy cluster merger Abell 2744. *MNRAS, 417*(1), 333–347. https://doi.org/10.1111/j.1365-2966.2011.19266.x. arXiv: 1103.2772 [astro-ph.CO]

Merten, J., Meneghetti, M., Postman, M., Umetsu, K., Zitrin, A., Medezinski, E., ... Zheng, W. (2015). CLASH: The concentration-mass relation of galaxy clusters. *ApJ, 806*(1), 4. https://doi.org/10.1088/0004-637X/806/1/4. arXiv: 1404.1376 [astro-ph.CO]

Metcalf, R. B., Meneghetti, M., Avestruz, C., Bellagamba, F., Bom, C. R., Bertin, E., ... Vernardos, G. (2019). The strong gravitational lens finding challenge. *A & A, 625*, A119. https://doi.org/10.1051/0004-6361/201832797. arXiv: 1802.03609 [astro-ph.GA]

Miller, L., Heymans, C., Kitching, T. D., vanWaerbeke, L., Erben, T., Hildebrandt, H., ... Velander, M. (2013). Bayesian galaxy shape measurement for weak lensing surveys—III. Application to the Canada-France-Hawaii Telescope Lensing Survey. *MNRAS, 429*(4), 2858–2880. https://doi.org/10.1093/mnras/sts454. arXiv: 1210.8201 [astro-ph.CO]

Millon, M., Courbin, F., Bonvin, V., Paic, E., Meylan, G., Tewes, M., ... Wyttenbach, A. (2020). COSMOGRAIL. XIX. Time delays in 18 strongly lensed quasars from 15 years of optical monitoring. *A & A, 640*, A105. https://doi.org/10.1051/0004-6361/202037740. arXiv: 2002.05736 [astro-ph.CO]

Miralda-Escudé, J. (2002). A test of the collisional dark matter hypothesis from cluster lensing. *ApJ, 564*(1), 60–64. https://doi.org/10.1086/324138. arXiv: astro-ph/0002050 [astro-ph]

Miyatake, H., Battaglia, N., Hilton, M., Medezinski, E., Nishizawa, A. J., More, S., . . . Wollack, E. J. (2019). Weak-lensing mass calibration of ACTPol Sunyaev-Zel'dovich clusters with the hyper suprime-cam survey. *ApJ, 875*(1), 63. https://doi.org/10.3847/1538-4357/ab0af0. arXiv: 1804.05873 [astro-ph.CO]

Myers, S. T., Jackson, N. J., Browne, I.W. A., de Bruyn, A. G., Pearson, T. J., Readhead, A. C. S., . . . Sykes, C. M. (2003). The cosmic lens all-sky survey—I. Source selection and observations. *MNRAS, 341*(1), 1–12. https://doi.org/10.1046/j.1365-8711.2003.06256.x. arXiv: astro-ph/0211073 [astro-ph]

Narayanan, D., Turk, M., Feldmann, R., Robitaille, T., Hopkins, P., Thompson, R., . . . Kereš, D. (2015). The formation of submillimetre-bright galaxies from gas infall over a billion years. *Nature, 525*(7570), 496–499. https://doi.org/10.1038/nature15383. arXiv: 1509.06377 [astro-ph.GA]

Natarajan, P., Chadayammuri, U., Jauzac, M., Richard, J., Kneib, J.-P., Ebeling, H., . . . Vogelsberger, M. (2017). Mapping substructure in the HST Frontier Fields cluster lenses and in cosmological simulations. *MNRAS, 468*(2), 1962–1980. https://doi.org/10.1093/mnras/stw3385. arXiv: 1702.04348 [astro-ph.GA]

Natarajan, P., De Lucia, G., & Springel, V. (2007). Substructure in lensing clusters and simulations. *MNRAS, 376*(17), 180–192. https://doi.org/10.1111/j.1365-2966.2007.11399.x. arXiv: astro-ph/0604414 [astro-ph]

Natarajan, P., & Kneib, J.-P. (1997). Lensing by galaxy haloes in clusters of galaxies. *MNRAS, 287*(4), 833–847. https://doi.org/10.1093/mnras/287.4.833. arXiv: astro-ph/9609008 [astro-ph]

Natarajan, P., Kneib, J.-P., & Smail, I. (2002). Evidence for tidal stripping of dark matter halos in massive cluster lenses. *ApJL, 580*(1), L11–L15. https://doi.org/10.1086/345399. arXiv: astroph/0207049 [astro-ph]

Natarajan, P., & Springel, V. (2004). Abundance of substructure in clusters of galaxies. *ApJL, 617*(1), L13–L16. https://doi.org/10.1086/427079. arXiv: astro-ph/0411515 [astro-ph]

Navarro, J. F., Frenk, C. S., & White, S. D. M. (1997). A universal density profile from hierarchical clustering. *ApJ, 490*(2), 493–508. https://doi.org/10.1086/304888. arXiv: astro-ph/9611107 [astro-ph]

Negrello, M., Amber, S., Amvrosiadis, A., Cai, Z. -Y., Lapi, A., Gonzalez-Nuevo, J., . . . van der Werf, P. (2017). The Herschel-ATLAS: A sample of 500 mm-selected lensed galaxies over 600 deg^2. *MNRAS, 465*(3), 3558–3580. https://doi.org/10.1093/mnras/stw2911. arXiv: 1611.03922 [astro-ph.GA]

Negrello, M., Hopwood, R., De Zotti, G., Cooray, A., Verma, A., Bock, J., . . . Zmuidzinas, J. (2010). The detection of a population of submillimeter-bright, strongly lensed galaxies. *Science, 330*(6005), 800. https://doi.org/10.1126/science.1193420. arXiv: 1011.1255 [astro-ph.CO]

Newman, A. B., Ellis, R. S., & Treu, T. (2015). Luminous and dark matter profiles from galaxies to clusters: Bridging the gap with group-scale lenses. *ApJ, 814*(1), 26. https://doi.org/10.1088/0004-637X/814/1/26. arXiv: 1503.05282 [astro-ph.GA]

Newman, A. B., Treu, T., Ellis, R. S., Sand, D. J., Nipoti, C., Richard, J., & Jullo, E. (2013). The density profiles of massive, relaxed galaxy clusters. I. The total density over three decades in radius. *ApJ, 765*(1), 24. https://doi.org/10.1088/0004-637X/765/1/24. arXiv: 1209.1391 [astro-ph.CO]

Nierenberg, A. M., Gilman, D., Treu, T., Brammer, G., Birrer, S., Moustakas, L., . . . Sluse, D. (2020). Double dark matter vision: Twice the number of compact-source lenses with narrowline lensing and the WFC3 grism. *MNRAS, 492*(4), 5314–5335. https://doi.org/10.1093/mnras/stz3588. arXiv: 1908.06344 [astro-ph.GA]

Penna-Lima, M., Bartlett, J. G., Rozo, E., Melin, J.-B., Merten, J., Evrard, A. E., . . . Rykoff, E. (2017). Calibrating the Planck cluster mass scale with CLASH. *A & A, 604*, A89. https://doi.org/10.1051/0004-6361/201629971. arXiv: 1608.05356 [astro-ph.CO]

Petrillo, C. E., Tortora, C., Chatterjee, S., Vernardos, G., Koopmans, L. V. E., Verdoes Kleijn, G., . . . McFarland, J. (2017). Finding strong gravitational lenses in the Kilo Degree Survey with Convolutional Neural Networks. *MNRAS, 472*(1), 1129–1150. https://doi.org/10.1093/mnras/stx2052. arXiv: 1702.07675 [astro-ph.GA]

Pillepich, A., Porciani, C., & Reiprich, T. H. (2012). The X-ray cluster survey with eRosita: forecasts for cosmology, cluster physics and primordial non-Gaussianity. *MNRAS, 422*(1), 44–69. https://doi.org/10.1111/j.1365-2966.2012.20443.x. arXiv: 1111.6587 [astro-ph.CO]

Planck Collaboration, Ade, P. A. R., Aghanim, N., Armitage-Caplan, C., Arnaud, M., Ashdown, M., . . . Zonca, A. (2014). Planck 2013 results. XX. Cosmology from Sunyaev-Zeldovich cluster counts. *A & A, 571*, A20. https://doi.org/10.1051/0004-6361/201321521. arXiv: 1303.5080 [astro-ph.CO]

Planck Collaboration, Aghanim, N., Akrami, Y., Ashdown, M., Aumont, J., Baccigalupi, C., . . . Zonca, A. (2020). Planck 2018 results. VI. Cosmological parameters. *A & A, 641*, A6. https://doi.org/10.1051/0004-6361/201833910. arXiv: 1807.06209 [astro-ph.CO]

Postman, M., Coe, D., Benítez, N., Bradley, L., Broadhurst, T., Donahue, M., . . . Van der Wel, A. (2012). The cluster lensing and supernova survey with hubble: An overview. *ApJS, 199*(2), 25. https://doi.org/10.1088/0067-0049/199/2/25. arXiv: 1106.3328 [astro-ph.CO]

Pourrahmani, M., Nayyeri, H., & Cooray, A. (2018). LensFlow: A convolutional neural network in search of strong gravitational lenses. *ApJ, 856*(1), 68. https://doi.org/10.3847/1538-4357/aaae6a. arXiv: 1705.05857 [astro-ph.IM]

Powell, D., Vegetti, S., McKean, J. P., & Spingola, C. (2020). A novel approach to visibility-space modelling of interferometric gravitational lens observations at high angular resolution. arXiv e-prints, arXiv:2005.03609. arXiv: 2005.03609 [astro-ph.IM]

Randall, S. W., Markevitch, M., Clowe, D., Gonzalez, A. H., & Bradač, M. (2008). Constraints on the self-interaction cross section of dark matter from numerical simulations of the merging galaxy cluster 1E 0657-56. *ApJ, 679*(2), 1173–1180. https://doi.org/10.1086/587859. arXiv: 0704.0261 [astro-ph]

Refregier, A., & Bacon, D. (2003). Shapelets—II. A method for weak lensing measurements. *MNRAS, 338*(1), 48–56. https://doi.org/10.1046/j.1365-8711.2003.05902.x. arXiv: astro-ph/0105179 [astro-ph]

Refsdal, S. (1964). On the possibility of determining Hubble's parameter and the masses of galaxies from the gravitational lens effect. *MNRAS, 128*, 307. https://doi.org/10.1093/mnras/128.4.307

Ribli, D., Dobos, L., & Csabai, I. (2019). Galaxy shape measurement with convolutional neural networks. *MNRAS, 489*(4), 4847–4859. https://doi.org/10.1093/mnras/stz2374. arXiv: 1902.08161 [astro-ph.CO]

Riess, A. G., Casertano, S., Yuan, W., Macri, L. M., & Scolnic, D. (2019). Large magellanic cloud cepheid standards provide a 1% foundation for the determination of the hubble constant and stronger evidence for physics beyond LCDM. *ApJ, 876*(1), 85. https://doi.org/10.3847/1538-4357/ab1422. arXiv: 1903.07603 [astro-ph.CO]

Riess, A. G., Casertano, S., Yuan, W., Macri, L., Bucciarelli, B., Lattanzi, M. G., . . . Anderson, R. I. (2018). Milky way cepheid standards for measuring cosmic distances and application to Gaia DR2: Implications for the hubble constant. *ApJ, 861*(2), 126. https://doi.org/10.3847/1538-4357/aac82e. arXiv: 1804.10655 [astro-ph.CO]

Riess, A. G., Macri, L. M., Hoffmann, S. L., Scolnic, D., Casertano, S., Filippenko, A. V., . . . Foley, R. J. (2016). A 2.4% determination of the local value of the hubble constant. *ApJ, 826*(1), 56. https://doi.org/10.3847/0004-637X/826/1/56. arXiv: 1604.01424 [astro-ph.CO]

Rigaut, F., & Neichel, B. (2020). Multiconjugate adaptive optics for astronomy. arXiv e-prints, arXiv:2003.03097. arXiv: 2003.03097 [astro-ph.IM]

Rivera-Thorsen, T. E., Dahle, H., Chisholm, J., Florian, M. K., Gronke, M., Rigby, J. R., . . . Bayliss, M. (2019). Gravitational lensing reveals ionizing ultraviolet photons escaping from a distant galaxy. *Science, 366*(6466), 738–741. https://doi.org/10.1126/science.aaw0978. arXiv: 1904.08186 [astro-ph.GA]

Rosati, P., Borgani, S., & Norman, C. (2002). The evolution of X-ray clusters of galaxies. *Annual Review of Astronomy and Astrophysics, 40*, 539–577. https://doi.org/10.1146/annurev.astro.40.120401.150547. arXiv: astro-ph/0209035 [astro-ph]

Rousset, G., Fontanella, J. C., Kern, P., Gigan, P., & Rigaut, F. (1990). First diffraction-limited astronomical images with adaptive optics. *A & A, 230*(2), L29–L32.

Rozo, E., Wechsler, R. H., Rykoff, E. S., Annis, J. T., Becker, M. R., Evrard, A. E., ... Weinberg, D. H. (2010). Cosmological constraints from the Sloan digital sky survey maxBCG cluster catalog. *ApJ, 708*(1), 645–660. https://doi.org/10.1088/0004-637X/708/1/645. arXiv: 0902.3702 [astro-ph.CO]

Saha, P., & Williams, L. L. R. (2004). A portable modeler of lensed quasars. *AJ, 127*(5), 2604–2616. https://doi.org/10.1086/383544. arXiv: astro-ph/0402135 [astro-ph]

Sand, D. J., Treu, T., Ellis, R. S., Smith, G. P., & Kneib, J.-P. (2008). Separating Baryons and dark matter in cluster cores: A full two-dimensional lensing and dynamic analysis of Abell 383 and MS 2137-23. *ApJ, 674*(2), 711–727. https://doi.org/10.1086/524652. arXiv: 0710.1069 [astro-ph]

Sandage, A., Tammann, G. A., Saha, A., Reindl, B., Macchetto, F. D., & Panagia, N. (2006). The hubble constant: A summary of the hubble space telescope program for the luminosity calibration of type Ia supernovae by means of cepheids. *ApJ, 653*(2), 843–860. https://doi.org/10.1086/508853. arXiv: astro-ph/0603647 [astro-ph]

Sarazin, C. L. (1988). *X-ray emission from clusters of galaxies.*

Schneider, P. (2006). Part 3: Weak gravitational lensing. In G. Meylan, P. Jetzer, P. North, P. Schneider, C. S. Kochanek, & J. Wambsganss (Eds.), *Saas-fee advanced course 33: Gravitational lensing: Strong, weak and micro* (pp. 269–451).

Schneider, P., & Seitz, C. (1995). Steps towards nonlinear cluster inversion through gravitational distortions. I. Basic considerations and circular clusters. *A & A, 294*, 411–431. arXiv: astroph/9407032 [astro-ph]

Sebesta, K., Williams, L. L. R., Mohammed, I., Saha, P., & Liesenborgs, J. (2016). Testing lighttraces-mass in Hubble Frontier Fields Cluster MACS-J0416.1-2403. *MNRAS, 461*(2), 2126–2134. https://doi.org/10.1093/mnras/stw1433. arXiv: 1507.08960 [astro-ph.CO]

Seidel, G., & Bartelmann, M. (2007). Arcfinder: an algorithm for the automatic detection of gravitational arcs. *A & A, 472*(1), 341–352. https://doi.org/10.1051/0004-6361:20066097. arXiv: astro-ph/0607547 [astro-ph]

Seitz, S., & Schneider, P. (1996). Cluster lens reconstruction using only observed local data: An improved finite-field inversion technique. *A & A, 305*, 383. arXiv: astro-ph/9503096 [astro-ph]

Sereno, M., Covone, G., Izzo, L., Ettori, S., Coupon, J., & Lieu, M. (2017). PSZ2LenS. Weak lensing analysis of the Planck clusters in the CFHTLenS and in the RCSLenS. *MNRAS, 472*(2), 1946–1971. https://doi.org/10.1093/mnras/stx2085. arXiv: 1703.06886 [astro-ph.CO]

Shu, Y., Brownstein, J. R., Bolton, A. S., Koopmans, L. V. E., Treu, T., Montero-Dorta, A. D., ... Moustakas, L. A. (2017). The Sloan lens ACS survey. XIII. Discovery of 40 new galaxyscale strong lenses. *ApJ, 851*(1), 48. https://doi.org/10.3847/1538-4357/aa9794. arXiv: 1711.00072 [astro-ph.GA]

Smith, G. P., Mazzotta, P., Okabe, N., Ziparo, F., Mulroy, S. L., Babul, A., ... Umetsu, K. (2016). LoCuSS: Testing hydrostatic equilibrium in galaxy clusters. *MNRAS, 456*(1), L74–L78. https://doi.org/10.1093/mnrasl/slv175. arXiv: 1511.01919 [astro-ph.CO]

Sonnenfeld, A., Chan, J. H. H., Shu, Y., More, A., Oguri, M., Suyu, S. H., ... Komiyama, Y. (2018). Survey of gravitationally-lensed objects in HSC imaging (SuGOHI). I. Automatic search for galaxy-scale strong lenses. *Publications of the ASJ, 70*, S29. https://doi.org/10.1093/pasj/psx062. arXiv: 1704.01585 [astro-ph.GA]

Spergel, D., Gehrels, N., Baltay, C., Bennett, D., Breckinridge, J., Donahue, M., ... Zhao, F. (2015). Wide-field infrared survey telescope-astrophysics focused telescope assets WFIRST AFTA 2015 report. arXiv e-prints, arXiv:1503.03757. arXiv: 1503.03757 [astro-ph.IM]

Spilker, J. S., Marrone, D. P., Aravena, M., Béthermin, M., Bothwell, M. S., Carlstrom, J. E., ... Welikala, N. (2016). ALMA imaging and gravitational lens models of south pole telescopeselected dusty, star-forming galaxies at high redshifts. *ApJ, 826*(2), 112. https://doi.org/10.3847/0004-637X/826/2/112. arXiv: 1604.05723 [astro-ph.GA]

Spingola, C., McKean, J. P., Auger, M. W., Fassnacht, C. D., Koopmans, L. V. E., Lagattuta, D. J., & Vegetti, S. (2018). SHARP—V. Modelling gravitationally lensed radio arcs imaged with global VLBI observations. *MNRAS, 478*(4), 4816–4829. https://doi.org/10.1093/mnras/sty1326. arXiv: 1807.05566 [astro-ph.GA]

Stapelberg, S., Carrasco, M., & Maturi, M. (2019). EasyCritics—I. Efficient detection of strongly lensing galaxy groups and clusters in wide-field surveys. *MNRAS, 482*(2), 1824–1839. https://doi.org/10.1093/mnras/sty2784. arXiv: 1709.09758 [astro-ph.CO]

Suyu, S. H., & Blandford, R. D. (2006). The anatomy of a quadruply imaged gravitational lens system. *MNRAS, 366*(1), 39–48. https://doi.org/10.1111/j.1365-2966.2005.09854.x. arXiv: astroph/0506629 [astro-ph]

Suyu, S. H., Marshall, P. J., Blandford, R. D., Fassnacht, C. D., Koopmans, L. V. E., McKean, J. P., & Treu, T. (2009). Dissecting the gravitational lens B1608+656. I. Lens potential reconstruction. *ApJ, 691*(1), 277–298. https://doi.org/10.1088/0004-637X/691/1/277. arXiv: 0804.2827 [astro-ph]

Suyu, S. H., Marshall, P. J., Hobson, M. P., & Blandford, R. D. (2006). A Bayesian analysis of regularized source inversions in gravitational lensing. *MNRAS, 371*(2), 983–998. https://doi.org/10.1111/j.1365-2966.2006.10733.x. arXiv: astro-ph/0601493 [astro-ph]

Tewes, M., Courbin, F., & Meylan, G. (2013). COSMOGRAIL: The cosmological monitoring of gravitational lenses. XI. Techniques for time delay measurement in presence of microlensing. *A & A, 553*, A120. https://doi.org/10.1051/0004-6361/201220123. arXiv: 1208.5598 [astro-ph.CO]

The Dark Energy Survey Collaboration. (2005). The dark energy survey. arXiv e-prints, astro-ph/0510346. arXiv: astro-ph/0510346 [astro-ph]

Treu, T. (2010a). Strong lensing by galaxies. *ARAA, 48*, 87–125. https://doi.org/10.1146/annurev-astro-081309-130924. arXiv: 1003.5567

Treu, T. (2010b). Strong lensing by galaxies. *Annual Review of Astronomy and Astrophysics, 48*, 87–125. https://doi.org/10.1146/annurev-astro-081309-130924. arXiv: 1003.5567 [astro-ph.CO]

Treu, T., Brammer, G., Diego, J. M., Grillo, C., Kelly, P. L., Oguri, M., . . . Patel, B. (2016). "Refsdal" meets popper: comparing predictions of the re-appearance of the multiply imaged supernova behind MACSJ1149.5+2223. *ApJ, 817*(1), 60. https://doi.org/10.3847/0004-637X/817/1/60. arXiv: 1510.05750 [astro-ph.CO]

Treu, T., Dutton, A. A., Auger, M. W., Marshall, P. J., Bolton, A. S., Brewer, B. J., . . . Koopmans, L. V. E. (2011). The SWELLS survey—I. A large spectroscopically selected sample of edge-on late-type lens galaxies. *MNRAS, 417*(3), 1601–1620. https://doi.org/10.1111/j.1365-2966.2011.19378.x. arXiv: 1104.5663 [astro-ph.CO]

Treu, T., & Marshall, P. J. (2016). Time delay cosmography. *The Astronomy and Astrophysics Review, 24*(1), 11. https://doi.org/10.1007/s00159-016-0096-8. arXiv: 1605.05333 [astro-ph.CO]

Umetsu, K. (2020). Cluster-galaxy weak lensing. *The Astronomy and Astrophysics Review, 28*(1), 7. https://doi.org/10.1007/s00159-020-00129-w. arXiv: 2007.00506 [astro-ph.CO]

Umetsu, K., Medezinski, E., Nonino, M., Merten, J., Postman, M., Meneghetti, M., . . . Zitrin, A. (2014). CLASH: Weak-lensing Shear-and-magnification analysis of 20 galaxy clusters. *ApJ, 795*(2), 163. https://doi.org/10.1088/0004-637X/795/2/163. arXiv: 1404.1375 [astro-ph.CO]

Vanzella, E., Calura, F., Meneghetti, M., Mercurio, A., Castellano, M., Caminha, G. B., . . . Coe, D. (2017a). Paving the way for the JWST: Witnessing globular cluster formation at z > 3. *MNRAS, 467*(4), 4304–4321. https://doi.org/10.1093/mnras/stx351. arXiv: 1612.01526 [astro-ph.GA]

Vanzella, E., Castellano, M., Meneghetti, M., Mercurio, A., Caminha, G. B., Cupani, G., . . . Tozzi, P. (2017b). Magnifying the early episodes of star formation: Super star clusters at cosmological distances. *ApJ, 842*(1), 47. https://doi.org/10.3847/1538-4357/aa74ae. arXiv: 1703.02044 [astro-ph.GA]

Vanzella, E., Meneghetti, M., Caminha, G. B., Castellano, M., Calura, F., Rosati, P., . . . Balestra, I. (2020). Candidate population III stellar complex at z = 6.629 in the MUSE deep lensed field. *MNRAS, 494*(1), L81–L85. https://doi.org/10.1093/mnrasl/slaa041. arXiv: 2001.03619 [astro-ph.GA]

Vegetti, S., & Koopmans, L. V. E. (2009). Bayesian strong gravitational-lens modelling on adaptive grids: Objective detection of mass substructure in Galaxies. *MNRAS, 392*(3), 945–963. https://doi.org/10.1111/j.1365-2966.2008.14005.x. arXiv: 0805.0201 [astro-ph]

Vegetti, S., Koopmans, L. V. E., Bolton, A., Treu, T., & Gavazzi, R. (2010). Detection of a dark substructure through gravitational imaging. *MNRAS, 408*(4), 1969–1981. https://doi.org/10.1111/j.1365-2966.2010.16865.x. arXiv: 0910.0760 [astro-ph.CO]

Vegetti, S., Lagattuta, D. J., McKean, J. P., Auger, M. W., Fassnacht, C. D., & Koopmans, L. V. E. (2012). Gravitational detection of a low-mass dark satellite galaxy at cosmological distance. *Nature, 481*(7381), 341–343. https://doi.org/10.1038/nature10669. arXiv: 1201.3643 [astro-ph.CO]

Vieira, J. D., Crawford, T. M., Switzer, E. R., Ade, P. A. R., Aird, K. A., Ashby, M. L. N., ... Zenteno, A. (2010). Extragalactic millimeter-wave sources in south pole telescope survey data: Source counts, catalog, and statistics for an 87 square-degree field. *ApJ, 719*(1), 763–783. https://doi.org/10.1088/0004-637X/719/1/763. arXiv: 0912.2338 [astro-ph.CO]

Vikhlinin, A., Kravtsov, A. V., Burenin, R. A., Ebeling, H., Forman, W. R., Hornstrup, A., ... Voevodkin, A. (2009). Chandra cluster cosmology project III: Cosmological parameter constraints. *ApJ, 692*(2), 1060–1074. https://doi.org/10.1088/0004-637X/692/2/1060. arXiv: 0812.2720 [astro-ph]

von der Linden, A., Mantz, A., Allen, S. W., Applegate, D. E., Kelly, P. L., Morris, R. G., ... Ebeling, H. (2014). Robust weak-lensing mass calibration of Planck galaxy clusters. *MNRAS, 443*(3), 1973–1978. https://doi.org/10.1093/mnras/stu1423. arXiv: 1402.2670 [astro-ph.CO]

Wang, J., Bose, S., Frenk, C. S., Gao, L., Jenkins, A., Springel, V., & White, S. D. M. (2020). Universal structure of dark matter haloes over a mass range of 20 orders of magnitude. *Nature, 585*(7823), 39–42. https://doi.org/10.1038/s41586-020-2642-9

Warren, S. J., & Dye, S. (2003). Semilinear gravitational lens inversion. *ApJ, 590*(2), 673–682. https://doi.org/10.1086/375132. arXiv: astro-ph/0302587 [astro-ph]

Wong, K. C., Suyu, S. H., Chen, G. C.-F., Rusu, C. E., Millon, M., Sluse, D., ... Meylan, G. (2020). H0LiCOW XIII. A 2.4% measurement of H0 from lensed quasars: 5.3s tension between early and late-Universe probes. *MNRAS*. https://doi.org/10.1093/mnras/stz3094. arXiv: 1907.04869 [astro-ph.CO]

Xu, B., Postman, M., Meneghetti, M., Seitz, S., Zitrin, A., Merten, J., ... Koekemoer, A. (2016). The detection and statistics of giant arcs behind CLASH clusters. *ApJ, 817*(2), 85. https://doi.org/10.3847/0004-637X/817/2/85. arXiv: 1511.04002 [astro-ph.CO]

Zaroubi, S. (2013). The epoch of reionization. In T. Wiklind, B. Mobasher, & V. Bromm (Eds.), *The first galaxies* (Vol. 396, p. 45). Astrophysics and Space Science Library. https://doi.org/10.1007/978-3-642-32362-1_2

Zheng, W., Postman, M., Zitrin, A., Moustakas, J., Shu, X., Jouvel, S., ... van derWel, A. (2012). A magnified young galaxy from about 500 million years after the Big Bang. *Nature, 489*(7416), 406–408. https://doi.org/10.1038/nature11446. arXiv: 1204.2305 [astro-ph.CO]

Lensing by Large-Scale Structure

7

In a universe dominated by cold dark matter (CDM), structure formation occurs hierarchically. Collapsed objects grow via the merging of smaller objects until the most massive galaxy clusters form. Cosmological simulations show that these clusters are located at the intersections of filaments, which form a web-like structure called "Cosmic Web." This filament structure acts as a huge gravitational lens, inducing tiny distortions on distant galaxies' shapes. It also causes multiple deflections of photons that emerged from the last scattering surface, constituting the Cosmic Microwave Background (CMB). In this chapter, we discuss the propagation of light through an in-homogeneous universe. We introduce cosmic shear, and we illustrate its application as a probe of cosmology. Finally, we describe how gravitational lensing by large-scale structure affects the appearance of the CMB temperature fluctuations and polarization.

7.1 Light Propagation Through an In-homogeneous Universe

7.1.1 Deflection of Light

In unperturbed space–time, light travels along null geodesic lines of the symmetric, homogeneous, and isotropic Friedmann–Lemaitre–Robertson–Walker (FLRW) space–time (see Sect. 9.1).

In contrast to the earlier treatment, we have to take into account that lenses can now be comparable in size to the curvature scale of the universe, thus we need to refine the picture of straight light paths which are instantly deflected by sheet-like, thin lenses.

© Springer Nature Switzerland AG 2021
M. Meneghetti, *Introduction to Gravitational Lensing*, Lecture Notes
in Physics 956, https://doi.org/10.1007/978-3-030-73582-1_7

The propagation equation of light rays in arbitrary space–times can be derived using the theory of geometrical optics in General Relativity. The rigorous derivation of such equation can be found elsewhere (e.g. Misner et al. 1973, Schneider et al. 1992). Here, we will start from the following result: as shown by Bartelmann and Schneider (2001), light rays propagate through the unperturbed FLRW space–time such that the comoving separation vector $\vec{x}$ between them changes with the radial coordinate w as

$$\frac{\mathrm{d}^2 \vec{x}}{\mathrm{d}w^2} + K\vec{x} = 0 \,, \tag{7.1}$$

where $K = (H_0/c)^2(\Omega_{m,0} + \Omega_{\Lambda,0} - 1)$ is the curvature parameter of the universe. The physical separation, $\vec{r}$, between the rays is related to $\vec{x}$ through the scale factor a,

$$\vec{x} = \frac{\vec{r}}{a} \,. \tag{7.2}$$

The propagation equation is easily solved. It has the form of a simple harmonic oscillator equation, so its solutions are trigonometric or hyperbolic functions depending on the sign of K. For example, for $K > 0$ its general solution is

$$\vec{x} = \vec{A} \cos \sqrt{K}w + \vec{B} \sin \sqrt{K}w \,. \tag{7.3}$$

With the boundary conditions $\vec{x}(w = 0) = 0$ and $\mathrm{d}\vec{x}/\mathrm{d}w|_{w=0} = \vec{\theta}$, where $\vec{\theta}$ is the angle between the two rays at the observer position, we find

$$\vec{x}(w) = \vec{\theta}\,\frac{1}{\sqrt{K}} \sin \sqrt{K}w \,. \tag{7.4}$$

More generally, for negative and vanishing K, we find

$$\vec{x}(w) = \vec{\theta} f_K(w) \,. \tag{7.5}$$

These solutions have a very simple interpretation: obviously, for $K = 0$, $\vec{x} = \vec{\theta}w$, as we know in Euclidean space. For positive or negative curvature, the light rays approach each other or diverge from each other faster than in the flat case.

Adding perturbations is simple considering that the lensing masses are typically much smaller than the Hubble radius. Then, space–time can be considered flat in their surroundings, and we can use our earlier result on the deflection angle (cfr Sect. 2.2) in the form

$$\frac{d^2\vec{x}}{dw^2} = -\frac{2}{c^2}\vec{\nabla}_\perp \Phi \,, \tag{7.6}$$

where it must now be noted that the perpendicular gradient of Φ must be taken with respect to the comoving coordinates as well. Thus, the relation between the gradient in comoving and in angular units is

$$\vec{\nabla}_\perp \phi = \frac{1}{f_K(w)}\vec{\nabla}_{\vec{\theta}}\Phi \,. \tag{7.7}$$

Equation 7.6 describes how the actual light ray deviates from a straight line in an unperturbed Minkowski's space. In an expanding universe, the propagation equation changes to

$$\frac{d^2\vec{x}}{dw^2} + K\vec{x} = -\frac{2}{c^2}\vec{\nabla}_\perp \Phi \,, \tag{7.8}$$

which now incorporates overall space–time curvature and local perturbations caused by a potential Φ.

This inhomogeneous oscillator equation can be solved by constructing a Green's function $G(w, w')$, which is defined on the square $0 \leq w \leq w_s,\, 0 \leq w' \leq w_s$, where w_s is the coordinate distance to the source (see Figure on the side). Even in this case, the boundary conditions $\vec{x}(w = 0) = 0$ and $d\vec{x}/dw_{w=0} = \vec{\theta}$ apply.

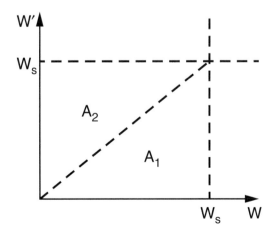

According to the definition of a Green's function, $G(w, w')$ must satisfy the following conditions:

- $G(w, w')$ is continuously differentiable in both triangles $A_{1,2}$ and satisfies the homogeneous differential equation

$$\frac{\mathrm{d}^2}{\mathrm{d}w^2} G(w, w') + K G(w, w') = 0 \qquad (7.9)$$

  for $w \neq w'$.
- $G(w, w')$ is continuous on the entire square.
- The derivative of $G(w, w')$ with respect to w jumps by 1 on the boundary between A_1 and A_2.
- As a function of w, $G(w, w')$ satisfies the homogeneous boundary conditions on the solution.

Accordingly, we set up

$$G(w, w') = \begin{cases} A(w') \cos \sqrt{K}w + B(w') \sin \sqrt{K}w & \text{on } A_1 \\ C(w') \cos \sqrt{K}w + D(w') \sin \sqrt{K}w & \text{on } A_2 \end{cases} . \qquad (7.10)$$

The homogeneous boundary conditions demand $A = B = 0$. Ensuring continuity at $w = w'$ requires

$$C \cos \sqrt{K}w' + D \sin \sqrt{K}w' = 0 , \qquad (7.11)$$

and the jump in the derivative implies

$$- C \sin \sqrt{K}w' + D \cos \sqrt{K}w' = \frac{1}{\sqrt{K}} . \qquad (7.12)$$

Thus,

$$C = -\frac{1}{\sqrt{K}} \sin \sqrt{K}w' \qquad (7.13)$$

$$D = \frac{1}{\sqrt{K}} \cos \sqrt{K}w' , \qquad (7.14)$$

and we obtain that

$$G(w, w') = \begin{cases} 0 & (w < w') \\ \frac{1}{\sqrt{K}} \sin \sqrt{K}(w - w') & (w > w') \end{cases} . \tag{7.15}$$

More generally, i.e. for arbitrary sign of K, we find

$$G(w, w') = \begin{cases} 0 & (w < w') \\ f_K(w - w') & (w > w') \end{cases} . \tag{7.16}$$

Therefore the general solution of the propagation equation reads

$$\vec{x} = f_K(w)\vec{\theta} - \frac{2}{c^2} \int_0^w dw' f_K(w - w')\vec{\nabla}_\perp \Phi . \tag{7.17}$$

As in the single-lens plane approach, we evaluate this integral along the unperturbed path, $f_K(w)\vec{\theta}$.

The deflection angle is defined as the difference between the perturbed and the unperturbed path,

$$\vec{\alpha}(\vec{\theta}, w) = \frac{f_K(w)\vec{\theta} - \vec{x}}{f_K(w)} = \frac{2}{c^2} \int_0^w dw' \frac{f_K(w - w')}{f_K(w)} \vec{\nabla}_\perp \Phi[f_K(w')\vec{\theta}, w'] . \tag{7.18}$$

This is now the deflection angle accumulated along a light path propagating into direction $\vec{\theta}$ out to the coordinate distance w. Hence, we denote it as $\vec{\alpha}(\vec{\theta}, w)$.

For a spatially flat universe, $K = 0$ and $f_K(w) = w$. Then,

$$\begin{aligned} \vec{\alpha}(\vec{\theta}, w) &= \frac{2}{c^2} \int_0^w dw' \left(1 - \frac{w'}{w}\right) \vec{\nabla}_\perp \Phi(w'\vec{\theta}, w') \\ &= \frac{2w}{c^2} \int_0^1 dy(1 - y)\vec{\nabla}_\perp \Phi(wy\vec{\theta}, wy) . \end{aligned} \tag{7.19}$$

7.1.2 Effective Convergence

In the single-lens plane case, the convergence is one half the divergence of $\vec{\alpha}$. Analogously, we define here an effective convergence for the large-scale structure,

$$\begin{aligned} \kappa_{\text{eff}}(\vec{\theta}, w) &= \frac{1}{2}\vec{\nabla}_{\vec{\theta}}\vec{\alpha}(\vec{\theta}, w) \\ &= \frac{1}{c^2} \int dw' \frac{f_K(w')f_K(w - w')}{f_K(w)} \Delta^{(2)}\phi[f_K(w')\vec{\theta}', w'] , \end{aligned} \tag{7.20}$$

where $\triangle^{(2)}$ is the two-dimensional Laplacian with respect to comoving coordinates,

$$\triangle^{(2)} = \vec{\nabla}_\perp^2 = \frac{\partial^2}{\partial x^2} + \frac{\partial^2}{\partial y^2} \; . \tag{7.21}$$

We now do the same as we did when we introduced the lensing potential: we replace

$$\triangle^{(2)} \to \triangle = \frac{\partial^2}{\partial x^2} + \frac{\partial^2}{\partial y^2} + \frac{\partial^2}{\partial z^2} \; , \tag{7.22}$$

and assume that $\partial\phi/\partial z = 0$ at the boundaries of the perturbations. Then, we can write

$$\kappa_{\text{eff}} = \frac{1}{c^2} \int_0^w \mathrm{d}w' \frac{f_K(w') f_K(w - w')}{f_K(w)} \triangle\Phi(f_K(w')\vec{\theta}, w') \; . \tag{7.23}$$

Now, we use Poisson's equation,

$$\triangle_r \Phi = 4\pi G\rho \; , \tag{7.24}$$

where the Laplacian is now taken with respect to physical coordinates. Using comoving coordinates, the same equation reads

$$\triangle\Phi = 4\pi G\rho a^2 \; . \tag{7.25}$$

Introducing the density contrast

$$\delta \equiv \frac{\rho - \overline{\rho}}{\overline{\rho}} \; , \tag{7.26}$$

we can write

$$\triangle\Phi = 4\pi G\overline{\rho}(1 + \delta)a^2 = 4\pi G\overline{\rho}_0 a^{-1}(1 + \delta) \; , \tag{7.27}$$

where we have inserted $\overline{\rho}a^{-3}$ as for ordinary (non-relativistic) matter. Decoupling the potential into a background potential

$$\triangle\overline{\Phi} = 4\pi G\overline{\rho}_0 a^{-1} \; , \tag{7.28}$$

and a peculiar (perturbing) potential ϕ, we have

$$\Delta\Phi = 4\pi G \bar{\rho}_0 a^{-1}\delta .$$

(7.29)

Using further

$$\bar{\rho}_0 = \Omega_{m,0}\frac{3H_0^2}{8\pi G}$$

(7.30)

yields the Poisson equation that we need,

$$\Delta\Phi = \frac{3}{2}H_0^2\Omega_{m,0}\frac{\delta}{a} .$$

(7.31)

The effective convergence can then be written as

$$\kappa_{\text{eff}}(\vec{\theta}, w) = \frac{3\Omega_{m,0}}{2}\left(\frac{H_0}{c}\right)^2 \int_0^w dw' \frac{f_K(w')f_K(w - w')}{f_K(w)} \frac{\delta[f_K(w')\vec{\theta}, w']}{a(w')} .$$

(7.32)

Notice the similarity of the distance factor with the factor $D_L D_{LS}/D_S$ that we had in the single-lens case.

If the sources are distributed in redshift or, equivalently, in the radial coordinate w, the mean effective convergence up to distance w_H is

$$\langle\kappa_{\text{eff}}\rangle(\vec{\theta}) = \int_0^{w_H} dw\, G(w)\kappa_{\text{eff}}(\vec{\theta}, w) ,$$

(7.33)

where $G(w)dw$ is the probability to find a source within dw of w. Then we can write

$$\langle\kappa_{\text{eff}}\rangle(\vec{\theta}) = \frac{3H_0^2\Omega_{m,0}}{2c^2} \int_0^{w_H} dw\, W(w) f_K(w)\frac{\delta[f_K(w)\vec{\theta}, w]}{a(w)} ,$$

(7.34)

with the effective weight function

$$W(w) = \int_w^{w_H} dw'\, G(w')\frac{f_K(w' - w)}{f_K(w')} .$$

(7.35)

7.1.3   Limber's Equation and the Convergence Correlation Function

We consider now the correlation function of the effective convergence,

$$\langle \kappa(\vec{\theta})\kappa(\vec{\theta} + \vec{\phi})\rangle_{\vec{\theta}} = \xi_{\kappa}(\phi) , \tag{7.36}$$

where the average extends over the positions $\vec{\theta}$ on the sky, and over the directions of the separation vector $\vec{\phi}$. Due to isotropy, the result cannot depend on the direction of $\vec{\phi}$. We can demonstrate that this quantity is related to the density contrast's three-dimensional correlation function via the so-called *Limber equation*. Similarly, a relation exists between the power spectra of the effective convergence and the density contrast. For this purpose, we follow Bartelmann and Schneider (2001) and use the flat-sky approximation.

Consider a generic homogeneous isotropic random field $g(\vec{x})$ (either real or complex) defined in a n-dimensional space. Its two point correlation function is

$$\xi_{gg}(|\vec{x} - \vec{y}|) = \langle g(\vec{x})g^*(\vec{y}))\rangle . \tag{7.37}$$

The Fourier transform of $g(\vec{x})$ is

$$\hat{g}(\vec{k}) = \int d^n x g(\vec{x}) \exp(i\vec{k} \cdot \vec{x}) . \tag{7.38}$$

Thus, the correlation function in Fourier space can be computed as,

$$\langle \hat{g}(\vec{k})\hat{g}^*(\vec{k}')\rangle = \left\langle \int d^n x g(x) \exp(i\vec{k} \cdot \vec{x}) \int d^n x' g(\vec{x}') \exp(-i\vec{k}' \cdot \vec{x}') \right\rangle$$

$$= \int d^n x \exp(i\vec{k} \cdot \vec{x}) \int d^n x' \exp(-i\vec{k}' \cdot \vec{x}')\langle g(\vec{x})g(\vec{x}')\rangle. \tag{7.39}$$

Inserting $\vec{y} + \vec{x} = \vec{x}'$ and using Eq. 7.37, we obtain

$$\langle \hat{g}(\vec{k})\hat{g}^*(\vec{k}')\rangle = \int d^n x \exp[i(\vec{k} - \vec{k}') \cdot \vec{x}] \int d^n y \exp(-i\vec{k}' \cdot \vec{y})\xi_{gg}(y)$$

$$= (2\pi)^n \delta_D^{(n)}(\vec{k} - \vec{k}') P_g(k) , \tag{7.40}$$

where we have introduced the power spectrum

$$P_g(k) \equiv \int d^n y \exp(-i\vec{k} \cdot \vec{y})\xi_{gg}(y) \tag{7.41}$$

as the Fourier transform of the correlation function. The function $\delta_D^{(n)}$ is the Dirac delta function in n dimensions.

Now, we consider the homogeneous isotropic random field $\delta(\vec{u})$, where $\vec{u}$ is a position vector in the three-dimensional space. Specifically, $\delta(\vec{u})$ is the density contrast defined in Sect. 9.7.2. The vector $\vec{u}$ can be decomposed into the comoving components perpendicular and parallel to the radial direction with respect to the origin, $\vec{u} = (f_K(w)\vec{\theta}, w)$. We define two weighted projections of δ along w as

$$g_i(\vec{\theta}) = \int_0^{w_H} dw q_i(w)\delta[f_K(w)\vec{\theta}, w] , \tag{7.42}$$

where $q_i(w)$ is the weight function and $i \in [1, 2]$. The integral extends from $w = 0$ to $w = w_H$, where w_H is the radial scale of the horizon.

If $\vec{\theta} - \vec{\theta}' = \vec{\phi}$, we can define the correlation function

$$\xi_{12}(\phi) = \langle g_1(\vec{\theta})g_2^*(\vec{\theta}')\rangle$$

$$= \int q_1(w)dw \int q_2(w')dw' \langle \delta[f_K(w)\vec{\theta}, w]\delta[f_K(w')\vec{\theta}', w']\rangle . \tag{7.43}$$

In the so-called *Limber approximation*, we assume that the correlation function ξ_{12} vanishes when $|w - w'|$ is larger than a coherence length $w_c \ll w_H$. We further assume that the weight functions $q_i(w)$ do not vary significantly over a scale smaller than w_c. Consequently, over the scales where ξ_{12} is not vanishing, we can set $q_2(w') \approx q_2(w)$ and $f_K(w') \approx f_K(w)$. This allows us to write the previous equation as

$$\xi_{12}(\phi) = dw \int q_1(w)q_2(w) \int d\Delta w \xi_{\delta\delta}(\sqrt{f_K^2(w)\theta^2 + \Delta w^2}, w) , \tag{7.44}$$

where $\Delta w = |w - w'|$. This *Limber equation* shows how the 2-point correlation function of the projected field is related to that of the three-dimensional field (Limber 1953).

We can now derive the relation between the corresponding power spectra. Writing $\delta(\vec{u})$ and $\delta(\vec{u}')$ in terms of their Fourier transforms, $\hat{\delta}(\vec{k})$ and $\hat{\delta}(\vec{k}')$, in Eq. 7.43, we obtain

$$\xi_{12}(\phi) = \int q_1(w)dw \int q_2(w')dw' \int \frac{d^3k}{(2\pi)^3} \int \frac{d^3k'}{(2\pi)^3} \langle \hat{\delta}(\vec{k})\hat{\delta}^*(\vec{k}')\rangle \exp(-if_K(w)\vec{\theta}\cdot\vec{k}_\perp)$$

$$\times \exp(if_K(w')\vec{\theta}'\cdot\vec{k}'_\perp) \exp(-ik_s w) \exp(ik'_s w') , \tag{7.45}$$

where we split the wave vector $\vec{k}$ into the components perpendicular and parallel to the line-of-sight, $\vec{k}_\perp$ and k_s, respectively. Then, we use Limber approximation and we replace $\langle \hat{\delta}(\vec{k})\hat{\delta}^*(\vec{k}')\rangle$ with $(2\pi)^3\delta_D(\vec{k} - \vec{k}')P_\delta(k, w)$ (as shown in Eq. 7.40). Carrying out the $\vec{k}'$ integration, we obtain

$$\xi_{12}(\phi) = \int dw q_1(w)q_2(w) \int \frac{d^3k}{(2\pi)^3} P_\delta(k, w) \exp[-if_K(w)\vec{k}_\perp \cdot (\vec{\theta} - \vec{\theta}')] \exp(-ik_s w)$$

$$\times \int dw' \exp(ik_s w') . \tag{7.46}$$

Since $\int dw' \exp(ik_s w') = 2\pi \delta_D(k_s)$, the latest integral implies that the only contribution to the correlation function comes from the modes perpendicular to the line-of-sight, i.e. with $\vec{k} = (\vec{k}_\perp, 0)$. Thus, we obtain

$$\xi_{12}(\phi) = \int dw q_1(w) q_2(w) \int \frac{d^2 k_\perp}{(2\pi)^2} P_\delta(k_\perp, w) \exp\left[-i f_K(w) \vec{k}_\perp \cdot \vec{\phi}\right]. \quad (7.47)$$

Using Eq. 7.41, the correlation function can be used to compute the power spectrum

$$P_{12}(l) = \int d^2 \phi \xi_{12}(\phi) \exp(i\vec{l} \cdot \vec{\phi})$$

$$= \int dw q_1(w) q_2(w) \int \frac{d^2 k_\perp}{(2\pi)^2} P_\delta(k_\perp, w) \delta_D[\vec{l} - f_K(w)\vec{k}_\perp]. \quad (7.48)$$

The latest δ_D function allows to carry out the $\vec{k}_\perp$-integral:

$$P_{12}(l) = \int dw \frac{q_1(w) q_2(w)}{f_K^2(w)} P_\delta\left(\frac{l}{f_K(w)}, w\right). \quad (7.49)$$

This is Limber's equation in Fourier space (Kaiser 1992, Kaiser et al. 1998).

Now, we specialize this result to the effective convergence in Eq. 7.34. Comparing Eqs. 7.42 and 7.34, we see that

$$q_1(w) = q_2(w) = \frac{3H_0^2 \Omega_{m,0}}{2c^2} W(w) \frac{f_K(w)}{a(w)}. \quad (7.50)$$

Thus, the convergence power spectrum is

$$P_\kappa(l) = \frac{9H_0^4 \Omega_{m,0}^2}{4c^4} \int_0^{w_H} \frac{W^2(w)}{a^2(w)} P_\delta\left(\frac{l}{f_K(w)}, w\right) dw. \quad (7.51)$$

Panel A of Fig. 7.1 show the power spectra of the effective convergence for several choices of the cosmological parameters and of the source redshift distribution. The curves $l^2 P_\kappa(l)$ are derived using the fitting function of $P_\delta(l, w)$ provided by Peacock and Dodds (1996), including non-linear evolution. They are characterized by a peak at $l \sim 10^4$, corresponding to an angular scale of ~ 1 arcmin. The non-linear evolution of the power spectrum of the density contrast is very important at $l \gtrsim 200$ (see e.g. Fig. 32 of Schneider 2006).

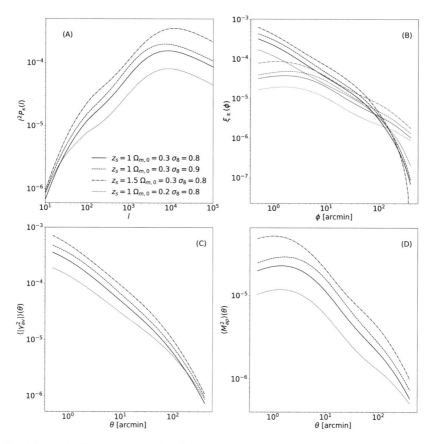

Fig. 7.1 Panel **a**: Power spectra of the effective convergence. Panel **b**: Shear correlation functions $\xi_+(\phi)$ and $\xi_-(\phi)$ (black and red curves, respectively). Panel **c**: Shear variance in apertures of different sizes. Panel **d**: Mass variance in apertures of different sizes. In each panel we show the curves corresponding to cosmological models with different parameters. The reference model is given by the solid lines. It is a flat ΛCDM cosmological model with $\Omega_{m,0} = 0.3$ and $\sigma_8 = 0.8$. The sources are assumed to lay on a single plane at $z_S = 1$. Then, the results are shown for other models with increased values of $\sigma_8 = 0.9$ (dashed lines) and lower $\Omega_{m,0} = 0.2$ (dotted lines). Finally, the dot-dashed lines show how the results change for a larger source redshift of 1.5

7.1.4 Effective Lensing Potential, Lensing Jacobian, Shear

In analogy to what was found in the limit of a single-lens plane, we can define an effective lensing potential, $\Psi(\vec{\theta}, w)$, related to the effective convergence through the equation

$$\kappa_{\text{eff}}(\vec{\theta}, w) = \frac{1}{2}\triangle\Psi(\vec{\theta}, w) \; . \tag{7.52}$$

Similarly, we can define the effective lensing Jacobian

$$A(\vec{\theta}, w) = I - \frac{\partial \vec{\alpha}(\vec{\theta}, w)}{\partial \vec{\theta}} = \frac{1}{f_K(w)} \frac{\partial \vec{x}(\vec{\theta}, w)}{\partial \vec{\theta}} . \tag{7.53}$$

This is still symmetric to high accuracy, except for light rays that encounter more than one compact deflector along their path to the observer (Jain et al. 2000). Therefore, we can still describe the an-isotropic distortions of sources lensed by the large-scale structure of the universe using the traceless part of this matrix, i.e. by means of an effective shear, $\gamma((\vec{\theta}, w) = \gamma_1(\vec{\theta}, w) + i\gamma_2(\vec{\theta}, w)$. As seen earlier in Chap. 3, circular sources are mapped onto elliptical images. Given that in this regime of lensing the amplitude of the effective convergence is $\kappa_{\text{eff}} \ll 1$, the image ellipticity can be used to estimate γ.

Since both κ_{eff} and γ are related to the effective potential, the power spectra of these quantities are also related. For example, as seen in Sect. 6.2.5, by taking the Fourier transform of both sides of Eq. 7.52, we obtain that

$$\tilde{\kappa}_{\text{eff}}(\vec{l}) = -\frac{l^2}{2} \tilde{\Psi}(\vec{l}) . \tag{7.54}$$

Thus, the power spectra of κ_{eff} and Ψ are related through

$$P_\kappa(l) = \frac{l^4}{4} P_\Psi(l) . \tag{7.55}$$

Similarly, since

$$\tilde{\gamma}_1 = -\frac{1}{2}(l_1^2 - l_2^2)\tilde{\Psi} , \tag{7.56}$$

$$\tilde{\gamma}_2 = -l_1 l_2 \tilde{\Psi} , \tag{7.57}$$

we obtain that

$$P_\gamma(l) = \frac{l^4}{4} P_\Psi(l) = P_\kappa(l) . \tag{7.58}$$

Thus, the convergence and the shear power spectra are identical, suggesting that it is possible to measure the power spectrum of the projected density contrast from source ellipticity correlations. We generally refer to the shear produced by the universe's large-scale structure as *cosmic shear*.

7.2 Cosmic Shear

7.2.1 Shear Correlation Functions

We consider a pair of galaxies at $\vec{\theta}$, separated by $\vec{\phi}$. We can use the polar angle φ of the separation vector $\vec{\phi}$ to define the tangential and the cross components of the shear (cfr. Sect. 6.2.3) as

$$\gamma_t = \gamma \cos(2\varphi) \tag{7.59}$$

$$\gamma_\times = \gamma \sin(2\varphi). \tag{7.60}$$

Using these two shear components, we can further define the correlation functions $\xi_{tt}(\phi) = \langle \gamma_t(\vec{\theta})\gamma_t(\vec{\theta}+\vec{\phi})\rangle$ and $\xi_{\times\times}(\phi) = \langle \gamma_\times(\vec{\theta})\gamma_\times(\vec{\theta}+\vec{\phi})\rangle$, as well as the mixed correlator $\xi_{t\times}(\phi) = \langle \gamma_t(\vec{\theta})\gamma_\times(\vec{\theta}+\vec{\phi})\rangle = \xi_{\times t}(\phi)$.

We begin with the correlation function of the tangential shear, which can be written in terms of the tangential shear power spectrum as

$$\xi_{tt}(\phi) = \int \frac{d^2l}{(2\pi)^2} P_{\gamma_t}(l) \exp\left(-i\vec{l}\cdot\vec{\phi}\right). \tag{7.61}$$

Since the Fourier transform of the tangential shear is

$$\tilde{\gamma}_t = -\frac{l^2}{2}(\cos^2\varphi - \sin^2\varphi)\tilde{\Psi}, \tag{7.62}$$

the power spectrum is

$$P_{\gamma_t}(l) = \frac{l^4}{4}(\cos^2\varphi - \sin^2\varphi)^2 P_\Psi(l) = (\cos^2\varphi - \sin^2\varphi)^2 P_\kappa(l). \tag{7.63}$$

Inserting this result into Eq. 7.61, we obtain that the correlation function $\xi_{tt}(\phi)$ can be written as

$$\xi_{tt}(\phi) = \int \frac{l\,dl}{2\pi} P_\kappa(l)[J_0(l\phi) + J_4(l\phi)], \tag{7.64}$$

where $J_n(x)$ are the n-order Bessel functions of the first kind[1].

---

[1]The integral representation of the Bessel functions of the first kind is given by

$$J_n(x) = \frac{1}{\pi}\int_0^\pi \cos(n\varphi - x\sin\varphi)d\varphi = \frac{(-i)^n}{\pi}\int_0^\pi e^{ix\varphi}\cos(n\varphi)d\varphi. \tag{7.65}$$

Similar considerations for the $\xi_{\times\times}$ correlation function lead to the result that

$$\xi_{\times\times}(\phi) = \int \frac{l\,dl}{2\pi} P_\kappa(l)[J_0(l\phi) - J_4(l\phi)] \,, \tag{7.66}$$

given that

$$P_{\gamma_\times}(l) = 4\cos^2\varphi\sin^2\varphi\, P_\kappa(l) \,. \tag{7.67}$$

Finally, for the mixed correlation functions, due to parity symmetry we find that

$$\xi_\times(\phi) = \xi_{t\times}(\phi) = \xi_{\times t}(\phi) = 0 \,. \tag{7.68}$$

For any measurement of cosmic shear, $\xi_\times(\phi) = 0$ provides a test for the reliability of the measurement, while $\xi_\times(\phi) \neq 0$ points to systematic errors.

We can now define the combinations

$$\xi_\pm(\phi) \equiv \langle\gamma_t\gamma_t'\rangle \pm \langle\gamma_\times\gamma_\times'\rangle \,, \tag{7.69}$$

and

$$\xi_\times(\phi) \equiv \langle\gamma_t\gamma_\times'\rangle \,. \tag{7.70}$$

such that

$$\xi_+(\phi) = \int \frac{l\,dl}{2\pi} P_\kappa(l)\,J_0(l\phi) \,, \tag{7.71}$$

$$\xi_-(\phi) = \int \frac{l\,dl}{2\pi} P_\kappa(l)\,J_4(l\phi) \,. \tag{7.72}$$

Measuring ξ_+ and ξ_- we can thus constrain $P_\kappa(l)$.

Note that the orthonormality relation of Bessel functions implies that

$$P_\kappa(l) = 2\pi \int_0^\infty d\phi\,\phi\,\xi_+(\phi)\,J_0(l\phi) = 2\pi \int_0^\infty d\phi\,\phi\,\xi_-(\phi)\,J_4(l\phi) \,, \tag{7.73}$$

i.e. we can express the power spectrum in terms of the observable correlation functions. However, we cannot use these equations to measure $P_\kappa(l)$, because they require to know the correlation functions for all angles ϕ. This is not possible due to both the limited spatial resolution of the shear measurements and the finite field-of-view of the observations: we can measure the correlation functions over a limited range of angular separations.

The correlation functions $\xi_\pm(\phi)$ for the same cosmological models and source redshift distributions used in panel A of Fig. 7.1 are shown in panel B of the same figure.

7.2.2 Shear in Apertures and Aperture Mass

Other convenient second order shear statistics are given by the shear dispersion and by the aperture mass.

The mean shear in a (circular) aperture of radius θ is given by

$$\gamma_{\mathrm{av}}(\theta) = \int_0^\theta \frac{d^2\Theta}{\pi\Theta^2}\gamma(\vec{\Theta}) . \tag{7.74}$$

Its dispersion is related to the convergence correlation function and thus to the convergence power spectrum:

$$
\begin{aligned}
\langle|\gamma_{\mathrm{av}}|^2\rangle(\theta) &= \left\langle \int_0^\theta \frac{d^2\Theta}{\pi\Theta^2} \int_0^\theta \frac{d^2\Theta'}{\pi\Theta'^2}[\gamma_1(\vec{\Theta})\gamma_1(\vec{\Theta}') + \gamma_2(\vec{\Theta})\gamma_2(\vec{\Theta}')] \right\rangle \\
&= \int_0^\theta \frac{d^2\Theta}{\pi\Theta^2} \int_0^\theta \frac{d^2\Theta'}{\pi\Theta'^2}\xi_\kappa(|\vec{\Theta}' - \vec{\Theta}|) \\
&= \int_0^\theta \frac{d^2\Theta}{\pi\Theta^2} \int_0^\theta \frac{d^2\Theta'}{\pi\Theta'^2} \int \frac{d^2l}{(2\pi)^2} P_\kappa(l) \exp[-i\vec{l}(\vec{\Theta} - \vec{\Theta}')] \\
&= 4\pi^2 \int \frac{ldl}{2\pi} P_\kappa(l) \left[\frac{J_1(l\theta)}{\pi l\theta}\right]^2 \\
&= \frac{1}{2\pi} \int ldl\, P_\kappa(l) W_{\mathrm{TH}}(l\theta) ,
\end{aligned}
\tag{7.75}
$$

where

$$W_{\mathrm{TH}}(x) = \frac{4J_1^2(x)}{x^2} \tag{7.76}$$

is a top-hat filter function[2].

The aperture mass is a weighted integral of the (effective) convergence within a (circular) aperture,

$$M_{\mathrm{ap}}(\theta) = \int d^2\Theta\, U(\vec{\Theta})\kappa_{\mathrm{eff}}(\vec{\Theta}) . \tag{7.77}$$

If the weight function satisfies the condition

$$\int_0^\theta \Theta d\Theta\, U(\Theta) = 0 , \tag{7.78}$$

---

[2]In Eq. 7.75, we have used $\int_0^1 x dx\, J_0(ax) = \frac{1}{a}J_1(a)$.

i.e. if it is compensated, the aperture mass can also be written as

$$M_{\mathrm{ap}}(\theta) = \int \mathrm{d}^2\Theta \, Q(\Theta) \gamma_t(\vec{\Theta}) \, , \tag{7.79}$$

where γ_t is the tangential shear with respect to the aperture center. Q is related to U through

$$Q(x) = \frac{2}{x^2} \int_0^x \mathrm{d}x' x' U(x') - U(x) \, . \tag{7.80}$$

A common choice (but not a necessary one) is

$$U(\Theta) = \frac{9}{\pi\Theta^2}(1 - x^2)\left(\frac{1}{3} - x^2\right) \, , \tag{7.81}$$

with $x \equiv \theta/\Theta$. This implies

$$Q(\Theta) = \frac{6}{\pi\Theta^2}x^2(1 - x^2) \, . \tag{7.82}$$

With this choice, the variance of the aperture mass turns out to be

$$\begin{aligned}
\langle M_{\mathrm{ap}}^2(\theta) \rangle &= \left\langle \int_0^\theta \mathrm{d}^2\Theta \int_0^\theta \mathrm{d}^2\Theta' U(\Theta)U(\Theta')\kappa_{\mathrm{eff}}(\vec{\Theta})\kappa_{\mathrm{eff}}(\vec{\Theta}') \right\rangle \\
&= \int \mathrm{d}^2\Theta \int \mathrm{d}^2\Theta' U(\Theta)U(\Theta')\xi_\kappa(|\vec{\Theta}' - \vec{\Theta}|) \\
&= \int \mathrm{d}^2\Theta \int \mathrm{d}^2\Theta' U(\Theta)U(\Theta') \int \frac{\mathrm{d}^2 l}{(2\pi)^2} P_\kappa(l) \exp[-i\vec{l}(\vec{\Theta} - \vec{\Theta}')] \\
&= 4 \int \frac{l\mathrm{d}l}{2\pi} P_\kappa(l) J^2(l\theta) \, ,
\end{aligned} \tag{7.83}$$

where

$$J(l\theta) \equiv \frac{12}{(l\theta)^2} J_4(l\theta) \, . \tag{7.84}$$

Some examples of $\langle |\gamma_{\mathrm{av}}|^2 \rangle(\theta)$ and $\langle M_{\mathrm{ap}}^2 \rangle(\theta)$ are shown in panels C and D of Fig. 7.1.

7.2.3 E- and B-modes

As we have seen, the shear is related to the convergence by a convolution (see Sect. 6.2.5). Thus, the shear components originate from a single scalar field κ, implying that they are not independent. Consequently, not all combinations of γ_1 and γ_2 are allowed. For example, a mass over-density can only yield tangential or radial alignments of the shear, while it cannot produce other distortions. Those shear modes that are allowed are called E-modes, while the others are called B-modes.

E- and B-modes can be separated using aperture measures. For example, the variance of the aperture mass, $\langle M_{ap}^2 \rangle(\theta)$ is sensitive only to the E-mode. For analogy, one can define

$$M_{\perp}(\theta) = \int d^2\Theta\, Q(\Theta)\gamma_{\times}(\vec{\Theta}) , \qquad (7.85)$$

obtaining a quantity, $\langle M_{\perp}^2 \rangle(\theta)$ which is sensitive only to B-modes. A pure E-mode signal would yield $M_{\perp} = 0$.

The observed shear field does not necessarily contain only E-modes. Indeed, B-modes can be generated by noise, by the fact that the Born approximation used to derive the previous equations is not strictly valid (Jain et al. 2000), by the clustering of galaxies (Schneider et al. 2002b), or by the intrinsic alignment of the galaxies mimicking a shear signal (see, e.g. Crittenden et al. 2001, Troxel & Ishak 2015). The tidal interactions between galaxies during structure formation can produce correlations between the galaxy ellipticities that can mimic a shear signal. Thus, the detection of a B-mode signals the presence of systematics affecting the measurements.

We may describe the E- and the B-modes with two different potentials, Ψ^E and Ψ^B, and define the complex potential

$$\Psi(\vec{\theta}) = \Psi^E(\vec{\theta}) + i\Psi^B(\vec{\theta}) . \qquad (7.86)$$

Using this potential, we can define the complex convergence

$$\kappa(\vec{\theta}) = \kappa^E(\vec{\theta}) + i\kappa^B(\vec{\theta}) , \qquad (7.87)$$

and shear,

$$\gamma(\vec{\theta}) = \left[\frac{1}{2}(\Psi_{11}^E - \Psi_{22}^E) - \Psi_{12}^B \right] + i\left[\Psi_{12}^E + \frac{1}{2}(\Psi_{11}^B - \Psi_{22}^B) \right] . \qquad (7.88)$$

We can compute the power spectra of the E- and B-mode components of the convergence by

$$\langle \hat{\kappa}^E(\vec{l})\hat{\kappa}^{E*}(\vec{l}) \rangle = (2\pi)^2\delta_D(\vec{l} - \vec{l}')P_E(l) , \qquad (7.89)$$

$$\langle \hat{\kappa}^B(\vec{l})\hat{\kappa}^{B*}(\vec{l}) \rangle = (2\pi)^2\delta_D(\vec{l} - \vec{l}')P_B(l) . \qquad (7.90)$$

Finally, from Eq. 7.83 we obtain

$$\langle M_{\mathrm{ap}}^2 \rangle (\theta) = 4 \int \frac{l \, dl}{2\pi} P_E(l) J^2(l\theta) \,, \tag{7.91}$$

and

$$\langle M_{\perp}^2 \rangle (\theta) = 4 \int \frac{l \, dl}{2\pi} P_B(l) J^2(l\theta) \,. \tag{7.92}$$

7.2.4 Cosmic Shear as a Cosmological Probe

Eq. 7.51 has very important implications. It shows that the convergence power spectrum (and all the second order statistics defined earlier) depends on cosmology in several ways. First of all, it is sensitive to the growth of structures, described by the density contrast's power spectrum. It is proportional to the square of the matter density, $\Omega_{m,0}$. Finally, it depends on the geometry of the universe through the distances appearing in the factor $f_K(w') f_K(w - w')/f_K(w)$. This explains why, after it was first detected in the early 2000s (Bacon et al. 2000, Van Waerbeke et al. 2000, Wittman et al. 2000), *cosmic shear*, i.e. gravitational lensing by the large-scale structure of the universe, became one of the key probes in several cosmological experiments. These include surveys like the ongoing Dark Energy Survey (DES; Abbott et al. 2020), the Kilo-Degree Survey (KiDS; de Jong et al. 2017), and the Hyper-Suprime-Cam Subaru Strategic Program (HSC-SSP; Aihara et al. 2018), which are measuring cosmic shear over thousands of square degrees of the sky. In the next years, these measurements will be extended to nearly the whole sky thanks to the surveys that will be conducted by the Vera Rubin Observatory (*The Legacy Survey of Space and Time* LSST; Ivezić et al. 2019) and by the space missions Euclid (Laureijs et al. 2011) and Nancy Grace Roman Space Telescope (Akeson et al. 2019).

The dependence of the power spectrum of the effective convergence and the second order statistics introduced in the previous sections on some cosmological parameters and source redshift are illustrated in the four panels of Fig. 7.1. The solid lines refer to a flat ΛCDM cosmological model with $\Omega_{m,0} = 0.3$ and $\sigma_8 = 0.8$. The source redshift is $z_S = 1$. The other lines show how the results change by increasing σ_8 and the source redshift (dashed and dot-dashed lines, respectively) or reducing the value of $\Omega_{m,0}$. As we can see, the power spectrum of the effective convergence grows as a function of σ_8, $\Omega_{m,0}$, and z_s, and so do the second order statistics $\xi_\pm$, $\langle |\gamma_{\mathrm{av}}| \rangle$, and M_{ap}^2

Cosmological parameters can be measured as follows. We assume that the shear correlation functions (or any other second order shear statistics) have been measured in n angular bins. These measurements are the elements of our data vector $\vec{\xi}_{obs}$. We can define the likelihood of the data given a set of cosmological parameters $\vec{p}$ as $\mathcal{L}(\vec{\xi}_{obs} | \vec{p})$. Assuming that it is Gaussian in the multi-dimensional parameter space, the likelihood is given by

$$\mathcal{L}(\vec{\xi}_{obs}|\vec{p}|) = \frac{1}{(2\pi)^n|\mathbf{S}|} \exp\left[-\frac{1}{2}(\vec{\xi}_{obs} - \vec{\xi}_{model})^T \mathbf{S}^{-1}(\vec{\xi}_{obs} - \vec{\xi}_{model})\right],$$
(7.93)

where $\vec{\xi}_{model}$ is the vector of theoretical predictions for the shear correlations computed for the same angular bins with the fiducial cosmological model characterized by the parameters $\vec{p}$. $\mathbf{S}$ is the covariance matrix, defined as

$$\mathbf{S} = \langle(\vec{\xi}_{obs} - \vec{\xi}_{model})^T (\vec{\xi}_{obs} - \vec{\xi}_{model})\rangle.$$
(7.94)

The average is over multiple realizations of the cosmic shear survey under consideration.

To determine the cosmological parameters $\vec{p}$, we can sample the likelihood and use it to maximize a posterior probability obtained by multiplying the likelihood by the priors on the cosmological parameters, $\mathcal{P}(\vec{p})$,

$$\mathcal{P}(\vec{p}|\vec{\xi}_{obs}) \propto \mathcal{L}(\vec{\xi}_{obs}|\vec{p}) \cdot \mathcal{P}(\vec{p}).$$
(7.95)

It is important to note that the signal depends on the source redshift (or distance) distribution, described by the function $G(w)$. Suppose we divide the sources into two groups, closer and farther than a certain comoving distance. The first group sources will be lensed only by the cosmic structure closest to us, while more distant matter distributions will also distort those in the second group. Thus, by measuring the lensing signal in different redshift bins, we can trace the cosmic structure's growth. This is particularly useful to constrain dark energy, whose effect is to push the universe through an accelerated expansion, thus suppressing the growth of cosmic structures below a certain redshift. This technique is called *lensing tomography* (Hu 1999).

Unfortunately, extracting the cosmological parameters from the lensing signal is not trivial. First of all, the signal amplitude is tiny (at the level of ~ 0.01). Second, the cosmological parameters are degenerate. Third, the measurements can be affected by several systematics due, e.g. to biased shape measurements, intrinsic alignments, etc., each of which needs to be adequately modeled when fitting the data (Hildebrandt et al. 2017, Troxel et al. 2018). Fourth, the theoretical calculation of the power spectrum of the convergence, necessary to compare to the data, requires to know P_δ to high accuracy. Unfortunately, no analytical methods exist to calculate the non-linear evolution of the density contrast's power spectrum, which needs to be estimated empirically from numerical simulations. Fitting formulae were proposed for P_δ (Peacock & Dodds 1996, Smith et al. 2003, Takahashi et al. 2012), but they have only a finite accuracy. Finally, the inverse of the covariance matrix $\mathbf{S}^{-1}$ is crucial for calculating the errors on the cosmological parameters. It can be estimated using several methods. The first possibility is to evaluate it from extensive, high-resolution cosmological simulations (see, e.g. Harnois-Déraps et al. 2018). The second is to derive it from the data by dividing the survey area into several regions (Friedrich et al. 2016, Norberg et al. 2009). The third is through analytical modeling (Joachimi et al. 2008, Krause and Eifler 2017, Schneider et al. 2002a, Takada and Jain 2009). Each method has its advantages and disadvantages, but the latest is preferred in the most recent cosmic shear analyses because of its lower computational cost and noise-free estimates (Barreira et al. 2018, Hildebrandt et al. 2017, Krause et al. 2017, Troxel et al. 2018). However, its accuracy is not well established, and comparisons with the simulation ensemble approach yield discrepancies at the level of 0.5σ in the estimate of the cosmological parameters.

7.3   Lensing of Cosmic Microwave Background

To conclude this chapter, we consider the weak lensing effects of the LSS on the Cosmic Microwave Background (CMB).

The CMB is relic radiation observable in all directions with a nearly perfect black body spectrum corresponding to a temperature of about 2.7 Kelvin. According to the Big-Bang Theory, the universe was extremely hot at early times. During these epochs, the matter and radiation were in a state of thermal equilibrium. Indeed, the temperature was so high that ordinary matter was fully ionized and capable of efficiently scattering photons. At a redshift of $z \approx 1100$, the universe's temperature dropped to nearly 3000 Kelvin, allowing protons and electrons to recombine into atomic hydrogen nuclei. When this recombination was complete, the photons ceased to scatter at all. They started to propagate freely throughout the universe, originating the CMB radiation that we see today, redshifted by the universe's expansion.

Observations show that the CMB temperature is only nearly uniform. Indeed, the CMB shows anisotropies at the level of $\mathcal{O}(10^{-5})$. These anisotropies arise from acoustic oscillations in the primordial baryon-photon fluid that remains imprinted on the last scattering's surface. They can be accurately modeled and predicted using the linear perturbation theory.

During the recombination, it is also expected that Thomson scattering of photons gives rise to the CMB radiation's polarization. This polarization is expected to be linear to a degree directly related to the quadrupole an-isotropy in the photons when they last scatter.

The power spectrum of the fluctuations of the CMB and its polarization properties provide constraints to cosmological parameters and structures' formation. In the next sections, we will show that both CMB temperature fluctuations and polarization are affected by gravitational lensing.

7.3.1 Lensing of the CMB Temperature

Compared to the lensing effects discussed in the first part of this chapter, the CMB can be considered a single source as large as the whole sky at the last scattering surface's redshift.

We consider the relative temperature fluctuation

$$\tau(\vec{\theta}) = \frac{T(\vec{\theta}) - \overline{T}}{\overline{T}} = \frac{\Delta T}{\overline{T}}. \tag{7.96}$$

Gravitational lensing has the effect that the relative temperature fluctuation τ_{obs} observed in the direction $\vec{\theta}$ corresponds to the intrinsic relative fluctuation τ in the direction $\vec{\theta} - \vec{\alpha}(\vec{\theta})$, where $\vec{\alpha}(\vec{\theta})$ is the effective deflection angle in the direction $\vec{\theta}$:

$$\tau_{obs}(\vec{\theta}) = \tau[\vec{\theta} - \vec{\alpha}(\vec{\theta})]. \tag{7.97}$$

To illustrate this effect, in Fig. 7.2, we show a small patch of the CMB temperature map. The map on the left shows the unlensed temperature map, characterized by a slight temperature gradient. The map on the right shows the imprint on the CMB of a mass over-density. The brightness of the CMB temperature is left unchanged, but the direction where its fluctuations are seen is changed. Note that no effect would be visible if the CMB temperature were perfectly uniform.

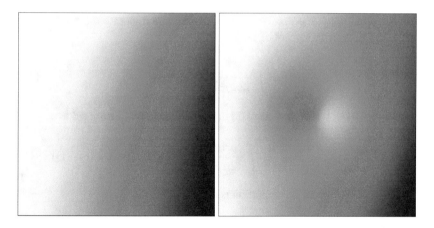

Fig. 7.2 Simulation of the lensing effects of a spherical mass over-density on the CMB temperature. The left image shows a mock CMB temperature map without lensing. An essential aspect of this simulation is that the temperature is not constant but varies across the map (blue is colder, white is hotter). The right panel shows the lensed temperature map. Lensing does not change the brightness (the color scale in the two images is the same). It changes the direction where the temperature fluctuations are seen

We wish to determine what is the effect of lensing on the statistical properties of the CMB temperature. In particular, we calculate, in the flat-sky approximation, how the spectrum of the CMB temperature fluctuations changes because of lensing.

Remark 7.1. We perform the calculations in the limit of flat-sky, i.e. assuming a small section of the sky, $\theta \lesssim 60$ deg, for which Fourier techniques can replace spherical harmonic analysis. For the generalization to the curved sky, we refer the reader to the review paper on CMB lensing by Lewis and Challinor (2006).

We can safely assume that the temperature of the CMB changes on scales that are substantially larger than those over which the deflection angle changes. Therefore, Eq. 7.97 can be Taylor-expanded as

$$\tau_{obs}(\vec{\theta}) \simeq \tau(\vec{\theta}) - \vec{\alpha}(\vec{\theta}) \cdot \vec{\nabla}\tau(\vec{\theta}) + \frac{1}{2} \frac{\partial \tau(\vec{\theta})}{\partial \theta_i \partial \theta_j} \alpha_i \alpha_j. \tag{7.98}$$

Since our purpose is to calculate the power spectrum of τ_{obs}, we need to calculate the Fourier transform of Eq. 7.98. Since the deflection angle is $\vec{\alpha}(\vec{\theta}) = \vec{\nabla}\Psi(\vec{\theta})$, the relation between the deflection angle and the Fourier transform of the potential, $\hat{\Psi}(\vec{l})$, is

$$\vec{\alpha}(\vec{\theta}) = \int \frac{d^2l}{(2\pi)^2} [i\vec{l}\hat{\Psi}(\vec{l})] \exp(i\vec{l} \cdot \vec{\theta}) . \tag{7.99}$$

On the other hand, the gradient of the temperature fluctuation can be written in terms of the Fourier transform of $\tau(\vec{\theta})$ as

$$\vec{\nabla}\tau(\vec{\theta}) = \int \frac{d^2l}{(2\pi)^2} [i\vec{l}\hat{\tau}(\vec{l})] \exp(i\vec{l} \cdot \vec{\theta}) . \tag{7.100}$$

Thus, we can write the second term on the right-hand side of Eq. 7.98 as

$$-\vec{\alpha}(\vec{\theta}) \cdot \vec{\nabla}\tau(\vec{\theta}) = \int \frac{d^2l_1}{(2\pi)^2} \int \frac{d^2l_2}{(2\pi)^2} \vec{l}_1 \cdot \vec{l}_2 \hat{\Psi}(\vec{l}_1) \hat{\tau}(\vec{l}_2) \exp[i(\vec{l}_1 + \vec{l}_2) \cdot \vec{\theta}] . \tag{7.101}$$

The Fourier transform of this equation is easily calculated:

$$\begin{aligned}
\hat{\mathcal{T}}_1(\vec{l}) &= \int \frac{d^2l_1}{(2\pi)^2} \int \frac{d^2l_2}{(2\pi)^2} \vec{l}_1 \cdot \vec{l}_2 \hat{\Psi}(\vec{l}_1) \hat{\tau}(\vec{l}_2) \int d^2\theta \exp[i(\vec{l}_1 + \vec{l}_2 - \vec{l}) \cdot \vec{\theta}] \\
&= \int \frac{d^2l_1}{(2\pi)^2} \int \frac{d^2l_2}{(2\pi)^2} \vec{l}_1 \cdot \vec{l}_2 \hat{\Psi}(\vec{l}_1) \hat{\tau}(\vec{l}_2) (2\pi)^2 \delta_D(\vec{l}_1 + \vec{l}_2 - \vec{l}) \\
&= \int \frac{d^2l_1}{(2\pi)^2} \vec{l}_1 \cdot (\vec{l} - \vec{l}_1) \hat{\Psi}(\vec{l}_1) \hat{\tau}(\vec{l} - \vec{l}_1).
\end{aligned} \tag{7.102}$$

Likewise, we obtain that

$$\frac{1}{2}\frac{\partial^2 \tau(\vec{\theta})}{\partial \theta_i \partial \theta_j}\alpha_i(\vec{\theta})\alpha_j(\vec{\theta}) = \frac{1}{2}\int \frac{d^2 l_1}{(2\pi)^2} l_{1i} l_{1j}\hat{\tau}(\vec{l}_1)$$

$$\times \int \frac{d^2 l_2}{(2\pi)^2} l_{2i}\hat{\Psi}(\vec{l}_2)$$

$$\times \int \frac{d^2 l_3}{(2\pi)^2} l_{3j}\hat{\Psi}(\vec{l}_3) \exp\left[i(\vec{l}_1 + \vec{l}_2 + \vec{l}_3)\cdot\vec{\theta}\right], \quad (7.103)$$

whose Fourier transform is

$$\hat{\mathcal{F}}_2(\vec{l}) = \frac{1}{2}\int \frac{d^2 l_1}{(2\pi)^2}\int \frac{d^2 l_2}{(2\pi)^2}(\vec{l}_1\cdot\vec{l}_2)[\vec{l}_1\cdot(\vec{l}-\vec{l}_1-\vec{l}_2)]\hat{\tau}(\vec{l}_1)\hat{\Psi}(\vec{l}_2)\hat{\Psi}(\vec{l}-\vec{l}_1-\vec{l}_2). \quad (7.104)$$

Thus, using the results in Eqs. 7.102 and 7.103, from Eq. 7.98, we obtain that

$$\hat{\tau}_{obs}(\vec{l}) = \hat{\tau}(\vec{l}) + \hat{\mathcal{F}}_1(\vec{l}) + \hat{\mathcal{F}}_2(\vec{l}). \quad (7.105)$$

The power spectrum of the observed temperature fluctuation can now be computed from Eq. 7.105, since

$$\langle\hat{\tau}_{obs}(\vec{l})\hat{\tau}_{obs}(\vec{l}')\rangle = (2\pi)^2\delta_D(\vec{l}-\vec{l}')P_{\tau_{obs}}(l). \quad (7.106)$$

Keeping only the terms up to first order in P_Ψ, we find that

$$P_{\tau_{obs}}(l) = P_\tau(l)(1-l^2 R_\Psi) + \int \frac{d^2 l_1}{(2\pi)^2}[\vec{l}_1\cdot(\vec{l}-\vec{l}_1)]^2 P_\Psi(l_1)P_\tau(|\vec{l}-\vec{l}_1|), \quad (7.107)$$

where

$$R_\Psi = \frac{1}{4\pi}\int l_1^3 dl_1 P_\Psi(l_1). \quad (7.108)$$

The first term in the equation above shows that the lensed power spectrum differs from the unlensed one by a factor that scales as l^2. The second term represents the convolution of the unlensed power spectrum with the power spectrum of the lensing potential.

The comparison between the unlensed and the lensed power spectra of the CMB temperature fluctuations is shown in the upper panels of Fig. 7.3 for a *Planck*-like cosmological model (Planck Collaboration et al. 2020). In short, the effects of lensing on the CMB temperature power spectrum are (1) to lower the amplitude of the peaks, (2) to broaden them, and (3) to create power at the small scales or high multipoles.

7.3.2 Lensing of the CMB Polarization

Since Thomson scattering does not produce circular polarization, the polarization of the CMB can be described by a symmetric, trace-free, and rank-2 tensor field, whose components in the coordinate basis are the Stokes parameters Q and U:

$$\mathscr{P} = \frac{1}{2} \begin{pmatrix} Q & U \\ U & -Q \end{pmatrix} . \tag{7.109}$$

An alternative way to describe the polarization is by using the complex representation of $\mathscr{P}$,

$$\mathscr{P}_\pm = Q \pm iU . \tag{7.110}$$

Any polarization pattern can be decomposed into parity conserving "electric" $\mathscr{E}$ and parity flipping "magnetic" $\mathscr{B}$ modes. In flat-sky approximation, this decomposition can be written in Fourier space as

$$\hat{\mathscr{E}} \pm i\hat{\mathscr{B}} = \exp(\pm 2i\varphi)(\hat{U} \pm i\hat{Q}) . \tag{7.111}$$

For the unlensed case, since the CMB polarization at the last scattering surface is expected to be a pure E-mode (except for the polarization possibly produced by gravitational waves), we have that

$$\hat{U} \pm i\hat{Q} = \exp(\mp 2i\varphi)\hat{\mathscr{E}} . \tag{7.112}$$

Now, we compute the E- and B-mode polarization power spectra for the lensed CMB. After substituting $\tau(\vec{\theta})$ with $\mathscr{P}_\pm(\vec{\theta})$ in Eq. 7.98, we can repeat the calculations shown in Sect. 7.3.1. This implies substituting $\hat{\tau}(\vec{l}_1)$ with $\exp(\mp 2i\varphi_1)\hat{\mathscr{E}}$ in Eqs. 7.102 and 7.103. By doing so, and by multiplying the result by $\exp \pm 2i\varphi$, we obtain the observed (or lensed) combinations

$$\hat{\mathscr{E}}_{obs}(\vec{l}) \pm i\hat{\mathscr{B}}_{obs}(\vec{l}) \approx \hat{\mathscr{E}}(\vec{l}) + \mathscr{T}_1^{\mathscr{E}}(\vec{l}) + \mathscr{T}_2^{\mathscr{E}}(\vec{l}) , \tag{7.113}$$

where

$$\mathscr{T}_1^{\mathscr{E}}(\vec{l}) = \int \frac{d^2 l_1}{(2\pi)^2} \vec{l}_1 \cdot (\vec{l} - \vec{l}_1) \exp[\pm 2i(\varphi_1 - \varphi)]\hat{\Psi}(\vec{l}_1)\hat{\mathscr{E}}(\vec{l} - \vec{l}_1) , \tag{7.114}$$

and

$$\mathscr{T}_2^{\mathscr{E}}(\vec{l}) = \frac{1}{2} \int \frac{d^2 l_1}{(2\pi)^2} \int \frac{d^2 l_2}{(2\pi)^2} (\vec{l}_1 \cdot \vec{l}_2)[\vec{l}_1 \cdot (\vec{l} - \vec{l}_1 - \vec{l}_2)]$$
$$\times \exp[\pm 2i(\varphi_1 - \varphi)]\hat{\mathscr{E}}(\vec{l}_1)\hat{\Psi}(\vec{l}_2)\hat{\Psi}(\vec{l} - \vec{l}_1 - \vec{l}_2) . \tag{7.115}$$

We can use them to form the power spectra

$$\langle (\hat{\mathscr{E}} + i\hat{\mathscr{B}})(\hat{\mathscr{E}} + i\hat{\mathscr{B}}) \rangle = P_E(l) + P_B(l) \,, \tag{7.116}$$

and

$$\langle (\hat{\mathscr{E}} + i\hat{\mathscr{B}})(\hat{\mathscr{E}} - i\hat{\mathscr{B}}) \rangle = P_E(l) - P_B(l) \,. \tag{7.117}$$

Their sum and difference yield the separate power spectra $P_E(l)$ and $P_B(l)$, which turn out to be

$$P_{E,obs}(l) = (1 - l^2 R_\Psi) P_E(l)$$
$$+ \int \frac{\mathrm{d}l_1}{(2\pi)^2} [\vec{l}_1 \cdot (\vec{l} - \vec{l}_1)]^2 \cos^2 2(\varphi_1 - \varphi) P_\Psi(|\vec{l} - \vec{l}_1|) P_E(l_1), \tag{7.118}$$

$$P_{B,obs}(l) = \int \frac{\mathrm{d}l_1}{(2\pi)^2} [\vec{l}_1 \cdot (\vec{l} - \vec{l}_1)]^2 \sin^2 2(\varphi_1 - \varphi) P_\Psi(|\vec{l} - \vec{l}_1|) P_E(l_1). \tag{7.119}$$

These equations give the power spectra to the lowest order in $P_\Psi(l)$, and they show some interesting results:

- First of all, the observed E-mode polarization power spectrum differs from the temperature fluctuation power spectrum by the phase factor $\cos^2 2(\varphi_1 - \varphi)$ in the integral. The angles φ and φ_1 are the angles that $\vec{l}$ and $\vec{l}_1$ form with the θ_1-axis. If the Fourier modes of the lensing potential and the unlensed E-mode polarization are not aligned, then the phase factor is different from unity.
- In this case, $\sin^2 2(\varphi_1 - \varphi) \neq 0$ and lensing creates B-mode from E-mode polarization. Thus, lensing introduces a B-mode, even if the unlensed polarization is a pure E-mode.

The lensed E- and B-mode polarization power spectra are shown in the bottom panels of Fig. 7.3, where they are compared to their unlensed counterparts. The effects of lensing on the shape of the E-mode polarization power spectrum are similar to those shown in the upper panels for the temperature fluctuations power spectrum.

7.3.3   Reconstruction of the Lensing Potential

A mass over-density produces a characteristic pattern on the CMB temperature (see Fig 7.2). This pattern can be used to reveal the presence of such over-density. The idea is as follows. Because of lensing, the observed temperature distribution around the over-density will not be isotropic, as if the CMB was unlensed. On the contrary,

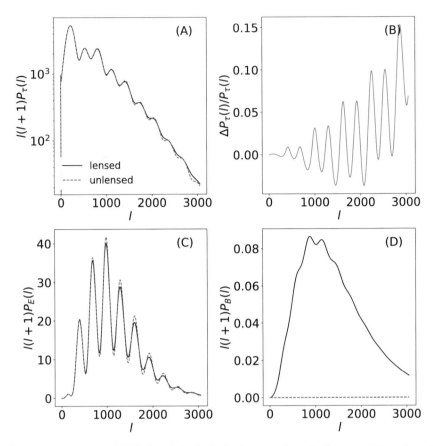

Fig. 7.3 Panel **a**: the solid black and the dashed red curves show the lensed and the unlensed power spectra of the CMB temperature fluctuations. Panel **b**: the solid blue curve shows the relative difference between the lensed and the unlensed power spectra in Panel **a**. Panel **c**: same as panel **a** but for the E-mode polarization power spectra. Panel **d**: same as panel **c**, but for the B-mode polarization power spectra

the correlations between many temperature anisotropies will provide information about the over-density's lensing potential and allow its recovery.

One possibility to achieve this recovery is through the construction of a temperature quadratic estimator (Hu 2001). As seen in the previous sections, lensing mixes Fourier modes, introducing correlations between formerly uncorrelated modes, $\hat{\tau}(\vec{l})$ and $\hat{\tau}(\vec{l} - \vec{L})$. Considering only the first order terms in $\hat{\Psi}$ in Eq. 7.105, we find

$$\hat{\tau}_{obs}(\vec{l}) = \hat{\tau}(\vec{l}) + \int \frac{d^2 l_1}{(2\pi)^2} \vec{l}_1 \cdot (\vec{l} - \vec{l}_1) \hat{\Psi}(\vec{l}_1) \hat{\tau}(\vec{l} - \vec{l}_1) . \qquad (7.120)$$

Thus, the expected correlation between different modes is

$$\langle \hat{\tau}_{obs}(\vec{l}) \hat{\tau}^*_{obs}(\vec{l} - \vec{L}) \rangle = (2\pi)^2 \delta_D(\vec{L}) P_\tau(l) - \left[\vec{L} \cdot (\vec{L} - \vec{l}) P_\tau(|\vec{l} - \vec{L}|) + \vec{L} \cdot \vec{l} P_\tau(l) \right] \hat{\Psi}(\vec{L}) .$$
(7.121)

Let us define now the filter $w(\vec{l}, \vec{L})$ such that

$$N(\vec{L}) \int \frac{d^2 l}{(2\pi)^2} \left[\hat{\tau}_{obs}(\vec{l}) \hat{\tau}^*_{obs}(\vec{l} - \vec{L}) \right] w(\vec{l}, \vec{L}) = \hat{\Psi}_{est}(\vec{L}) .$$
(7.122)

The quantity on the right-hand side of the equation is the estimate of the Fourier transform of lensing potential. We want this estimate to be unbiased,

$$\langle \hat{\Psi}_{est} \rangle = \hat{\Psi} .$$
(7.123)

Given Eq. 7.121, this constraint implies that the normalization factor is

$$N(\vec{L})^{-1} = - \int \frac{d^2 l}{(2\pi)^2} \left[\vec{L} \cdot (\vec{L} - \vec{l}) P_\tau(|\vec{l} - \vec{L}|) + \vec{L} \cdot \vec{l} P_\tau(l) \right] w(\vec{l}, \vec{L}) .$$
(7.124)

To find $w(\vec{l}, \vec{L})$, we require that the variance,

$$\langle (\hat{\Psi}_{est} - \hat{\Psi})^2 \rangle \approx \langle |\Psi_{est}|^2 \rangle ,$$
(7.125)

is minimized. From Eq. 7.122, we obtain

$$\langle \Psi_{est}(\vec{L}) \Psi^*_{est}(\vec{L}') \rangle = N(\vec{L})^2 \delta_D(\vec{L} - \vec{L}') \int \frac{d^2 l}{(2\pi)^2} P^{tot}_\tau(l) P^{tot}_\tau(|\vec{l} - \vec{L}|) w^2(\vec{l}, \vec{L}) ,$$
(7.126)

where we have included the noise contribution to the total, lensed power spectrum, $P^{tot}_\tau(l) = P_{\tau_{obs}}(l) + P_{noise}(l)$. By taking the derivative of the variance with respect to w and setting it to zero, under the constraint given by Eq. 7.123, we obtain

$$w(\vec{l}, \vec{L}) = \frac{\vec{L} \cdot (\vec{L} - \vec{l}) P_\tau(|\vec{l} - \vec{L}|) + \vec{L} \cdot \vec{l} P_\tau(l)}{2 P^{tot}_\tau(|\vec{l} - \vec{L}|) P^{tot}_\tau(l)} .$$
(7.127)

Thus, applying this filter to the squared temperature field will return an unbiased estimate of the lensing potential. Note that, however, the estimator is only valid to the lowest order in the lensing potential. This means that, in reality, the estimator is biased due to the significance of higher-order terms. Note also that this is only one of the possible ways to recover $\hat{\Psi}$ from the CMB (Lewis & Challinor 2006). For example, quadratic estimators can also be constructed for the polarization (Hu & Okamoto 2002).

7.4   Python Applications

7.4.1   Effective Shear and Potential

We consider an effective convergence map resulting from the propagation of bundles of light rays through the matter distribution of a cosmological simulation. More specifically, the map used in these examples was created with the code MAPSIM (Giocoli et al. 2015) by ray-tracing through 32 lens planes extracted from an N-body simulation representative of a flat universe with cosmological parameters $\Omega_{m,0} = 0.32$, $\Omega_{b,0} = 0.049$, $\sigma_8 = 0.83$, $n_i = 0.96$, and $H_0 = 67$ km s^{-1} Mpc^{-1}. The same simulation was analyzed by Hilbert et al. (2020) for testing the accuracy of several weak lensing simulation codes, including MAPSIM. The effective convergence map is computed for a source redshift $z_S = 1$. Its side-length is 10 degrees. The map is shown in the upper left panel of Fig. 7.4.

In this example, we use the effective convergence to generate maps of effective shear and potential. As done earlier, we use the relations between these quantities in Fourier space (see, e.g., Sect. 6.2.5).

We begin with the effective shear. The procedure is the inverse of that described in Sect. 6.4.3. We begin by computing the Fourier transform of the convergence map, $\hat{\kappa}$. The two components of the shear in Fourier space are given by

$$\hat{\gamma}_1 = \frac{k_1^2 - k_2^2}{k^2}\hat{\kappa} \ ,$$

$$\hat{\gamma}_2 = \frac{2k_1k_2}{k^2}\hat{\kappa} \ , \tag{7.128}$$

where $\vec{k} = (k_1, k_2)$ is the wave vector, as usual. The inverse Fourier transform of $\hat{\gamma}_1$ and $\hat{\gamma}_2$ returns the maps of γ_1 and γ_2. The procedure is implemented in the function shearFromConvergence below. Note that we do not perform any zero-padding in this application. Indeed, given the size of the effective convergence map we can safely assume periodic conditions on the boundary.

```
import scipy.fftpack as fftengine
import numpy as np
import astropy.io.fits as pyfits


def shearFromConvergence(kappa):
    """

    Computes the components of the shear from the
    input map of the effective convergence

    :param kappa: input convergence map (assumed to be a n x n array)
    :return: shear1 and shear2 - maps of the two shear components
    """

    kfreq = fftengine.fftfreq(kappa.shape[0])
    kx,ky = np.meshgrid(kfreq, kfreq)
```

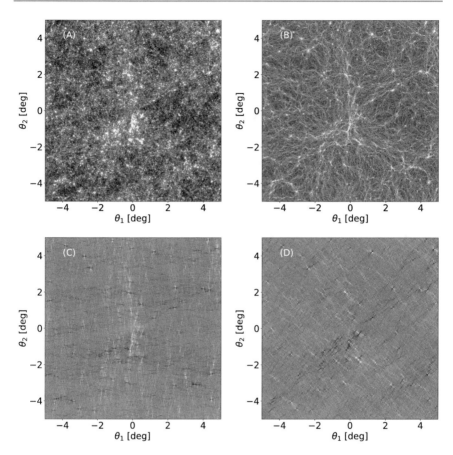

Fig. 7.4 Maps of effective convergence (panel **a**), shear modulus γ (panel **b**), and shear components (panels **c** and **d**)

```
k2 = kx**2+ky**2
k2[0,0] = 1

# Fourier transform the convergence map
fourier_kappa = fftengine.fftn(kappa)

# compute the shear components in Fourier space
# and inverse transform


shear1_ft = (kx**2-ky**2)*fourier_kappa/k2
shear2_ft = 2.0*kx*ky*fourier_kappa/k2
shear1 = fftengine.ifftn(shear1_ft).real
shear2 = fftengine.ifftn(shear2_ft).real
```

```
    return shear1, shear2

# read convergence map from file:
hdl=pyfits.open('test_kappaBApp_2.fits')
kappa=hdl[0].data

shear1, shear2 = shearFromConvergence(kappa)
```

The resulting maps of the shear components γ_1 and γ_2 are shown in panels C and D of Fig. 7.4. The map of the shear modulus, $\gamma = \sqrt{\gamma_1^2 + \gamma_2^2}$, is shown in panel B.

We can now compute the map of the effective lensing potential corresponding to the effective convergence κ_{eff}. As seen in Sect. 3.7.2, the calculation is also done in Fourier space:

```
def PotentialFromConvergence(kappa):
    # wave vector components:
    npix=kappa.shape[0]
    kfreq = fftengine.fftfreq(npix,1.0/npix)*2.0*np.pi
    kx,ky = np.meshgrid(kfreq, kfreq)
    k2 = kx**2+ky**2
    k2[0,0] = 1

    # Fourier transform the convergence map
    fourier_kappa = fftengine.fftn(kappa)
    # Apply laplacian in Fourier space
    fourier_pot = -fourier_kappa*2.0/k2

    # inverse Fourier transform and return potential map
    pot = fftengine.ifftn(fourier_pot).real/2.0/np.pi

    return pot
```

The map of the effective potential is shown in Fig. 7.5.

7.4.2 Power Spectrum

In this example, we illustrate how to compute the power spectrum of a 2D-map. We assume a flat-sky, and we calculate the power spectrum of the effective convergence and of the shear.

The procedure is as follows. First, we compute the Fourier transform of the input map. Note that in this example we assume that the map is a square numpy array:

```
# set the number of pixels and the unit conversion factor
npix = input_map.shape[0]
factor = 2.0*np.pi/(BoxSize*np.pi/180.0)

# take the Fourier transform of the input map:
fourier_map = fftengine.fftn(input_map)/npix**2
```

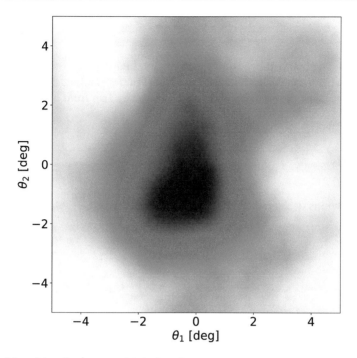

Fig. 7.5 Map of the effective potential obtained from the convergence map in panel **a** of Fig. 7.4

Then, we compute the Fourier amplitudes and store them in a one-dimensional array:

```
# compute the Fourier amplitudes
fourier_amplitudes = np.abs(fourier_map)**2
fourier_amplitudes = fourier_amplitudes.flatten()
```

We need to associate a wave number with each Fourier amplitude:

```
# compute the wave vectors
kfreq = fftengine.fftfreq(input_map.shape[0])*input_map.shape[0]
kfreq2D = np.meshgrid(kfreq, kfreq)
# take the norm of the wave vectors
knrm = np.sqrt(kfreq2D[0]**2 + kfreq2D[1]**2)
knrm = knrm.flatten()
```

We create bins of the wave number k. We will evaluate the power spectrum in these bins:

```
# set up k bins. The PS will be evaluated in these bins
half = npix/2
rbins = int(np.sqrt(2*half**2))+1
kbins = np.linspace(0.0,rbins,(rbins+1))
# use the middle points in each bin to define the values of k
```

```
# where the PS is evaluated
kvals = 0.5 * (kbins[1:] + kbins[:-1])*factor
```

Finally, we compute the power spectrum by binning the wave numbers and calculating the mean value of the corresponding Fourier amplitudes in each bin. This can be done using the function `binned_statistics` from `scipy.stats`:

```
# now compute the PS: calculate the mean of the
# Fourier amplitudes in each kbin
Pbins, _, _ = stats.binned_statistic(knrm, fourier_amplitudes,
                                      statistic = "mean",
                                      bins = kbins)
```

All the previous steps are included in the function `compute_PS`:

```
def compute_PS(input_map,FieldSize):
    """
    Compute the angular power spectrum of input_map.
    :param input_map: input map (n x n numpy array)
    :param FieldSize: the side-length of the input map in degrees

    :return: l, Pl - the power-spectrum at l
    """
    # set the number of pixels and the unit conversion factor
    npix = input_map.shape[0]
    factor = 2.0*np.pi/(FieldSize*np.pi/180.0)


    # take the Fourier transform of the input map:
    fourier_map = fftengine.fftn(input_map)/npix**2

    # compute the Fourier amplitudes
    fourier_amplitudes = np.abs(fourier_map)**2
    fourier_amplitudes = fourier_amplitudes.flatten()

    # compute the wave vectors
    kfreq = fftengine.fftfreq(input_map.shape[0])*input_map.shape[0]
    kfreq2D = np.meshgrid(kfreq, kfreq)

    # take the norm of the wave vectors
    knrm = np.sqrt(kfreq2D[0]**2 + kfreq2D[1]**2)
    knrm = knrm.flatten()

    # set up k bins. The PS will be evaluated in these bins
    half = npix/2
    rbins = int(np.sqrt(2*half**2))+1
    kbins = np.linspace(0.0,rbins,(rbins+1))

    # use the middle points in each bin to define the values of k
    # where the PS is evaluated
    kvals = 0.5 * (kbins[1:] + kbins[:-1])*factor
```

```
# now compute the PS: calculate the mean of the
# Fourier amplitudes in each kbin
Pbins, _, _ = stats.binned_statistic(knrm, fourier_amplitudes,
                                     statistic = "mean",
                                     bins = kbins)

# return kvals and PS
l=kvals[1:]
Pl=Pbins[1:]/factor**2

return l, Pl
```

Now, we can use the function `compute_PS` to compute the power spectra of the maps of the effective convergence and the shear described in the previous example:

```
FieldSize=10

# compute the power spectra of convergence and of the shear
# components:
lk,Plk=compute_PS(kappa,FieldSize)
lg,Pg1k=compute_PS(shear1,FieldSize)
lg,Pg2k=compute_PS(shear2,FieldSize)
# the shear power spectrum is the sum of the power spectra of the
# two shear components:
Pgk=Pg1k+Pg2k
```

The power spectra of the convergence and the shear are shown in Fig. 7.6 (blue dashed and red dotted lines, respectively). Not surprisingly, they are (almost) identical (see Eq. 7.58)! We also show the convergence power spectrum calculated for the same cosmological model by using Eq. 7.51 with the non-linear power spectrum $P_\delta(l)$ as given by the fitting formulae of Peacock and Dodds (1996) (solid black line). As we can see, the spectra measured from the maps agree well with the theoretical expectations, except for high l-values. At these small angular scales, we approach the resolution limit of the input maps. For this reason, we measure a power lower than expected at these scales.

7.4.3 Correlation Functions

We use the maps from the previous examples to generate a catalog of reduced shear measurements. Using this catalog, we compute second order shear statistics. Note that, in this illustrative example, we ignore important sources of noise affecting the shear measurements (including the intrinsic shape noise).

We assume a random distribution of 6,000,000 sources in our field-of-view of 100 sq. degrees (i.e. ~17 sources per sq. arcmin). We use the `rand` function of the module `numpy.random` to generate their positions x and y:

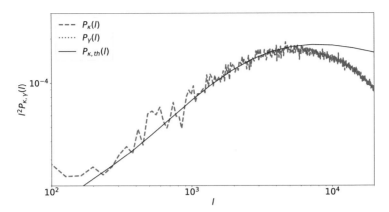

Fig. 7.6 Power spectra of the effective convergence (blue dashed line) and the effective shear (red dotted line) measured from the maps in panels **a**, **c**, and **d** of Fig. 7.4. The solid black line shows the power spectrum calculated using Eq. 7.51

```
x = np.random.rand(6000000)*kappa.shape[0]
y = np.random.rand(6000000)*kappa.shape[0]
x_deg=(x/kappa_in.shape[0]*10.0-5.0)
y_deg=(y/kappa_in.shape[0]*10.0-5.0)
```

Then, by means of bi-linear interpolation of the shear and convergence maps at (x,y), we assign a lensing "ellipticity" to each source. For this purpose, we use the map_coordinates function of scipy.ndimage:

```
from scipy.ndimage import map_coordinates
# interpolate the convergence and shear maps
kappa_ = map_coordinates(kappa, [[y], [x]], order=1,
                         prefilter=True).flatten()
shear1_ = map_coordinates(shear1, [[y], [x]], order=1,
                          prefilter=True).flatten()
shear2_ = map_coordinates(shear2, [[y], [x]], order=1,
                          prefilter=True).flatten()
# compute the real and imaginary parts of the reduced
# shear as an estimate of the ellipticity
e1=shear1_/(1.0-kappa_)
e2=shear2_/(1.0-kappa_)
```

Having assigned the ellipticity $\epsilon = \epsilon_1 + i\epsilon_2$ to each source, we can proceed with the calculation of the correlation functions $\xi_+(\phi)$ and $\xi_-(\phi)$. For this purpose we can do as follows. We select all pairs of sources that have an angular separation within $\Delta\phi$ of ϕ. For each pair, we compute the tangential and the cross component of the ellipticity, ϵ_t and $\epsilon_\times$ (cfr. Eqs. 7.60). Then, we take the averages $\langle \epsilon_{ti}\epsilon_{tj}\rangle$ and $\langle \epsilon_{\times i}\epsilon_{\times j}\rangle$ over all the pairs. As said, we did not assign intrinsic ellipticities to the sources. However, even if we did it, since $\epsilon \approx \epsilon_s + \gamma$, the expectation values of $\langle \epsilon_{ti}\epsilon_{tj}\rangle$ and $\langle \epsilon_{\times i}\epsilon_{\times j}\rangle$ are $\langle \gamma_{ti}\gamma_{tj}\rangle$ and $\langle \gamma_{\times i}\gamma_{\times j}\rangle$ assuming that $\kappa \ll 1$. More generally, we could define the estimators

$$\xi_{tt}(\phi) = \frac{\sum_{ij} w_i w_j \epsilon_{ti}(\vec{\theta}_i) \epsilon_{tj}(\vec{\theta}_j)}{\sum_{ij} w_i w_j} , \tag{7.129}$$

$$\xi_{\times\times}(\phi) = \frac{\sum_{ij} w_i w_j \epsilon_{\times i}(\vec{\theta}_i) \epsilon_{\times j}(\vec{\theta}_j)}{\sum_{ij} w_i w_j} , \tag{7.130}$$

where $\phi = |\vec{\theta}_i - \vec{\theta}_j|$ and w_i are weights assigned to each source. The correlation functions $\xi_+(\phi)$ and $\xi_-(\phi)$ are then obtained from Eqs. 7.69 and 7.70.

Performing this calculation can be expensive in terms of computation time and memory usage. However, algorithms exist that can make it more efficient. For example, one can use space-partitioning data structures, like, e.g. k-d trees or ball-trees (Friedman et al. 1145, Omohundro 1989, "Orthogonal Range Searching" 2008). This approach is used in the python package TREECORR (Jarvis et al. 2004), which we employ in this example.

First of all, we save the source positions and ellipticities into a TREECORR catalog:

```
import treecorr
mycat = treecorr.Catalog(x=x_deg, y=y_deg, g1=g1, g2=g2,
                         x_units='deg', y_units='deg')
```

Then, we specify the angular scales and binning for the computation of the correlation functions, and we process the catalog:

```
# we choose do compute the correlation functions
# in 50 bins between 0.4 and 100 arcmin
corrf = treecorr.GGCorrelation(min_sep=0.4, max_sep=100,
                        nbins=50,sep_units='arcmin')
corrf.process(mycat)
```

The resulting correlation functions can be extracted from `corrf` as follows:

```
phi = numpy.exp(corrf.meanlogr)  # the angles where the
                                 # correlation functions are
                                 # computed
xip = corrf.xip                  # the function xi_+
xim = corrf.xim                  # the function xi_-
```

The results of the computation are shown in Fig. 7.7. The measured correlation functions $\xi_\pm(\phi)$ are given by the solid blue and green lines, respectively. The theoretical models using the cosmological parameters of the simulation correspond to the dashed lines.

Fig. 7.7 Shear 2-point correlation functions measured from the maps in Fig. 7.4. The measured ξ_+ and ξ_- functions are given by the solid blue and green lines, respectively. The dashed lines show the theoretical expectations given the cosmological parameters used in the simulation

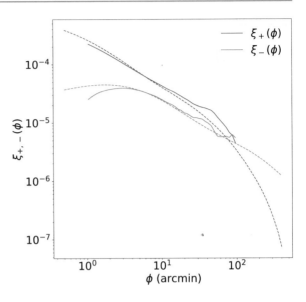

Note that TREECORR can compute the other second order statistics introduced in this chapter.

References

Abbott, T. M. C., Aguena, M., Alarcon, A., Allam, S., Allen, S., Annis, J., ... DES Collaboration. (2020). Dark energy survey year 1 results: cosmological constraints from cluster abundances and weak lensing. *Physical Review, 102*(2), 023509. https://doi.org/10.1103/PhysRevD.102. 023509. arXiv: 2002.11124 [astro-ph.CO]

Aihara, H., Arimoto, N., Armstrong, R., Arnouts, S., Bahcall, N. A., Bickerton, S., ... Yuma, S. (2018). The hyper suprime-cam SSP survey: overview and survey design. *Publications of the ASJ, 70*, S4. https://doi.org/10.1093/pasj/psx066. arXiv: 1704.05858 [astro-ph.IM]

Akeson, R., Armus, L., Bachelet, E., Bailey, V., Bartusek, L., Bellini, A., ... Zimmerman, N. (2019). The wide field infrared survey telescope: 100 hubbles for the 2020s. arXiv e-prints, arXiv:1902.05569. arXiv: 1902.05569 [astro-ph.IM]

Bacon, D. J., Refregier, A. R., & Ellis, R. S. (2000). Detection of weak gravitational lensing by large-scale structure. *MNRAS, 318*(2), 625–640. https://doi.org/10.1046/j.1365-8711.2000. 03851.x. arXiv: astro-ph/0003008 [astro-ph]

Barreira, A., Krause, E., & Schmidt, F. (2018). Accurate cosmic shear errors: Do we need ensembles of simulations? *JCAP, 2018*(10), 053. https://doi.org/10.1088/1475-7516/2018/10/ 053. arXiv: 1807.04266 [astro-ph.CO]

Bartelmann, M., & Schneider, P. (2001). Weak gravitational lensing. *Physics Reports, 340*, 291–472. https://doi.org/10.1016/S0370-1573(00)00082-X. eprint: astro-ph/9912508

Crittenden, R. G., Natarajan, P., Pen, U.-L., & Theuns, T. (2001). Spin-induced galaxy alignments and their implications for weak-lensing measurements. *ApJ, 559*(2), 552–571. https://doi.org/ 10.1086/322370. arXiv: astro-ph/0009052 [astro-ph]

de Jong, J. T. A., Verdoes Kleijn, G. A., Erben, T., Hildebrandt, H., Kuijken, K., Sikkema, G., ... Viola, M. (2017). The third data release of the Kilo-Degree Survey and associated data products. *A & A, 604*, A134. https://doi.org/10.1051/0004-6361/201730747. arXiv: 1703.02991 [astro-ph.GA]

Friedman, J. H., Bentley, J. L., & Finkel, R. A. (1977). An algorithm for finding best matches in logarithmic expected time. *ACM Transactions on Mathematical Software, 3*(3), 209–226. https://doi.org/10.1145/355744. 355745

Friedrich, O., Seitz, S., Eifler, T. F., & Gruen, D. (2016). Performance of internal covariance estimators for cosmic shear correlation functions. *MNRAS, 456*(3), 2662–2680. https://doi.org/10.1093/mnras/stv2833. arXiv: 1508.00895 [astro-ph.CO]

Giocoli, C., Metcalf, R. B., Baldi, M., Meneghetti, M., Moscardini, L., & Petkova, M. (2015). Disentangling dark sector models using weak lensing statistics. *MNRAS, 452*(3), 2757–2772. https://doi.org/10.1093/mnras/stv1473. arXiv: 1502.03442 [astro-ph.CO]

Harnois-Déraps, J., Amon, A., Choi, A., Demchenko, V., Heymans, C., Kannawadi, A., . . . Tröster, T. (2018). Cosmological simulations for combined-probe analyses: Covariance and neighbourexclusion bias. *MNRAS, 481*(1), 1337–1367. https://doi.org/10.1093/mnras/sty2319. arXiv: 1805.04511 [astro-ph.CO]

Hilbert, S., Barreira, A., Fabbian, G., Fosalba, P., Giocoli, C., Bose, S., . . . Monaco, P. (2020). The accuracy of weak lensing simulations. *Monthly Notices of the Royal Astronomical Society, 493*(1), 305–319. https://doi.org/10.1093/mnras/staa281. eprint: https://academic.oup.com/mnras/articlepdf/493/1/305/32468457/staa281.pdf

Hildebrandt, H., Viola, M., Heymans, C., Joudaki, S., Kuijken, K., Blake, C., . . . Van Waerbeke, L. (2017). KiDS-450: Cosmological parameter constraints from tomographic weak gravitational lensing. *MNRAS, 465*(2), 1454–1498. https://doi.org/10.1093/mnras/stw2805. arXiv: 1606.05338 [astro-ph.CO]

Hu, W. (1999). Power spectrum tomography with weak lensing. *ApJL, 522*(1), L21–L24. https://doi.org/10.1086/312210. arXiv: astro-ph/9904153 [astro-ph]

Hu, W. (2001). Mapping the dark matter through the cosmic microwave background damping tail. *ApJL, 557*(2), L79–L83. https://doi.org/10.1086/323253. arXiv: astro-ph/0105424 [astro-ph]

Hu, W., & Okamoto, T. (2002). Mass reconstruction with cosmic microwave background polarization. *ApJ, 574*(2), 566–574. https://doi.org/10.1086/341110. arXiv: astro-ph/0111606 [astro-ph]

Ivezić, Ž., Kahn, S. M., Tyson, J. A., Abel, B., Acosta, E., Allsman, R., . . . et al. (2019). LSST: From science drivers to reference design and anticipated data products. *ApJ, 873*, 111. https://doi.org/10.3847/1538-4357/ab042c. arXiv: 0805.2366

Jain, B., Seljak, U., & White, S. (2000). Ray-tracing simulations of weak lensing by large-scale structure. *ApJ, 530*(2), 547–577. https://doi.org/10.1086/308384. arXiv: astro-ph/9901191 [astro-ph]

Jarvis, M., Bernstein, G., & Jain, B. (2004). The skewness of the aperture mass statistic. *MNRAS, 352*(1), 338–352. https://doi.org/10.1111/j.1365-2966.2004.07926.x. arXiv: astro-ph/0307393 [astro-ph]

Joachimi, B., Schneider, P., & Eifler, T. (2008). Analysis of two-point statistics of cosmic shear. III. Covariances of shear measures made easy. *A & A, 477*(1), 43–54. https://doi.org/10.1051/0004-6361:20078400. arXiv: 0708.0387 [astro-ph]

Kaiser, N. (1992). Weak gravitational lensing of distant galaxies. *ApJ, 388*, 272. https://doi.org/10.1086/171151

Kaiser, N. (1998). Weak lensing and cosmology. *ApJ, 498*(1), 26–42. https://doi.org/10.1086/305515. arXiv: astro-ph/9610120 [astro-ph]

Krause, E., Eifler, T. F., Zuntz, J., Friedrich, O., Troxel, M. A., Dodelson, S., . . . Weller, J. (2017). Dark energy survey year 1 results: Multi-probe methodology and simulated likelihood analyses. arXiv e-prints, arXiv:1706.09359. arXiv: 1706.09359 [astro-ph.CO]

Krause, E., & Eifler, T. (2017). cosmolike—cosmological likelihood analyses for photometric galaxy surveys. *MNRAS, 470*(2), 2100–2112. https://doi.org/10.1093/mnras/stx1261. arXiv: 1601.05779 [astro-ph.CO]

Laureijs, R., Amiaux, J., Arduini, S., Auguères, J.-L., Brinchmann, J., Cole, R., . . . Zucca, E. (2011). Euclid definition study report. arXiv e-prints, arXiv:1110.3193. arXiv: 1110.3193 [astro-ph.CO]

Lewis, A., & Challinor, A. (2006). Weak gravitational lensing of the CMB. *Physics Reports, 429*(1), 1–65. https://doi.org/10.1016/j.physrep.2006.03.002

Limber, D. N. (1953). The analysis of counts of the extragalactic nebulae in terms of a fluctuating density field. *ApJ, 117*, 134. https://doi.org/10.1086/145672

Misner, C. W., Thorne, K. S., & Wheeler, J. A. (1973). *Gravitation*.

Norberg, P., Baugh, C. M., Gaztañaga, E., & Croton, D. J. (2009). Statistical analysis of galaxy surveys—I. Robust error estimation for two-point clustering statistics. *MNRAS, 396*(1), 19–38. https://doi.org/10.1111/j.1365-2966.2009.14389.x. arXiv: 0810.1885 [astro-ph]

Omohundro, S. M. (1989). *Five balltree construction algorithms*.

Orthogonal Range Searching. (2008). In *Computational geometry: Algorithms and applications* (pp. 95–120). https://doi.org/10.1007/978-3-540-77974-2_5

Peacock, J. A., & Dodds, S. J. (1996). Non-linear evolution of cosmological power spectra. *MNRAS, 280*(3), L19–L26. https://doi.org/10.1093/mnras/280.3.L19. arXiv: astro-ph/9603031 [astro-ph]

Planck Collaboration, Aghanim, N., Akrami, Y., Ashdown, M., Aumont, J., Baccigalupi, C., ... Zonca, A. (2020). Planck 2018 results. VI. Cosmological parameters. *A & A, 641*, A6. https://doi.org/10.1051/0004-6361/201833910. arXiv: 1807.06209 [astro-ph.CO]

Schneider, P. (2006). Part 3: Weak gravitational lensing. In G. Meylan, P. Jetzer, P. North, P. Schneider, C. S. Kochanek, & J. Wambsganss (Eds.), *Saas-fee advanced course 33: Gravitational lensing: Strong, weak and micro* (pp. 269–451).

Schneider, P., Ehlers, J., & Falco, E. E. (1992). Gravitational lenses. https://doi.org/10.1007/978-3-662-03758-4

Schneider, P., van Waerbeke, L., Kilbinger, M., & Mellier, Y. (2002a). Analysis of two-point statistics of cosmic shear. I. Estimators and covariances. *A & A, 396*, 1–19. https://doi.org/10.1051/0004-6361:20021341. arXiv: astro-ph/0206182 [astro-ph]

Schneider, P., van Waerbeke, L., & Mellier, Y. (2002b). B-modes in cosmic shear from source redshift clustering. *A & A, 389*, 729–741. https://doi.org/10.1051/0004-6361:20020626. arXiv: astroph/0112441 [astro-ph]

Smith, R. E., Peacock, J. A., Jenkins, A., White, S. D. M., Frenk, C. S., Pearce, F. R., ... Couchman, H. M. P. (2003). Stable clustering, the halo model and non-linear cosmological power spectra. *MNRAS, 341*(4), 1311–1332. https://doi.org/10.1046/j.1365-8711.2003.06503. x. arXiv: astro-ph/0207664 [astro-ph]

Takada, M., & Jain, B. (2009). The impact of non-Gaussian errors on weak lensing surveys. *MNRAS, 395*(4), 2065–2086. https://doi.org/10.1111/j.1365-2966.2009.14504.x. arXiv: 0810.4170 [astro-ph]

Takahashi, R., Sato, M., Nishimichi, T., Taruya, A., & Oguri, M. (2012). Revising the halofit model for the nonlinear matter power spectrum. *ApJ, 761*(2), 152. https://doi.org/10.1088/0004-637X/761/2/152. arXiv: 1208.2701 [astro-ph.CO]

Troxel, M. A., & Ishak, M. (2015). The intrinsic alignment of galaxies and its impact on weak gravitational lensing in an era of precision cosmology. *Physics Reports, 558*, 1–59. https://doi.org/10.1016/j.physrep.2014.11.001. arXiv: 1407.6990 [astro-ph.CO]

Troxel, M. A., MacCrann, N., Zuntz, J., Eifler, T. F., Krause, E., Dodelson, S., ... DES Collaboration. (2018). Dark energy survey year 1 results: Cosmological constraints from cosmic shear. *Physical Review, 98*(4), 043528. https://doi.org/10.1103/PhysRevD.98.043528. arXiv: 1708.01538 [astro-ph.CO]

Van Waerbeke, L., Mellier, Y., Erben, T., Cuilland re, J. C., Bernardeau, F., Maoli, R., ... Schneider, P. (2000). Detection of correlated galaxy ellipticities from CFHT data: First evidence for gravitational lensing by large-scale structures. *A & A, 358*, 30–44. arXiv: astro-ph/0002500 [astro-ph]

Wittman, D. M., Tyson, J. A., Kirkman, D., Dell'Antonio, I., & Bernstein, G. (2000). Detection of weak gravitational lensing distortions of distant galaxies by cosmic dark matter at large scales. *Nature, 405*(6783), 143–148. https://doi.org/10.1038/35012001. arXiv: astro-ph/0003014 [astro-ph]

Part III

Appendixes

Python Mini-Tutorial

8

8.1   Installation

For the Python examples discussed in this book, we used Anaconda Python 2.7 or 3.6. We expect that any other Python distribution should work equally well.

 If the reader opts for Anaconda Python, they can download the installer, available for Windows, Mac OSX, and Linux platforms, from https://www.anaconda.com. Following the installation instructions, Python should be ready to use within minutes.

8.2   Documentation

There are many books and resources online to learn how to program in Python. The list below is just a starting point and does not want to be complete:

- The official online documentation, as well as links to many other useful guides and books, can be found at this URL: http://www.python.org/doc.
- Several platforms for e-learning propose courses to learn Python. Some of these courses can be found at Codecademy, datacamp, and Coursera.
- Google also offers a Python class online.
- A Byte of Python is a free, online book for beginners.

8.3   Running Python

You can run Python programs in several ways:

- For using the interactive interpreter, launch "python" in a shell. At the Python prompt, type your commands. Quit the interpreter with Ctrl+D or type "exit()" when finished.

© Springer Nature Switzerland AG 2021   371
M. Meneghetti, *Introduction to Gravitational Lensing*, Lecture Notes
in Physics 956, https://doi.org/10.1007/978-3-030-73582-1_8

- Create your script with an extension ".py" and run it in a shell by typing "python <script name>.py."
- Use an interactive development environment (IDE). There are several options available (e.g., Spyder, Rodeo, PyCharm, etc.).
- Use a web-based interactive development environment like Jupyter Notebook, or Jupyter lab (if you use Python 3.x).

8.4   Your First Python Code

Try running the code:

```
# your first python code -- this is a comment
print ("Hello World!")
```

Congratulations! You have run your first Python code!

8.5   Variables

Variables are reserved memory locations to store values. Setting them in Python is extremely easy:

```
int_var = 4
float_var = 7.89778
boolean_var = True
string_var = "My name is Python"
obj_var=some_class_name(par1,par2)
```

8.6   Strings

String constants can be defined in three ways:

```
single_quotes = 'my name is Python'
double_quotes = "my name is Python"
triple_quotes = """my name is Python
and this is  a multiline  string.""" #This can contain line breaks!
```

Note that you can combine single and double quotes:

```
double_quotes1 = 'my name is "Python"'
double_quotes2 = "don't"
```

otherwise you have to use backslashes:

```
double_quotes3 = 'don\'t'
```

Strings can be sliced:

```
my_name='Massimo Meneghetti'
name = my_name[:7]
surname = my_name[8:]
a_piece_of_my_name=my_name[4:7]
```

You can make many operations with strings. Check out this URL to learn more: https://docs.python.org/2/library/stdtypes.html.
   Some examples:

- String concatenation:

    ```
    back_to_my_full_name=name+" "+surname
    ```

- Convert into upper case

    ```
    my_name_uppercase=back_to_my_full_name.upper()
    ```

The built-in function `str` converts numbers into strings:

```
my_int=2
my_float=2.0
str_int=str(my_int)
str_float=str(my_float)
```

Another way to include numbers in strings is

```
my_string1 = 'My integer is %d.' % my_int
my_string2 = 'My float is %f.' % my_float
my_string3 = 'My float is %3.1f (with only one decimal)' % my_float
```

With several variables, we need to use parentheses:

```
a = 2
b = 67
my_string4 = '%d + %d = %d' % (a, b, a+b)
```

```
a = 2
b = 67.3
my_string5 = '%d + %5.2f = %5.1f' % (a, b, a+b)
```

Not only you can convert numbers into string, but you can do the reverse operation:

```
s = '23'
i = int(s)
s = '23'
i = float(s)
```

To remove spaces at beginning and end of a string, you can do as follows:

```
stripped = a_string.strip()
```

The method to replace part of a string is instead:

```
newstring = a_string.replace('abc', 'def')
```

Important note: A Python string is "immutable." In other words, it is a constant that cannot be changed in place. All string operations create a new string.

8.7   Lists

A list is a dynamic array of any objects. It is declared with square brackets:

```
a_list = [1, 2, 3, 'abc', 'def']
```

Lists may contain lists:

```
another_list = [a_list, 'abc', a_list, [1, 2, 3]]
```

Note that a_list in this case is a pointer.
    You can access a specific element of the list using the index (index starts at zero):

```
elem = a_list[2]
elem2 = another_list[3][1]
```

It is easy to test if an object is in the list:

```
if 'abc' in a_list:
    print 'bingo!'
```

The operation of extracting part of a list is called slicing:

```
list2 = a_list[2:4]  # returns a list with items 2 and 3 (not 4)
```

Other examples of list operations are appending and removing operations:

```
a_list.append('ghi')
a_list.remove('abc')
```

You can learn about other list operations here: http://docs.python.org/lib/typesseq. html.

8.8   Tuples

The tuple data type is similar to a list, but it has a fixed size and is immutable. Once a tuple has been created, its elements may not be changed, removed, appended, or inserted.

You can define a tuple using round brackets and comma-separated values:

```
a_tuple = (1, 2, 3, 'abc', 'def')
```

but brackets are optional:

```
another_tuple = 1, 2, 3, 'abc', 'def'
```

Tip: A tuple containing only one item must be declared using a comma; otherwise, it is not considered as a tuple:

```
a_single_item_tuple = ('one value',)
```

Remark 8.1. Tuples are not constant lists—this is a common misconception. Lists are intended to be homogeneous sequences, while tuples are heterogeneous data structures.

In some sense, tuples may be regarded as simplified structures, in which position has a semantic value [e.g. (name,surname,age,height,weight)]. For this reason, they are immutable, contrary to lists.

8.9   Dictionaries

A dictionary (or "dict") is a way to store data just like a list. Dictionaries are initialized using curl brackets:

```
person = {'name': 'Massimo', 'surname': 'Meneghetti'}
```

You can access the elements of the dictionary by using keys:

```
person['name']
```

The keys can also be numbers:

```
person = {'name': 'Massimo', 'surname': 'Meneghetti', 1: 'new data'}
person[1]
```

8.10   Blocks and Indentation

Blocks of code are delimited using indentation, either spaces or tabs at the beginning of lines.

Tip: NEVER mix tabs and spaces in a script, as this could generate bugs that are very difficult to find.

8.11   IF/ELIF/ELSE

Here is an example of how to implement an IF/ELIF/ELSE loop:

```
if a == 3:
    print 'The value of a is:'
    print 'a=3'

if a == 'test':
    print 'The value of a is:'
    print 'a="test"'
    test_mode = True
else:
    print 'a!="test"'
    test_mode = False
    do_something_else()

if a == 1 or a == 2:
    pass # do nothing
elif a == 3 and b > 1:
    pass
elif a==3 and not b>1:
    pass
else:
    pass
```

8.12   While Loops

While loops can be implemented as follows:

```
a=1
while a<10:
    print a
    a += 1
```

8.13   For Loops

The following are examples of for loops. For running the index a between 0 and 9, you can do as follows:

```
for a in range(10):
    print a
```

For looping over the elements of a list, you can use this syntax:
```
my_list = [2, 4, 8, 16, 32]
for a in my_list:
    print a
```

8.14 Functions

Functions can be defined in Python as follows:

```
def compute_sum(arg1,arg2):
        # implement function to calculate the sum of two numbers
        res=arg1+arg2
    return(res)
```

The function declaration starts with the def statement followed by the function name and by the arguments within the round brackets. The arguments can be initialized to default values:

```
def compute_sum(arg1=1,arg2=1):
        # implement function to calculate the sum of two numbers
        res=arg1+arg2
    return res
```

They can be of any type (e.g., simple variables, lists, tuples, etc.).

The function can be called by typing the function name. If the function returns a value or object, this is assigned to a variable as follows:

```
summa=compute_sum(3.0,7.0)
```

It is possible to create global variables outside of a function. These variables can be used by everyone, both inside and outside the function.

```
c=3

def use_global_c(val):
        # this function uses a global variable
    print val+c
```

Variables created inside a function are local. They can be made global using the global keyword:

```
def global_c(val):
        # this function change the value of a global variable
    global c
    c=val

global_c(10)
print ( c )
```

8.15 Classes

Python is an object-oriented programming language. Almost everything in Python is an object with its properties and methods.

Classes are a way to group a set of functions and methods dedicated to specific objects. For example, we may want to define a class called "square," containing the methods to compute the square properties, such as the perimeter and the area. The object is created using a "constructor" function, which assigns values to the object properties. The name of this function is __init__. It is always executed when the class is initiated:

```python
class square:

    #the constructor:
    def __init__(self,side):
            self.side=side

    #area of the square:
    def area(self):
            return(self.side*self.side)

    #perimeter of the square:
    def perimeter(self):
            return(4.0*self.side)
```

The self-parameter is a reference to the current instance of the class and is used to access variables that belong to the class. It has to be the first parameter of any function in the class.

We can use the class to define a square object:

```python
s=square(3.0) # a square with side length 3
```

To use the object methods (or any other object attribute, like side, we use the "." operator:

```python
print s.area()
print s.perimeter()
print s.side
s.side=s.side+2 # here we are modifying the attribute side of s
```

8.16   Inheritance

As in other languages (e.g., C++), Python supports inheritance. Inheritance allows us to define a class that inherits all the methods and properties of another class. The class being inherited from is called *parent* or *base* class. The class that inherits from another class is called *child* or *derived* class.

For example:

```python
class geometricalFigure(object):

    def __init__(self,name):
        self.name=name
```

```
    def getName(self):
        print 'this is a %s' % self.name

class square(geometrical_figure):

    #the constructor:
    def __init__(self,side):
        geometricalFigure.__init__(self,'square')
        self.side=side

    #area of the square:
    def area(self):
        return(self.side*self.side)

    #perimeter of the square:
    def perimeter(self):
        return(4.0*self.side)

class circle(geometrical_figure):

    #the constructor:
    def __init__(self,radius):
        geometricalFigure.__init__(self,'circle')
        self.radius=radius

    #area of the square:
    def area(self):
        return(3.141592653*self.radius**2)

    #perimeter of the square:
    def perimeter(self):
        return(2.0*self.radius*3.141592653)


s=square(3.0)
c=circle(3.0)
s.getName()
c.getName()
```

In the example above, `square` and `circle` are two classes derived from the base class `geometricalFigure`. They have some specialized methods to compute the area and the perimeter, but both can access the method `getName`. They have inherited it from the parent class `geometricalFigure`.

8.17   Modules

A module is a sort of code library. It is a file containing Python definitions and statements (constants, functions, classes, etc.). The file name is the module name with the suffix .py appended.

Modules can be imported in another script by using the `import` statement:

```
import modulename
```

The classes, functions, and statements contained in the module can be accessed using the . operator.

Modules can import other modules. It is customary but not required to place all import statements at the beginning of a module (or script, for that matter).

A variant of the import statement imports names from a module directly into the importing module's symbol table:

```
from modulename import something
```

8.18   Importing Packages

Packages can be added to your Python distribution using either the `pip` or `easy_install` utilities. Anaconda has its utility for installing a set of supported packages, called `conda`. To learn more, check out https://packaging.python.org/installing/.

Packages can be used by importing modules and classes in the code, as discussed above.

Some packages that we use a lot are:

- NumPy: fundamental package for scientific computing with Python (powerful N-dimensional array object, sophisticated functions, tools for integrating C/C++ and Fortran code, useful linear algebra, Fourier transform, and random number capabilities)
- SciPy: provides many user-friendly and efficient numerical routines such as routines for numerical integration and optimization
- Matplotlib: a Python 2D plotting library that produces publication quality figures in a variety of hard-copy formats and interactive environments across platforms
- Astropy: a community effort to develop a single core package for Astronomy in Python and foster interoperability between Python astronomy packages

Other packages are introduced in the examples.

Cosmology Primer

We review in this appendix those aspects of the standard cosmological model that are relevant for understanding gravitational lensing and its applications.

9.1   The Friedmann–Lemaitre–Robertson–Walker Metric

The standard cosmological models are based on the assumption of the *"Cosmological Principle"*. This is the assertion that, on sufficiently large scales (beyond those traced by the large-scale structure of the galaxy distribution), the universe is homogeneous and isotropic. Originally introduced by Einstein and subsequent relativistic cosmologists without strict empirical justification, the Cosmological Principle is today accepted because it agrees with observations: data concerning radio galaxies, clusters of galaxies, quasars, and the microwave background all demonstrate that the level of an-isotropy of the universe on very large scales is about one part in 10^5.

The geometrical properties of the space–time are described by a metric. All events in the space–time have one time coordinate $x^0 = ct$, where c is the velocity of light and t is the proper time, and three space coordinates x^1, x^2, x^3. The interval between two events in the space–time can be written as

$$ds^2 = g_{ij} dx^i dx^j \,, \qquad (9.1)$$

where repeated suffixes imply summation and i, j both run from 0 to 3. The tensor g_{ij} is the metric tensor, which describes the space–time geometry.

© Springer Nature Switzerland AG 2021

M. Meneghetti, *Introduction to Gravitational Lensing*, Lecture Notes in Physics 956, https://doi.org/10.1007/978-3-030-73582-1_9

The most general space–time metric describing a universe in which the Cosmological Principle is obeyed is the *Friedmann–Lemaitre–Robertson–Walker* metric. Adopting this metric, Eq. 9.1 can be written as

$$ds^2 = (cdt)^2 - a(t)^2 \left[dw^2 + f_K(w)^2 (d\theta^2 + \sin\theta^2 d\phi^2) \right] . \tag{9.2}$$

Here, θ and ϕ are angles that uniquely identify positions on the unit sphere around an arbitrary point chosen as origin, and w is a radial coordinate. The parameter K determines the curvature of spatial hyper-surfaces. It is a constant that can be scaled to assume only the values $1, 0$ or -1. The case $K = 0$ corresponds to the flat, Euclidean space, whose properties are familiar. The other two cases, $K = 1$ and $K = -1$, correspond, respectively, to hyper-spheres and spaces of constant negative curvature: the first is a closed space, with finite volume and no boundary; the second is an open (hyperbolic) space, i.e. infinite. The radial function $f_K(w)$ is either a trigonometric, linear, or hyperbolic function of w, depending on whether K is positive, zero, or negative,

$$f_K(w) = \begin{cases} \frac{1}{\sqrt{K}} \sin(\sqrt{K}w) & (K > 0) \\ w & (K = 0) \\ \frac{1}{\sqrt{-K}} \sinh(\sqrt{-K}w) & (K < 0) \end{cases} \tag{9.3}$$

Because of isotropy, spatial surfaces of constant distance from an arbitrary point need to be two spheres. If we define the radius r of the two spheres by $f_K(w) \equiv r$, then the metric in Eq. 9.2 takes the form

$$ds^2 = (cdt)^2 - a(t)^2 \left[\frac{dr^2}{1 - Kr^2} + r^2(d\theta^2 + \sin\theta^2 d\phi^2) \right] . \tag{9.4}$$

As can be easily seen in Eq. 9.4, the distance between two points in space depends on time only through the scale factor $a(t)$, whose form will be shown later. Given that r, θ, and ϕ are time independent variables, these coordinates are called *comoving*.

9.2   Redshift

If the scale factor $a(t)$ changes with time, i.e. if the universe expands or shrinks, photons that are emitted by a source are redshifted or blueshifted while propagating to the observer.

Consider a luminous source at comoving distance r, emitting photons whose wavelength is λ_e at time t_e, and an observer placed at the origin of the coordinate system ($r = 0$). We define the *redshift* z of the source as

$$z = \frac{\lambda_o - \lambda_e}{\lambda_e} , \tag{9.5}$$

where λ_o is the wavelength of radiation from the source measured by the observer at time t_o.

Given that radiation travels along null geodesics in the space–time (i.e. $ds^2 = 0$), we obtain from Eq. 9.4:

$$\int_{t_e}^{t_o} \frac{c\,dt}{a(t)} = \int_0^r \frac{dr'}{(1 - Kr'^2)} = f(r) \,. \tag{9.6}$$

The same result of Eq. 9.6 can also be obtained for light that is emitted by the source at time $t'_e = t_e + \delta t_e$ and received by the observer at time $t'_o = t_o + \delta t_o$:

$$\int_{t'_e}^{t'_o} \frac{c\,dt}{a(t)} = f(r) \,. \tag{9.7}$$

Equations 9.6 and 9.7 imply that if δt_e and δt_o are small,

$$\frac{\delta t_o}{a_o} = \frac{\delta t_e}{a_e} \,, \tag{9.8}$$

where $a_o = a(t_o)$ and $a_e = a(t_e)$. Identifying the inverse of the time intervals t_o and t_e with the inverse frequencies of the observed and emitted radiations, ν_o and ν_e, Eq. 9.8 can be written as

$$\nu_e a_e = \nu_o a_o \,, \tag{9.9}$$

or equivalently,

$$\frac{a_e}{\lambda_e} = \frac{a_o}{\lambda_o} \,, \tag{9.10}$$

from which

$$1 + z = a_o/a_e \,. \tag{9.11}$$

This equation shows the relationship between the redshift z and the scale factor $a(t)$: if the universe expands and $a_o > a_e$, then the light observed from distant sources is redshifted ($z > 0$).

9.3   The Friedmann Equations

The geometry of space–time, expressed by the metric tensor g_{ij}, is related to the matter content of the universe, expressed by the energy–momentum tensor T_{ij}, through *Einstein's field equations*,

$$R_{ij} - \frac{1}{2}Rg_{ij} - \Lambda g_{ij} = \frac{8\pi G}{c^4}T_{ij} ,$$
(9.12)

where R_{ij} and R are the Ricci tensor and Ricci scalar, respectively.

The term Λ is called the *cosmological constant*. It was introduced by Einstein to enable for static cosmological solutions of the field equations. However, even after the expansion of the universe was observationally established, the cosmological constant refused to die. Its physical meaning can be easily understood by considering Eq. 9.12 in vacuum. In that case, the energy–momentum tensor is

$$T_{ij}^{vac} = -\frac{c^4 \Lambda}{8\pi G}g_{ij} .$$
(9.13)

In other words, if the cosmological constant differs from zero, the vacuum has non-zero energy density and pressure.

The energy–momentum tensor of the universe is that of an homogeneous perfect fluid, which is characterized by its density $\rho(t)$ and pressure $p(t)$. For the Robertson–Walker metric, the Einstein's equations simplify to the *Friedmann equations*,

$$\left(\frac{\dot{a}}{a}\right)^2 = \frac{8\pi G}{3}\rho - \frac{Kc^2}{a^2} + \frac{\Lambda c^2}{3}$$
(9.14)

$$\frac{\ddot{a}}{a} = -\frac{4}{3}\pi G\left(\rho + \frac{3p}{c^2}\right) + \frac{\Lambda c^2}{3} .$$
(9.15)

These two equations are not independent: the second can be recovered from the first if one takes the adiabatic expansion of the universe into account, i.e. if we assume that the change of internal energy equals minus the pressure times the change in proper volume,

$$\frac{d}{dt}\left[a^3(t)\rho(t)c^2\right] = -p\frac{da^3(t)}{dt} .$$
(9.16)

The time dependence of the scale factor $a(t)$ can then be determined by integrating these differential equations and by fixing its value at one instant of time. We choose $a = 1$ at the present epoch t_0.

9.4   Cosmological Parameters

In this section we give a summary of the relevant cosmological parameters. First of all, we introduce the parameter used to quantify the relative expansion rate of the universe,

$$H(t) \equiv \frac{\dot{a}(t)}{a(t)} \,, \tag{9.17}$$

which is called the *Hubble parameter*. Its value at the present epoch, $H_0 = H(t_0)$, is known as the *Hubble constant*. Current measurements of this quantity roughly fall in the range $H_0 = (60 \div 75)$ km s^{-1} Mpc^{-1} (Planck Collaboration 2020, Riess et al. 2016, Wong et al. 2020). The uncertainty in H_0 is commonly expressed as $H_0 = 100h$ km s^{-1} Mpc^{-1}, with $h = (0.6 \div 0.75)$.

We then define the present *critical density* of the universe as

$$\rho_{0,\mathrm{cr}} \equiv \frac{3H_0^2}{8\pi G} \approx 1.9 \times 10^{-29} h^2 \mathrm{g\ cm}^{-3} \,, \tag{9.18}$$

where G is the gravitational constant.

The density of the universe in units of $\rho_{0,\mathrm{cr}}$ is the *density parameter*,

$$\Omega_0 \equiv \frac{\rho_0}{\rho_{0,\mathrm{cr}}} \,. \tag{9.19}$$

The value of Ω_0 is crucial for cosmology.

All the components of the universe contribute to its density: baryonic and dark matter, radiation and relativistic particles, and vacuum. We call these contributions $\Omega_{b,0}$, $\Omega_{DM,0}$, $\Omega_{r,0}$, $\Omega_{\Lambda,0}$, respectively. The total contribution from the matter is $\Omega_{m,0} = \Omega_{b,0} + \Omega_{DM,0}$. The most recent observations of the Cosmic Microwave Background by the ESA's mission *Planck* suggest that the value of the total density parameter is very close to unity, $\Omega_0 \sim 1$, consistently with other cosmological probes such as baryon acoustic oscillation (BAO) measurements from galaxy redshift surveys (Alam et al. 2017), and observations of type Ia Supernovae (Scolnic et al. 2018). The contributions to the density parameter from dark and baryonic matter are $\Omega_{DM,0} \sim 0.27$ and $\Omega_{b,0} \sim 0.04$, respectively (Planck Collaboration 2020), i.e. the matter content of the universe is dominated by dark matter. The commonly accepted hypothesis is that dark matter is nearly collision-less and in the form of massive, weakly interacting particles. The relativistic component candidates today are photons and neutrinos. Their cosmic density contribution, $\Omega_{r,0} = \Omega_{\gamma,0} + \Omega_{\nu,0}$, is estimated to be $\sim 3.2 \times 10^{-5} h^{-2}$. This is four orders of magnitude smaller than the today estimated value of the matter density parameter of the universe and can thus be considered negligible.

As was pointed out in the previous section, a non-null cosmological constant means that the vacuum has a finite density. This density can be written in units of the critical density as

$$\Omega_{\Lambda,0} \equiv \frac{\Lambda c^2}{3H_0^2} \,.$$ (9.20)

Observations suggest that the largest contribution to the total density of universe comes from the cosmological constant or, more generally speaking, from the so-called *dark energy*, of which the cosmological constant is a particular form. The equation of state of dark energy can be written in the form

$$p_{DE} = w_{DE}(t)\rho_{DE}c^2,$$ (9.21)

where $w_{DE}(t)$ is a function of time. The case $w_{DE} = -1$ corresponds to the cosmological constant. Assuming that the energy of vacuum may not have the form of a cosmological constant, then its contribution to the density parameter at present time is generally indicated as $\Omega_{DE,0}$.

The values of $\Omega_{m,0}$ and $\Omega_{\Lambda,0}$ (or $\Omega_{DE,0}$) are related to the curvature of the spatial hyper-surfaces. Using the previous definitions, Eq. 9.14 can be written for $a = a_0 = 1$ as

$$H_0^2(1 - \Omega_{m,0} - \Omega_{\Lambda,0}) = -Kc^2 \,,$$ (9.22)

from which we see that if $K = 0$, then $\Omega_0 = \Omega_{m,0} + \Omega_{\Lambda,0} = 1$. Moreover, the sign of K is positive if $\Omega_0 > 1$, negative otherwise.

Finally, we define the *deceleration parameter*

$$q_0 = -\frac{\ddot{a}a}{\dot{a}^2}$$ (9.23)

at $t = t_0$. The sign of this parameter tells us if the universe is in decelerated expansion or not.

9.5   Cosmological Distances

Several types of distance between two points can be defined in a curved space–time. Indeed, distance definitions in terms of different measurement prescriptions generally lead to different distances.

The *proper distance* of P from P_0, which we can take as the origin of the polar coordinate system r, ϕ, θ, is the distance measured at time t by a chain of observers connecting P to P_0. From Eq. 9.4, this distance is

$$D_{\text{pr}} = \int_0^r \frac{a \, dr'}{(1 - Kr'^2)^{1/2}} = af(r) , \qquad (9.24)$$

where a is the scale factor at time t, $a = a(t)$, and the radial function $f(r)$ is

$$f(r) = \begin{cases} \arcsin r & K = 1 \\ r & K = 0 \\ \text{arcsinh } r & K = -1 \end{cases} . \qquad (9.25)$$

Given the time dependence of the scale factor, the proper distance also changes with time. Therefore, the point P has a radial velocity with respect to P_0 given by

$$v_r = \dot{a} f(r) = \frac{\dot{a}}{a} f(r) = H(t) D_{\text{pr}} . \qquad (9.26)$$

This equation is called the *Hubble's Law*.

The proper distance at the actual time t_0 defines the *comoving distance*

$$D_c = f(r) = a^{-1} D_{\text{pr}} . \qquad (9.27)$$

This is the distance on the spatial hyper-surface $t = t_0$ between the world lines of points P and P_0 comoving with the cosmic flow.

The *angular diameter distance* is defined in analogy to the relation in the Euclidean space between the physical diameter d_{pr} of a source and the angle $\Delta\theta$ that it subtends. From Eq. 9.4, it results in $d_{\text{pr}} = ar\Delta\theta$ and the angular diameter distance is then

$$D_A = \frac{d_{pr}}{\Delta\theta} = ar . \qquad (9.28)$$

The distance defined in such a way to preserve the Euclidean inverse-square law for the decrease of luminosity with distance from a source is called *luminosity distance*. If L is the amount of energy emitted by a source in P per unit time and l is the amount of energy received per unit time and per unit area by an observer in P_0, the luminosity distance between these two points is defined as

$$D_L = \left(\frac{L}{4\pi l} \right)^{1/2} . \qquad (9.29)$$

The area of a spherical surface centered on P and passing through P_0 at time t_0 is just $4\pi r^2$. The photons emitted by the source arrive at this surface having been redshifted by the expansion of the universe by a factor a. Also, photon arrival times are delayed by another factor a. Therefore, the flux l is given by

$$l = \frac{L}{4\pi r^2} a^2 , \qquad (9.30)$$

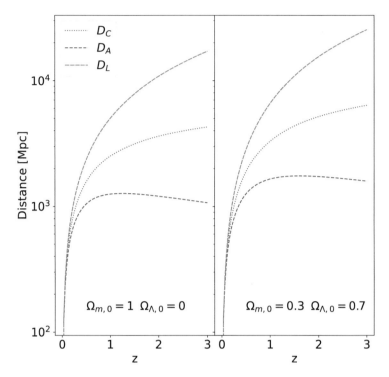

Fig. 9.1 Three distance measures are plotted as a function of redshift in two cosmological models: a model with $\Omega_{m,0} = 1$ and $\Omega_{\Lambda,0} = 0$ (left panel) and a model with $\Omega_{m,0} = 0.3$ and $\Omega_{\Lambda,0} = 0.7$ (right panel). These are the comoving distance D_c (red dotted line), the angular diameter distance D_A (blue short-dashed line), and the luminosity distance D_L (green log-dashed line)

from which

$$D_L = a^{-1}r = a^{-2}D_A . \tag{9.31}$$

We plot the four distances D_{pr}, D_c, D_A, and D_L as a function of redshift in Fig. 9.1. Given that they depend on the curvature of the space–time, distances change for different values of $\Omega_{m,0}$ and $\Omega_{\Lambda,0}$: they are larger for lower cosmic density and higher cosmological constant. The largest differences are at high redshift, while at low redshift all distances follow the Hubble's law,

$$\text{distance} = \frac{cz}{H_0} + O(z^2) . \tag{9.32}$$

9.6 The Friedmann Models

We now introduce the standard cosmological models described by Eqs. 9.14 and 9.15. They receive their name from A. Friedmann, who derived them in 1922. The main assumptions on which these cosmological models are based are:

- The universe can be approximated by a perfect fluid with some density ρ and pressure p.
- The appropriate equation of state, relating pressure to density, can be cast, either exactly or approximately, in the form

$$p = w\rho c^2. \tag{9.33}$$

The perfect fluid is, in fact, quite a realistic approximation in many situations. For example, if the scale of particle interaction is much smaller than the scale of physical interest, as is the case for the global properties of the universe, the fluid can be treated as perfect. Even the second assumption on the fluid equation of state is quite appropriate in many cases of physical interest.

Non-relativistic matter, which can be considered as pressure-less, is frequently called *dust* and corresponds to the case $w = 0$. This approximation is valid in general for any non-relativistic gas, which can then be treated as a fluid of dust.

On the other hand, relativistic matter and radiation correspond to the case $w = 1/3$.

Inserting these expressions into Eq. 9.16, we obtain:

$$\rho = \begin{cases} \rho_0 a^{-3} = \rho_0(1+z)^4 & \text{for dust, } p = 0 \\ \rho_0 a^{-4} = \rho_0(1+z)^3 & \text{for relativistic matter, } p = \frac{1}{3}\rho c^2 \end{cases}. \tag{9.34}$$

The energy density of relativistic matter therefore drops more rapidly than that of the non-relativistic one. This can be understood by considering a comoving box containing N particles. Let us assume that the box expands, with an expansion factor $a(t)$, and particles within it are neither created nor destroyed. In the case of non-relativistic matter, the density of particles simply changes as the inverse of the box volume, i.e. as a^{-3}. If the particles are relativistic, they behave like photons: not only their number density decreases as a^{-3}, but their wavelength λ is also increased by a factor a. Since the energy per particle is proportional to λ^{-1}, the total energy density decreases as a^{-4}.

At the epoch when the density of the relativistic matter component equaled that of the non-relativistic one, the expansion scale factor was

$$a_{\text{eq}} = \frac{\Omega_{r,0}}{\Omega_0} = 3.2 \times 10^{-5}\Omega_0^{-1}h^{-2}. \tag{9.35}$$

Much before that epoch, i.e. for $a \ll a_{\text{eq}}$, the expansion of the universe therefore was radiation-dominated.

An important property of the Friedmann models with no cosmological constant is that, if $w > -1/3$, they possess a point in time where $a = 0$, which is called the *Big-Bang singularity*. Indeed, inserting Eq. 9.33 into Eq. 9.14, we obtain

$$H^2(t) = \left(\frac{\dot{a}}{a}\right)^2 = H_0^2 a^{-2} \left[\Omega_{0w} a^{-(1+3w)} + (1 - \Omega_{0w})\right] , \qquad (9.36)$$

where $\Omega_{0w} = \rho_{0w}/\rho_{cr}$ is the actual density parameter of the fluid. Suppose that at some generic time t the universe is expanding, so that $\dot{a}(t) > 0$. If $w > -1/3$, from Eq. 9.14 we see that $\ddot{a} < 0$ for all t. The graph describing $a(t)$ therefore is necessarily concave and there must be a time, which we can set as $t = 0$, when $a(0) = 0$ and when the density diverges.

9.6.1 Single Component Models

We discuss in this section the solutions of Eq. 9.14 for flat, open, and closed models in the simple case of single component universes.

The solution appropriate to a flat universe, i.e. with $\Omega_{0w} = 1$, without cosmological constant is known as the *Einstein-de Sitter universe*. In this case, integrating Eq. 9.14, we obtain

$$a(t) = \begin{cases} \left(\frac{t}{t_0}\right)^{2/3} & \text{for the matter-dominated universe, } w = 0 \\ \left(\frac{t}{t_0}\right)^{1/2} & \text{for the radiation-dominated universe, } w = \frac{1}{3} \end{cases} . \qquad (9.37)$$

A general property of these models therefore is that the expansion parameter a grows indefinitely with time. Moreover, the deceleration parameter is constant and positive in both the cases of matter- and radiation-dominated universes. This means that the expansion in these models is decelerated.

The solutions for curved models are rather more complicated. At early times, they behave in a manner very similar to flat models, and the solutions seen for the Einstein-de Sitter model can be applied. Then, when $a(t) \gg a(t^\star) = a^\star$, where $a^\star$ is given by

$$a^\star = \left|\frac{\Omega_{0w}}{1 - \Omega_{0w}}\right|^{1/(1+3w)} , \qquad (9.38)$$

solutions differ. In models with $\Omega_{0w} < 1$ (open universes), for $t \gg t^\star$ the scale factor grows with time as

$$a(t) \simeq a^\star \frac{t}{t^\star} , \qquad (9.39)$$

and the deceleration parameter is approximately zero.

In models with $\Omega_w > 1$, at $t = t^\star$ the scale factor reaches a maximum $a_m = a_\star$. After that time $a(t)$ starts to decrease symmetrically around a_m. At time $t = 2t^\star$ there is another singularity in a symmetrical position with respect to the *Big-Bang*, called the *Big-Crunch*.

Analytical solutions exist for both matter- and radiation-dominated universes (e.g. Coles & Lucchin 2002).

9.6.2 Multiple Component Models

Considering a model universe made of matter, radiation, and cosmological constant (we consider this case of dark energy for the moment), Eq. 9.14 becomes

$$H^2(t) = H_0^2 \left[\frac{\Omega_{r,0}}{a^4} + \frac{\Omega_{m,0}}{a^3} + \frac{1 - \Omega_{m,0} - \Omega_{\Lambda,0}}{a^2} + \Omega_{\Lambda,0} \right]. \tag{9.40}$$

There is generally no simple solution to this equation. Qualitatively, we can say that at very early times the universe is flat and radiation-dominated. Even the term depending on the cosmological constant is negligible at that time. Therefore, solutions for the Einstein-de Sitter model are appropriate for describing the evolution of the scale factor during this period. Then, after equivalence, matter starts to dominate over radiation and the cosmological constant term becomes increasingly more significant.

Using Eq. 9.40, we can determine the dependence of Ω and Ω_Λ on the scale factor a. For a matter-dominated universe, we find

$$\Omega_m(a) = \frac{8\pi G}{3H^2(a)} \rho_{m,0} a^{-3} = \frac{\Omega_{m,0}}{a + \Omega_{m,0}(1-a) + \Omega_{\Lambda,0}(a^3 - a)}, \tag{9.41}$$

$$\Omega_\Lambda(a) = \frac{\Lambda c^2}{3H^2(a)} = \frac{\Omega_{\Lambda,0} a^3}{a + \Omega_{m,0}(1-a) + \Omega_{\Lambda,0}(a^3 - a)}. \tag{9.42}$$

Whatever the values of $\Omega_{m,0}$ and $\Omega_{\Lambda,0}$ are at the present epoch, Eqs. 9.41 and 9.42 show that $\Omega_m \to 1$ and $\Omega_\Lambda \to 0$ for $a \to 0$. On the other hand, if $\Omega_{m,0} + \Omega_{\Lambda,0} \le 1$, $\Omega_m \to 0$ and $\Omega_\Lambda \to 1$ monotonically for $a \to \infty$. Therefore, for this kind of models we can define a time t_Λ such that for $a \gg a_\Lambda = a(t_\Lambda)$, the dominant term in Eq. 9.40 is the cosmological constant one. Neglecting all the other terms, the solution of the Friedmann equations at that time is that of the so-called *de Sitter universe*, which is written in the form

$$a(t) \propto \exp \left[\left(\frac{1}{3}\Lambda \right)^{1/2} ct \right]. \tag{9.43}$$

This means that at a given time t_{acc}, the cosmological constant term starts to accelerate the expansion of the universe. This can be explained, if one remembers that the cosmological constant is proportional to the energy density and pressure of vacuum. In particular, from Eq. 9.13 (see also Eq. 9.21), we see that the vacuum has a negative-pressure equation of state:

$$p_{vac} = -\rho_{vac}c^2 . \tag{9.44}$$

This means that the pressure of vacuum can be interpreted as a source of gravitational repulsion. Therefore, we can assume that, when the matter density is sufficiently low, gravity loses its efficiency to decelerate the expansion of the universe and the vacuum pressure starts to accelerate it. This behavior characterizes also models with dark energy that differs from a cosmological constant.

9.7   Structure Formation

9.7.1   Linear Growth of Density Perturbations

The standard model assumes that structure in the universe formed via gravitational collapse from initial density fluctuations. The origin of these fluctuations is still unclear, but the general idea is that they were generated during an inflationary epoch from quantum fluctuations. They are assumed to be uncorrelated and the distribution of their amplitudes is assumed to be Gaussian.

Density perturbations are characterized by the density contrast

$$\delta(\vec{r}, a) = \frac{\rho(\vec{r}, a) - \overline{\rho}(a)}{\overline{\rho}(a)} , \tag{9.45}$$

where $\overline{\rho}$ is the average cosmic density and ρ is the density at the position given by $\vec{r}$.

Until $\delta \ll 1$, it is possible to study the evolution of the density perturbations using the linear approach. First, we consider a universe dominated by baryonic matter, i.e. gas having pressure p. The time evolution of a small density perturbation δ is determined by two opposite forces: the first is the pressure force, which acts such as to cancel the density fluctuation; the second is the gravitational force, which tends to amplify it. If the fluctuation is contained in a region of radius R, the pressure force per unit mass can be written as

$$F_p \approx \frac{pR^2}{M} = \frac{v_s^2}{R} , \tag{9.46}$$

where $M = \rho R^3$ is the mass of the fluctuation and $v_s \sim \sqrt{p/\rho}$ is the sound velocity in the fluid. On the other hand, the gravitational force per unit mass is

$$F_g \approx \frac{GM}{R^2} = G\rho R . \tag{9.47}$$

The density perturbation can grow only if $F_g \geq F_p$. Equaling Eqs. 9.46 and 9.47, we obtain that this condition is satisfied only if the size of the region where the fluctuation happens is larger than the *Jeans radius*,

$$R_J = \left(\frac{v_s^2}{G\rho}\right)^{1/2} . \tag{9.48}$$

Density perturbations on scales $\lambda \geq R_J$ therefore are amplified. Perturbations on scales $\lambda < R_J$ do not grow and propagate as acoustic waves.

Linear theory shows that perturbations on scales larger than the Jeans radius grow like

$$\delta(a) \propto \begin{cases} a^2 & \text{before } a_{\text{eq}} \\ a & \text{after } a_{\text{eq}} \end{cases} , \tag{9.49}$$

as long as the Einstein-de Sitter limit holds. If $\Omega_{m,0} \neq 1$ and $\Omega_{\Lambda,0} \neq 0$, this approximation, however, does not apply anymore for $a \gg a_{\text{eq}}$. Then, the linear growth of density perturbations is changed according to

$$\delta(a) = \delta_0 a \frac{g'(a)}{g'(1)} \equiv \delta_0 a g(a) , \tag{9.50}$$

where δ_0 is the density contrast linearly extrapolated to the present epoch and $g'(a)$ is the linear growth function, which depends on $\Omega_m(a)$ and $\Omega_\Lambda(a)$ as given by the fitting function

$$g'(a; \Omega_{m,0}, \Omega_{\Lambda,0}) = \frac{5}{2}\Omega_m(a) \left[\Omega_m^{4/7}(a) - \Omega_\Lambda(a) + \left(1 + \frac{\Omega_m(a)}{2}\right)\left(1 + \frac{\Omega_\Lambda(a)}{70}\right)\right]^{-1} . \tag{9.51}$$

The growth function $ag(a)$ is shown in Fig. 9.2 for a variety of parameters $\Omega_{m,0}$ and $\Omega_{\Lambda,0}$: the growth rate is constant for the Einstein-de Sitter model ($\Omega_{m,0} = 1$; $\Omega_{\Lambda,0} = 0$), while for low $\Omega_{m,0}$ models it is higher for $a \ll 1$ and lower for $a \approx 1$. This feature means that structures form earlier in low-density than in high-density models, which will turn out to be of great importance for the further discussion on arc statistics.

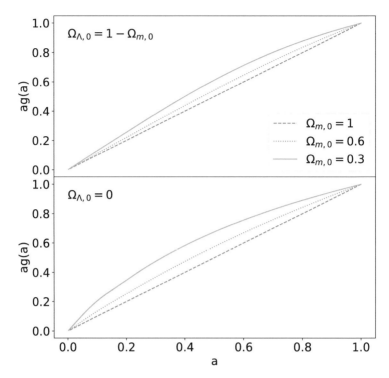

Fig. 9.2 The growth function $ag(a)$ given in Eqs. 9.50 and 9.51 for different values of $\Omega_{m,0}$ between $\Omega_{m,0} = 0.3$ and $\Omega_{m,0} = 1.0$: the solid, dotted, and dashed lines refer to $\Omega_{m,0} = 0.3, 0.1$, and 1, respectively. Curves are displayed for flat models with $\Omega_{\Lambda,0} = 1 - \Omega_{m,0}$ (top panel) and for open models with $\Omega_{\Lambda,0} = 0$ (bottom panel)

Instead of using the size of fluctuations, we can deal with their mass. For a perturbation of density ρ and size R, the mass is defined as

$$M = \frac{4}{3}\pi R^3 \rho \ . \tag{9.52}$$

This permits us to define the *Jeans mass*, $M_J \equiv M(R_J) = (4/3)\pi R_J^3 \rho$. Only perturbations of mass $M \geq M_J$ can be amplified by gravity.

Before equivalence, the Jeans mass M_J grows like $M_J \propto a^3$. Then, while the matter is coupled with radiation, it is constant, $M_J(a_{\mathrm{eq}}) \sim 10^{16} M_\odot/h$. When radiation and matter decouple, the Jeans mass drops down to $M_J \sim 10^5 M_\odot/h$. This is very important because the perturbations, which initially had mass $M < M_J$ and oscillated without being amplified, can now start to grow.

Density fluctuations can be on scales even larger than the size of the causally connected regions in the universe. This size is called the *cosmological horizon*. It is given by the distance by which a photon can travel in the time t since the Big-Bang. Since the appropriate time scale is given by the inverse Hubble parameter $H^{-1}(a)$, the horizon size is $R_H = cH^{-1}(a)$. The mass inside a sphere with radius R_H is called the *horizon mass*,

$$M_H = \frac{4}{3}\pi R_H^3 \rho . \tag{9.53}$$

Considering only the matter contribution to the density ρ, the horizon mass grows like $M_H \propto a^3$ before equivalence, being $M_H(a_{eq}) \sim 10^{15} M_\odot/h$, and like $M_H \propto a^{3/2}$ after equivalence. When a perturbation has $M \leq M_H$, it can experience all the physical processes, like dissipation, which happen in the expanding universe.

Before the epoch a_{rec} of the hydrogen recombination, when matter and radiation are still coupled, the baryonic matter in the universe can be considered as a plasma of photons, electrons, and protons. Dissipative effects arise in this plasma mainly due to the diffusion of photons. The size of the region that is affected by the dissipation effects grows with time like $R_S \propto (c^2 \tau_{\gamma e} t)^{1/2}$, where $\tau_{\gamma e}$ is the time scale for the collisions between photons and electron–proton pairs. The *dissipation mass* (or *Silk mass*) is then

$$M_S = \frac{4}{3}\pi R_S^3 \rho . \tag{9.54}$$

This mass grows like $M_S \propto a^{9/2}$ before equivalence, when $M_S(a_{eq}) \sim 10^{11} M_\odot/h$, and like $M_S \propto a^{15/4}$ after equivalence. After recombination, the dissipation mass drops to zero. It is important to note that dissipation obliterates the fluctuations on all scales smaller than the Silk mass almost immediately. In a collisional fluid, no structure will therefore be formed on all mass scales less than the Silk mass at recombination, $M_S(a_{rec}) \sim 10^{14} M_\odot/h$. Smaller structures can only form by fragmentation of the larger ones.

The main problem of this model for structure formation is that the cosmic microwave background reveals relative temperature fluctuations of order 10^{-5} on large scales. By the Sachs–Wolfe effect Sachs and Wolfe (1967), these temperature fluctuations reflect baryonic density fluctuations of the same order of magnitude at recombination. Given that recombination occurs at $a \sim 10^3$, Eq. 9.49 implies that density fluctuations today should only reach a level of 10^{-2}. Instead structure (e.g. galaxies) with $\delta \gg 1$ are observed.

This is one of the strongest arguments for the existence of an additional matter component that does not couple electromagnetically (probably only weakly), i.e. the *dark matter*. If this matter component exists, then fluctuations in that component can grow as soon as it decouples from the cosmic plasma, well before photons decouple from baryons to set the cosmic microwave background free.

We can consider the dark matter as a collision-less fluid. Previous definitions of Jeans mass, horizon mass, and dissipation mass can be extended even to this fluid by substituting the sound velocity with the average velocity $v_\star$ of the collision-less particles. In this case fluctuations on mass scales less than the Jeans mass do not propagate as acoustic waves but are damped by the velocity dispersion of particles. This dissipation process is called *free streaming*. Perturbations are completely canceled when their mass is equal to the *free streaming mass*,

$$M_{FS} = \frac{4}{3}\pi R_{FS}^3 \rho_{DM} , \tag{9.55}$$

where R_{FS} is the free streaming length,

$$R_{FS} = a \int_0^t \frac{v(t')}{a(t')} dt' . \tag{9.56}$$

If we assume that dark matter decoupled from the cosmic plasma when it was already non-relativistic (*cold dark matter*), the Jeans mass and the free streaming mass are almost identical and much lower than the masses of cosmological interest, being $M_J(a_{rec}) \simeq M_{FS}(a_{rec}) \sim 10^5 M_\odot/h$. Structures can thus form starting at the smallest scales. Baryons can then fall into the potential wells of the dark matter structures once they decouple from radiation.

Among the fluctuations of mass $M \leq M_J(a_{rec})$, those that enter the horizon at $a \leq a_{eq}$ cannot grow until radiation ceases dominating the expansion of the universe: before a_{eq}, the expansion time scale $t_{exp} \sim (G\rho_R)^{-1/2}$ is smaller than the collapse time scale $t_{DM} \sim (G\rho_{DM})^{-1/2}$ of dark matter fluctuations. In other words, the radiation-driven expansion of the universe prevents dark matter perturbations from collapsing. Since the time evolution of density perturbations before equivalence is $\delta \propto a^2$, fluctuations of mass $M < M_H(a_{eq})$ that enter the horizon at a_{enter} are suppressed at equivalence by a factor

$$f_{sup} = \left(\frac{a_{enter}}{a_{eq}}\right)^2 . \tag{9.57}$$

9.7.2 Density Power Spectrum

It is very convenient to think of the linear perturbations as of superposition of plane waves that evolve independently, while the fluctuations are still linear. In other words, we can decompose the density contrast into Fourier modes. This description should be valid only in flat space, but we can use it also for curved models because 1) at early times space can be considered flat in all cosmological models and 2) at late times the interesting scales of the density perturbations are much smaller than the curvature radius of the universe.

We therefore write the density contrast as

$$\delta(\vec{r}) = \int \frac{d^3k}{(2\pi)^3} \hat{\delta}(\vec{k}) e^{-i\vec{k}\vec{r}} , \tag{9.58}$$

where $\vec{k}$ is the wave vector and $\hat{\delta}$ denotes the Fourier transform of δ,

$$\hat{\delta}(\vec{k}) = \int d^3r \, \delta(\vec{r}) e^{i\vec{k}\vec{r}} \,. \tag{9.59}$$

The *power spectrum* of the density fluctuations is then defined as

$$P(k) \equiv \langle |\hat{\delta}^2(\vec{k})| \rangle \,. \tag{9.60}$$

It is commonly assumed that the primordial power spectrum has a power-law form,

$$P_i(k) = A_i k^{n_i} \,. \tag{9.61}$$

Moreover, if we require that the primordial power spectrum is *scale invariant*, i.e. the fluctuations in the gravitational potential are independent of the length scale, the power index must be $n_i = 1$. This power spectrum is then called the *Harrison–Zel'dovich spectrum*, which is compatible with predictions from inflationary models, although they tend to require $n_i \lesssim 1$.

The primordial power spectrum is later modified by all the physical processes that affect the growth of density perturbations. In particular, we must take into account the suppression of the fluctuations that enter the horizon before equivalence. Given that the time a_{eq} when a perturbation of comoving length λ or wave number $k = \lambda^{-1}$ enters the horizon is $a_{enter} \propto k^{-1}$ for $a_{enter} \ll a_{eq}$, the suppression factor 9.57 can be written as

$$f_{sup} = \left(\frac{k_0}{k}\right)^2 \,, \tag{9.62}$$

where k_0 is the wave number corresponding to a perturbation that enters the horizon at a_{eq}. Combining the primordial spectrum with this suppression of small-scale modes, we obtain

$$P(k) \propto \begin{cases} k & \text{for } k \ll k_0 \\ k^{-3} & \text{for } k \gg k_0 \end{cases} \,. \tag{9.63}$$

The normalization of the matter power spectrum is defined through the variance of mass within spheres of radius $8 \, h^{-1} \text{Mpc}$, σ_8. It has been measured using several methods, such as X-ray cluster counts, weak lensing, and the cosmic microwave background (CMB). The most recent results from *Planck* set this parameter to $\sigma_8 \sim 0.81$.

9.7.3 Non-linear Evolution

Once the non-linear regime is reached by the density perturbations, their evolution is separated from the universal expansion and they start to collapse to form virialized objects. It is very difficult to properly describe the collapse process unless one makes strong assumptions, for example on the symmetry of the perturbations.

The simplest assumption we can make is that the collapse is spherical. In the absence of the cosmological constant, the radius R of a mass shell in a spherically symmetric density perturbation evolves according to

$$\frac{d^2R}{dt^2} = -\frac{GM}{R^2} ,$$
(9.64)

where M is the mass within the mass shell. Integrating this equation, we obtain

$$\frac{1}{2}\left(\frac{dR}{dt}\right)^2 - \frac{GM}{R} = E .$$
(9.65)

If the energy $E < 0$, the shell collapses. Using this spherical approximation, the solution of Eq. 9.65 is characterized by:

- A maximum expansion, $R = R_{\max}$, at $t = t_{\max}$, when the perturbation density is $\rho_p(t_{\max}) = (3\pi/4)^2\overline{\rho}(t_{\max}) \approx 5.5\,\overline{\rho}(t_{\max})$, where $\overline{\rho}$ is the average density of the unperturbed background
- A singularity, $R = 0$, at the collapse time $t_c = 2t_{\max}$, when density ideally goes to infinity at the center

In fact, when the density is high, small departures from spherical symmetry will result in the formation of shocks and considerable pressure gradients. Heating of the material will occur due to the dissipation of shocks that converts the kinetic energy of collapse into heat. The final result will therefore be an equilibrium state that is not a singular point but some extended configuration with radius R_{vir} and mass M. This happens when the system reaches the virial equilibrium. From the virial theorem, $R_{\mathrm{vir}} = R_{\max}/2$. In an Einstein-de Sitter universe, the perturbation density in units of the critical density at t_c is

$$\Delta_c \equiv \frac{\rho_p(t_c)}{\rho_{\mathrm{cr}}(t_c)} = 18\pi^2 \approx 178 .$$
(9.66)

An extrapolation of linear theory would give a density contrast $\delta_c = \delta(t_c) = 3(12\pi)^{2/3}/20 \approx 1.687$ with a weak dependence on the cosmological parameters.

We show in Fig. 9.3 how Δ_c and δ_c for halos collapsing at $z_c = 0$ change for a variety of parameters $\Omega_{m,0}$ and $\Omega_{\Lambda,0}$. The critical over-density δ_c has only a weak dependence on $\Omega_{m,0}$ for both open models with $\Omega_{\Lambda,0} = 0$ and flat models with

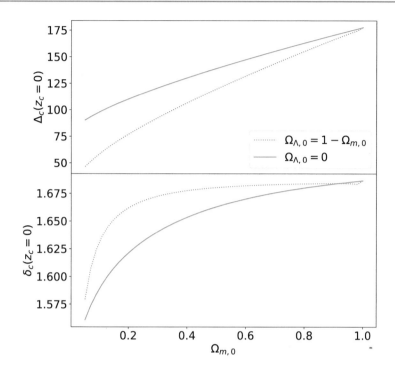

Fig. 9.3 Upper panel: virial over-density of collapsed objects in units of the critical density as a function of Ω_0. Results are plotted for open models with $\Omega_{\Lambda,0} = 0$ (solid line) and flat models with $\Omega_{m,0} + \Omega_{\Lambda,0} = 1$ (dotted line). Lower panel: linear extrapolation of the density contrast at the collapse time t_c. The solid and dotted lines are as in the upper panel

$\Omega_{m,0} + \Omega_{\Lambda,0} = 1$. The dependence of Δ_c on the cosmological parameters is much stronger.

Of course, non-linearity affects the shape of the power spectrum $P(k)$ in a very complicated way. Numerical methods are required for properly evaluating the non-linear power spectrum. Analytic formulae can be obtained only under some strong assumptions. For example, starting from the *ansatz* that the two point correlation functions in the linear and non-linear regimes are related by a general scaling relation Hamilton et al. (1991), analytic formulae describing the non-linear behavior of $P(k)$ have been derived Jain et al. (1995), Peacock and Dodds (1996). We show in Fig. 9.4 the CDM power spectrum corresponding to a ΛCDM model with $\Omega_{m,0} = 0.3$, $\Omega_{\Lambda,0} = 0.7$, and $h = 0.7$, and normalized to the local cluster abundance. The solid line shows the results obtained by assuming a primordial Harrison-Zel'dovich spectrum and linear evolution of the density perturbations on all scales. The dashed curve shows the non-linear evolution of the previous spectrum to $z = 0$. Non-linearity affects the small-scale (large k) part of the spectrum because small-scale perturbations are the first that enter the non-linear regime.

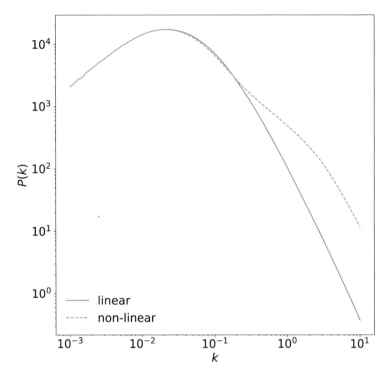

Fig. 9.4 CDM power spectrum, normalized to the local cluster abundance, for a ΛCDM model with $\Omega_{m,0} = 0.3$ and $\Omega_{\Lambda,0} = 0.7$ with $h = 0.7$. The solid curve shows the linear power spectrum, extrapolated to the present time; the dashed line shows its non-linear evolution

9.8   Mass Function

The mass distribution of dark matter halos undergoing spherical collapse in the framework of CDM models is described by the Press and Schechter function Press and Schechter (1974). The number density of collapsed lumps at redshift z with mass in the range $[M, M + dM]$ is

$$n(M, z)dM = \frac{\overline{\rho}}{M} f(v) \frac{dv}{dM} dM , \qquad (9.67)$$

where $\overline{\rho}$ is the universe mean density at redshift z. The function f depends only on the variable $v = \delta_c(z)\sigma_M$ and is normalized such that $\int f(v)dv = 1$. $\delta_c(z)$ is the linearly extrapolated density contrast of halos collapsed at redshift z (see previous section). The r.m.s. density fluctuation at the mass scale M, σ_M, is given by

$$\sigma_M = \frac{1}{2\pi^2} \int_0^\infty dk \, k^2 P(k) W^2(kR) , \qquad (9.68)$$

where $W(kR)$ is the Fourier transform of the window function, which describes the shape of the volume from which the collapsing object is accreting material. Finally, R is the comoving size of the fluctuation of mass M.

In their original derivation of the cosmological mass function, Press and Schechter obtained

$$f(v) = \frac{1}{\sqrt{2\pi}} \exp\left(-\frac{v^2}{2}\right). \tag{9.69}$$

The mass function has been used in this form for more than a decade. However, the last generation of N-body simulations revealed significant deviations of the original Press and Schechter function from the numerical description of the mass distribution of dark matter halos. Correcting the Press–Schechter approach by incorporating the effects of non-spherical collapse, Sheth and Tormen (1999) found

$$f(v) = \sqrt{\frac{2A}{\pi}} C \left(1 + \frac{1}{(Av^2)^q}\right) \exp\left(-\frac{Av^2}{2}\right), \tag{9.70}$$

where $A = 0.707$, $C = 0.3222$, and $q = 0.3$. This equation reduces to the Press and Schechter expression for $A = 1$, $C = 0.5$, and $q = 0$. Fitting the results of N-body simulations, Jenkins et al. (2001) found a formula that is statistically not distinguishable from the Sheth and Tormen one with $A = 0.75$ (see also Tinker et al. 2008).

We plot in Fig. 9.5 the three different mass functions by Press and Schechter, Sheth and Tormen, and Jenkins et al. The original Press and Schechter function underpredicts the abundance of large mass halos with respect to the Sheth and Tormen and the Jenkins et al. functions. The last two are very close to each other; there is only a small difference in the very high-mass tail.

9.9   Dark Energy Models

As seen in the previous sections, the cosmological constant is strongly connected to the energy density of vacuum. Actually most favored cosmological models predict that today this vacuum energy density is of the same order of magnitude as the amount of dark and baryonic matter energy density in the universe, $\epsilon \approx (10^{-3}\text{eV})^4$. This number is tiny in terms of the natural scale of primordial energy density given by the Planck mass $M_p = 1.22 \times 10^{19}$ GeV. A dominant radiation or matter energy density decreases as $\rho \sim M_p^2 \, t^{-2}$ and the present age of the universe is $t_0 \approx 1.5 \times 10^{10}$ yr. It is therefore very easy to explain the smallness of the actual matter energy density in terms of the long duration of the cosmological expansion. On the other hand, assuming that the vacuum energy density is constant, it is very difficult to understand why the cosmological constant is so small today.

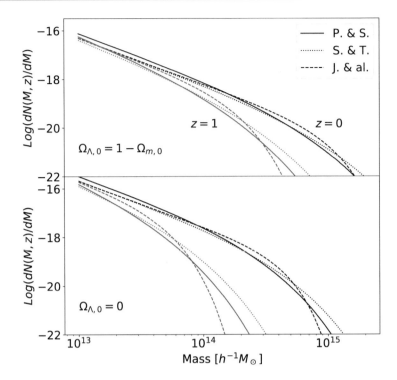

Fig. 9.5 Differential mass function of dark matter halos. Three different mass functions are showed: the original Press and Schechter function (solid lines), its modification obtained by including non-spherical collapse by Sheth and Tormen (dotted lines), and the mass function derived from numerical simulations by Jenkins et al. (dashed lines). Results are plotted for a flat model with $\Omega_{m,0} = 0.3$ and $\Omega_{\Lambda,0} = 0.7$ and for an Einstein-de Sitter model with $\Omega_{m,0} = 1$ and $\Omega_{\Lambda,0} = 0$. Black and red lines show the mass functions of halos at redshift $z = 0$ and $z = 1$, respectively

In order to solve this problem, it has been proposed that even Λ might be very small now because it has been rolling toward zero for a very long time. This idea that the universe contains nearly to homogeneous dark energy, called *Quintessence*, that approximates a time-variable cosmological "constant" arose also in particle physics, through the discussion of phase transition in the early universe and through the search for a dynamical cancellation of the vacuum energy density. We shall not present a rigorous treatment of quintessence models here but only summarize some basic concepts. For a more quantitative discussion, we refer to the excellent review by Peebles and Ratra (2003) and the book by Amendola and Tsujikawa (2010).

Dark energy is usually modeled as that of a homogeneous scalar field, Φ. In this case, if spatial curvature can be neglected, the field equation is

$$\ddot{\Phi} + 3\frac{\dot{a}}{a}\dot{\Phi} + \frac{dV}{d\Phi} = 0 \,, \tag{9.71}$$

where V is the potential energy density, which is a function of the field Φ.

The energy–momentum tensor of this homogeneous field is diagonal in the rest frame of an observer moving such that the universe appears isotropic, and its time and space parts along the diagonal define the energy density and pressure,

$$\rho_\Phi = \frac{1}{2}\dot{\Phi}^2 + V(\Phi) \,, \tag{9.72}$$

$$p_\Phi = \frac{1}{2}\dot{\Phi}^2 - V(\Phi) \,. \tag{9.73}$$

The general form of the equation of state relating ρ_Φ to p_Φ is again

$$p_\Phi = w\rho_\Phi c^2 \,, \tag{9.74}$$

with $w < -1/3$. If the scalar field varies slowly in time ($\dot{\Phi}^2 \ll V(\Phi)$), the field energy approximates the effect of the Einstein's cosmological constant with $p_\Phi \simeq -\rho_\Phi c^2$.

This condition is verified during the inflationary epoch but not at its the end, when $V \sim 0$. When this happens, Φ oscillates. The scalar field is supposed to vary rapidly enough to produce the entropy of our universe and the field or the entropy may produce the baryons, leaving ρ_Φ small or zero. If, after inflation, the time evolution of ρ_Φ starts to slow down and becomes slower than that of the matter density, there comes a time when ρ_Φ starts to dominate and the universe appears to have a cosmological constant.

There are many and well justified forms for the potential of this slowly evolving field (e.g. Lucchin & Matarrese 1985). A very simple model assumes a potential of the form

$$V = \frac{Q}{\Phi^\alpha} \,, \tag{9.75}$$

where the constant Q has dimensions of mass raised to the power $\alpha + 4$. Using this potential, the ratio of the mass densities in the scalar field and in the matter or radiation turns out to be

$$\frac{\rho_\Phi}{\rho} \propto t^{4/(2+\alpha)} \,. \tag{9.76}$$

In the limit where the parameter $\alpha = 0$, ρ_Φ is constant and this model is equivalent to the Einstein cosmological constant. Solutions as those obtained by using this potential for $\alpha > 0$ have two properties that seem desirable. First, they are said to be attractors Ratra and Peebles (1988) or trackers Steinhardt et al. (1999), meaning that they are asymptotic solutions for a broad range of initial conditions at high redshift. For example, energy distribution becomes nearly homogeneous even when gravity has collected the other matter into non-relativistic clumps. Second, the energy density in the attractor solution decreases less rapidly than that of matter and radiation.

References

Alam, S., Ata, M., Bailey, S., Beutler, F., Bizyaev, D., Blazek, J. A., … Zhao, G.-B. (2017). The clustering of galaxies in the completed SDSS-III Baryon Oscillation Spectroscopic Survey: cosmological analysis of the DR12 galaxy sample. *MNRAS, 470*(3), 2617–2652. https://doi.org/10.1093/mnras/stx721. arXiv: 1607.03155 [astro-ph.CO]

Amendola, L., & Tsujikawa, S. (2010). *Dark energy: Theory and observations.*

Coles, P., & Lucchin, F. (2002). *Cosmology: The origin and evolution of cosmic structure* (2nd ed.).

Hamilton, A. J. S., Kumar, P., Lu, E., & Matthews, A. (1991). Reconstructing the primordial spectrum of fluctuations of the universe from the observed nonlinear clustering of galaxies. *ApJL, 374*, L1. https://doi.org/10.1086/186057

Jain, B., Mo, H. J., & White, S. D. M. (1995). The evolution of correlation functions and power spectra in gravitational clustering. *MNRAS, 276*(1), L25–L29. https://doi.org/10.1093/mnras/276.1.L25. arXiv: astro-ph/9501047 [astro-ph]

Jenkins, A., Frenk, C. S., White, S. D. M., Colberg, J. M., Cole, S., Evrard, A. E., … Yoshida, N. (2001). The mass function of dark matter haloes. *MNRAS, 321*(2), 372–384. https://doi.org/10.1046/j.1365-8711.2001.04029.x. arXiv: astro-ph/0005260 [astro-ph]

Lucchin, F., & Matarrese, S. (1985). Power-law inflation. *Physical Review, 32*(6), 1316–1322. https://doi.org/10.1103/PhysRevD.32.1316

Peacock, J. A., & Dodds, S. J. (1996). Non-linear evolution of cosmological power spectra. *MNRAS, 280*(3), L19–L26. https://doi.org/10.1093/mnras/280.3.L19. arXiv: astro-ph/9603031 [astro-ph]

Peebles, P. J., & Ratra, B. (2003). The cosmological constant and dark energy. *Reviews of Modern Physics, 75*(2), 559–606. doi:10.1103/RevModPhys.75.559. arXiv: astro-ph/0207347 [astro-ph]

Planck Collaboration, Aghanim, N., Akrami, Y., Ashdown, M., Aumont, J., Baccigalupi, C., … Zonca, A. (2020). Planck 2018 results. VI. Cosmological parameters. *A & A, 641*, A6. https://doi.org/10.1051/0004-6361/201833910. arXiv: 1807.06209 [astro-ph.CO]

Press, W. H., & Schechter, P. (1974). Formation of galaxies and clusters of galaxies by self-similar gravitational condensation. *ApJ, 187*, 425–438. https://doi.org/10.1086/152650

Ratra, B., & Peebles, P. J. E. (1988). Cosmological consequences of a rolling homogeneous scalar field. *Physical Review, 37*(12), 3406–3427. https://doi.org/10.1103/PhysRevD.37.3406

Riess, A. G., Macri, L. M., Hoffmann, S. L., Scolnic, D., Casertano, S., Filippenko, A. V., … Foley, R. J. (2016). A 2.4% determination of the local value of the hubble constant. *ApJ, 826*(1), 56. https://doi.org/10.3847/0004-637X/826/1/56. arXiv: 1604.01424 [astro-ph.CO]

Sachs, R. K., & Wolfe, A. M. (1967). Perturbations of a cosmological model and angular variations of the microwave background. *ApJ, 147*, 73. https://doi.org/10.1086/148982

Scolnic, D. M., Jones, D. O., Rest, A., Pan, Y. C., Chornock, R., Foley, R. J., … Smith, K. W. (2018). The complete light-curve sample of spectroscopically confirmed SNe Ia from Pan-STARRS1 and cosmological constraints from the combined pantheon sample. *ApJ, 859*(2), 101. https://doi.org/10.3847/1538-4357/aab9bb. arXiv: 1710.00845 [astro-ph.CO]

Sheth, R. K., & Tormen, G. (1999). Large-scale bias and the peak background split. *MNRAS*, *308*(1), 119–126. https://doi.org/10.1046/j.1365-8711.1999.02692.x. arXiv: astro-ph/9901122 [astro-ph]

Steinhardt, P. J., Wang, L., & Zlatev, I. (1999). Cosmological tracking solutions. *Physical Review*, *59*(12), 123504. https://doi.org/10.1103/PhysRevD.59.123504. arXiv: astro-ph/9812313 [astro-ph]

Tinker, J., Kravtsov, A. V., Klypin, A., Abazajian, K., Warren, M., Yepes, G., ... Holz, D. E. (2008). Toward a halo mass function for precision cosmology: The limits of universality. *ApJ*, *688*(2), 709–728. https://doi.org/10.1086/591439. arXiv: 0803.2706 [astro-ph]

Wong, K. C., Suyu, S. H., Chen, G. C.-F., Rusu, C. E., Millon, M., Sluse, D., ... Meylan, G. (2020). H0LiCOW XIII. A 2.4% measurement of H0 from lensed quasars: 5.3σ tension between early and late-Universe probes. *MNRAS*. https://doi.org/10.1093/mnras/stz3094. arXiv: 1907.04869 [astro-ph.CO]

Index

© Springer Nature Switzerland AG 2021
M. Meneghetti, *Introduction to Gravitational Lensing*, Lecture Notes
in Physics 956, https://doi.org/10.1007/978-3-030-73582-1

Printed in the United States
by Baker & Taylor Publisher Services